THE MODERN TECHNIQUE OF

ROCK BLASTING

THE MODERN TECHNIQUE OF

ROCK BLASTING

THIRD EDITION

by

U. Langefors and B. Kihlström

A HALSTED PRESS BOOK

John Wiley & Sons

New York–London–Sydney–Toronto

© 1978 U. Langefors–B. Kihlström
Almqvist & Wiksell Förlag AB Stockholm

Third edition

Published by Halsted Press,
a Division of John Wiley & Sons, Inc.,
New York

Published in Scandinavia and Continental Europe
by Almqvist & Wiksell, Stockholm

Library of Congress Cataloging in Publication Data
Langefors, U.
The modern technique of rock blasting
"A Halsted Press book"
Bibliography: p. 398–407
1. Blasting
I. Kihlström, B., joint author
II. Title
TN279.L314 1977 624'.152 77-23895
ISBN 0-470-99282-4

Printed in Sweden by
Almqvist & Wiksell
Uppsala 1978

TO C. H. JOHANSSON

PREFACE

The technique of rock blasting has within recent years developed from a mere manual occupation in which long-time experience, personal skill and intuition have played a dominating part, into a technical science in which the fundamental concepts can be taught to students, engineers and workers.

The aim of the present book is to provide an introduction to this technology as it has been built up in mining and civil engineering in the post-war years through research, theoretical investigations, field tests and full scale application.

These years have seen a revolution in the technique as a whole. Drills with tungsten carbide bits and modern drilling machines have greatly improved drilling efficiency, adequate methods of calculating the charge have been worked out, development in electric ignition has increased accuracy and safety in operations. Methods for blasting multiple-row rounds with short delays have created new possibilities for mechanization of the heavy parts of the job and for blasting big rounds. New tools and equipment for charging the drill holes are about to change completely the procedure of loading. The use of AN-explosives in powder or slurries has reduced the costs of benching. The main principles of tunnel blasting have been established and indicate a further development towards increased mechanization, better fragmentation, higher driving speed and a more extensive use of methods for smooth blasting. Such smoothing will also become of ever increasing importance in most other forms of rock blasting.

An essential factor in the safety of the work and in the possibility of blasting big rounds is the wider knowledge of ground vibrations and of the ways of reducing them to safe levels. In underwater blasting the fundamental problems as well as the solutions of the practical difficulties have been studied. At the same time it has been shown how covering material on the rock face affects the calculations of water, as well as of heavier, more or less liquid overburden. The results indicate the conditions for blasting without uncovering the rock.

These various subjects in the modern technique of rock blasting—except where the tools and technique of drilling are concerned—are presented in the respective chapters of this monograph. The first chapter deals also with the mechanism of breakage, as a knowledge of this has proved to be essential in practical application.

Part of the material of the book has been given as lectures at the Colorado School of Mines, the University of Minnesota and the Pennsylvania State University, U.S.A., in 1959.

The headlines of the contents are based on original works carried out at Nitroglycerin AB (NAB), Sweden, in close collaboration with miners and contractors. The company and its President Mr T. Edlund have supported a generous research policy with an open exchange of results, within the country as well as with other research institutions all over the world. The company has in this respect followed its old traditions from Nobel and Nauckhoff of scientific approach to technical problems. There has been trustful and wide-ranging collaboration with all involved in rock blasting, and problems have been attacked with enthusiasm and decisiveness. This has made it possible to convey questions and ideas from research laboratories to the working fronts and for a reversed exchange of problems from these sites into the laboratories for analysis, study and suggestions for solutions. In this way research and development activities have been coordinated so as to include impressive combined resources right through from theory to experiment. Extensive full scale tests under working conditions in various kinds of rock have always been regarded as the final and decisive step in the investigations.

Experience from blasting operations has been carefully examined from reports and literature as well as in my activities as a consultant for projects where new methods of blasting have been called for and adopted. In this way it has been possible to check the fundamental calculations in the book from current rock blasting in most parts of the world. Even if no major changes have been necessary during that procedure there have been contributions and suggestions of importance, such as blasting with AN-explosives and presplit blasting. Practical experience has deeply influenced the presentation of the material as shown in tables, diagrams and recommendations. I hope this has made the book easier to read, easier to use and more complete in practical details.

*

I have dedicated the book to Professor C. H. Johansson in respect and admiration for his pioneering work in detonics research and with warmest thanks for inspiring years of collaboration and friendship. His influence will be felt throughout the book. He and Professor Luth Parkinsson, C.S.M., U.S.A., stressed the great need of and demand for a text book on rock blasting and they are largely responsible for my decision to start writing.

It has been a privilege for me to work together with Mr B. A. Kihlström, whose experience in the field of commercial blasting is outstanding and

7

whose experimental skill has been invaluable in the work. By naming him as a co-author I desire to recognize the far-reaching importance of his collaboration in the practical work connected with the book.

I am also indebted to our former mathematician Professor H. L. Selberg, to Dr A. Wetterholm and Mr N. Lundborg, Mr S. Ljungberg and Mr T. Sjölin of our Physical and Chemical Research Departments and of the Swedish Detonic Research Foundation for their invaluable help at many important stages of the research work. H. L. Selberg has revised the first chapter and he and C. H. Johansson have made essential contributions to the present knowledge of the mechanics of breakage.

I must make special reference to the original contributions made by S. Brännfors and O. Bännmark to tunnelling and underwater blasting, to S. Wikström, Å. Kallin and K.-F. Lautmann on tunnel blasting and coherent problems, to I. Janelid, J. Törnqvist and C. E. Wennerfeldt for their fresh and unorthodox views on mining methods. They have all with courage, clear vision and a sense of technical and economic realities linked the technique of today with that of to-morrow. The list of names could be extended to many of the members of the Swedish Rock Blasting Committee and to those in the Swedish Mining Association.

Outstanding contributions to the knowledge and safety of electric ignition and the practice of its application have been made by Y. Hagerman, I. Olsson and N. Lundborg.

The Du Ponts Explosives Department, U.S.A., has given me generous support and help. I owe personal thanks to their director of research Dr R. M. Cavanaugh and to their technical specialist N. G. Johnson both of whom have encouraged me in my work with the book.

Throughout the years I have received from the Project Department of Atlas Copco and its head Mr G. Fogelström, Sweden, reports and technical information from their international activities. Material for the book has also been supplied by the manufacturers of explosives: Hercules Powder Company, and Atlas Chemical Industries Inc., U.S.A.; Canadian Industries Limited (CIL), Canada; Imperial Chemical Industries Limited (ICI), Great Britain, and Dynamit Nobel AG, Germany.

For valuable material and permission to include it in the book I want to thank Mr D. K. Holmes, U.S.A. (Niagara Project); Dr MacGregor and Mr P. Harder, Georg Wimpey and Co., Great Britain (Rio Furnas, Brazil); Mr W. L. Kok of Ackerman and von Haaren, Belgium (Port of Thema, Ghana); Sr M. Castillo, Rubio, Hydroelectrica Espanola S.A., Spain, and Sr A. Barocio, Ingenieros Civiles Asociados, Mexico.

Important contributions by Dr T. D. Northwood, National Research Council, Canada, and A. T. Edwards of the Hydro-Electric Power Commission, Canada, have been made on the question of the destructiveness of ground vibrations and by Edwards on the reduction of water shock waves. I thank them for permission to publish material from their investigations.

I would also like to refer with appreciation to our scientific contact with Dr S. G. A. Bergman and Dr B. Broberg of the Research Section of the Royal Swedish Fortifications Service concerning shock waves and theories of cracking and fragmentation. The investigations of water shock waves by Dr E. Enhamre have been of great influence in underwater rock blasting.

Critical revision of the English text of the book has been made by my friend George Urquhart, whom I thank for valuable suggestions and for all his good and unselfish aid. The texts of Chapters 3 and 6 have been read by Dr Robert Stefanko, Penn. State University, U.S.A.

For permission to use material from the periodicals Mine and Quarry Engineering and Water Power I thank their editors D. R. Patey and G. R. Bamber.

My best thanks are due to Mrs G. Runwall who has typed the manuscript and to Mr R. Edberg for making drawings and diagrams with great care and accuracy.

Finally, I also wish to thank the following Companies and Institutions in Sweden for participation in the investigations: The State Power Board, The Board of Highways and Waterways, Sandvikens Jernverks AB, AB Skånska Cementgjuteriet and AB Widmark and Platzer.

Ulf Langefors

Preface to the second edition

The first edition of The Modern Technique of Rock Blasting was rapidly sold out and this new edition is therefore of immediate interest. As the general technical principles have remained unchanged only minor alterations have been required. Some of these will be found in Chapter 9 Ground Vibrations and serve to show some important results that have been attained in developments within the past two years.

Ulf Langefors

Preface to the revised third edition

With the book "Rock Blasting" first published in 1963 the intention was to show how the old handicraft of dealing with rock had been transformed into a technological science where the basic laws and general relations were established.

They remain and have still to be known better by those who have to "sculpture" in the hard crust of our Earth by the aid of explosives. As always new ideas and new methods will take time to be fully adopted—it is a mental process, but it is also a matter of requirements for the development of new equipment before full advantage of a new knowledge can be drawn.

Our modern society should be entirely different without the use of commercial explosives as a tool in constructional work and in the production of raw materials. The buildings you work and live in, the car you drive, subways, roads, drainage systems, water supply, electricity from water power plants, they all started their life with a round of explosives.

The technique of rock blasting opens new opportunities for a wider and better use of underground space in cities, in communication and in various kinds of large size rock store. It opens the third dimension for our planners and the years to come will certainly see a notable development in the field of application of rock blasting.

As the book has been a required reference in all these connections a new edition has been asked for. Changes and additions that have been done from the first and second editions concern such revolutionary development as the nonelectrical firing system, but also such ordinary changes as the data for electric detonators, blasting machines and various equipments for charging of drill holes, equipment for monitoring of ground vibrations and also the extended experience from the effects from ground vibrations on surrounding constructions. The information given already in the original edition could also be used for the formulation of new and more adequate criteria in seismology for the classification of the strength and damage capacity of earth quakes. Those hitherto used are shown not to be adequate as regards the relation to the risk for damage.

A chapter on nonelectric blasting and another illustrating the use of underground space has been added.

My friends Duri Prader and Per-Anders Persson have with their great experience and talent given valuable contributions to this third edition. Björn Kihlström's outstanding mastery of the art has of course had a great influence also on the new sections of the book.

Ulf Langefors

CONTENTS

13

SYMBOLS USED

A, A_m	advance of a round, mean value	m (ft)
A	amplitude in ground vibrations	mm (in)
a	acceleration (1 g = 9.8 m/s^2)	g
α	relative advance, $\alpha = A_m/H$	–
B	width of a tunnel or of a bench round	m
c	velocity of sound in a material	m/s (ft/s)
c	rock factor, c_0 for limit charge, $\bar{c}$ approximation	kg/m^3 (lb/cu.yd)
d	diameter of drill-bit	mm (in)
D	height of a tunnel from bottom to roof	m
E	spacing, distance between holes in a row	m (ft)
e	energy	
F	loss of energy	
F, F_1 F_2	forces	
f	degree of fixation, $f = 1$ for straight angle of breakage at the bottom of a bench hole	–
f	frequency in ground vibrations, cycles per sec.	c/s
f_0	proper frequency	c/s
ϕ	diameter of uncharged hole	mm (in)
G	total drilled length in a round	m/round (ft/round)
g	drilled length per m^3	m/m^3 (ft/cu.yd)
γ	reduction factor, exponent	
H	depth of drill holes	m (ft)
h	height of charge	m (ft)
i	impulse	kgm/sec (lb ft/sec)
K	height of bench (difference in level between top and bottom)	m (ft)
K'	height of bench measured along the slope	m (ft)
k	constant	–
$\varkappa$	ratio	–
L	length	m (ft)
l	charge per meter of hole	kg/m (lb/ft)
l_b, l_p	charge per meter of bottom charge, of column charge	kg/m (lb/ft)
λ	wave length	m (ft)
m	mass	kg (lb)
N	number of holes in a round	–
ν	relative vibration velocity v/c	mm/m (‰)
p	pressure	kg/cm^2 (lb/sq in)
P	degree of packing, $P = 1.27 \cdot 10^3\ l/d^2$	kg/dm^3
Q	quantity of explosive in a hole	kg (lb)
Q_b, Q_p	bottom charge, column charge	kg (lb)
Q_0	bottom charge, when concentrated ($h < 0.3\ V$)	kg (lb)

16

q	spec. charge	kg/m³ (lb/cu.yd)
q_0, q_1	spec. limit charges	kg/m³ (lb/cu.yd)
ϱ	density	kg/dm³ (lb/cu.yd)
R	distance	m (ft)
R	deviation for 5/6 of the holes	cm
r_e	electrical resistance per detonator	ohm
r	scaled distance, $r = R/Q^{1/3}$	m/kg$^{1/3}$ (ft/lb$^{1/3}$)
S	section of a tunnel	m² (sq.ft)
$S\ (\alpha, \beta, \gamma)$	damage criterion	
s	weight strength of explosive ($s = 1.30$ for blasting gelatine)	
σ	average deviation, $\sigma = R/H$	cm/m (%)
T	time for a cycle in vibration	ms, sec
t	time	ms
$\circ$	temperature degree	°C $= 5/9$ (°F $- 32$)
τ	interval time	ms, sec
$\Delta\tau$	spread in interval time	ms, sec
u	propagation velocity	m/s (ft/sec)
V	burden, maximum burden	m (ft)
V_1	practical burden (used in the drilling pattern)	m (ft)
v	vibration velocity	mm/s
w	angular frequency $w = 2\pi f$	sec^{-1}
$>$	greater than	
$<$	smaller than	
$\approx$	approximately equal to	
$\geqslant$	greater than or equal to	

1. THE MECHANICS OF BREAKAGE

Within some thousandths of a second after the initiation of the explosive there occurs in a charged hole a series of events which, in drama and violence, have few equivalents in civil technology. The chemical energy of the explosive is liberated and the compact explosive becomes transformed into a glowing gas with an enormous pressure which, in a densely packed hole, can amount to and exceed 100,000 atmospheres. The amount of energy developed per unit of time is even in a tiny hole drilled with a hand-held machine, of the order of magnitude of 25,000 MW, that is to say it exceeds the power of most of the world's present largest power stations. This is not due to the fact that the amount of energy latent in the explosive is extremely large but to the rapidity of the reaction (2500–6000 m/s). What is especially characteristic of the explosive as a tool in rock blasting is its ability to provide concentrated power in a limited part of the rock.

The high pressure to which the rock is exposed shatters the area adjacent to the drill hole and exposes the space beyond that to vast tangential strains and stresses. These take place under the influence of the outgoing shock wave which travels in the rock at a velocity of 3000–5000 m/s. Round a hole with a diameter of 40 mm in primitive rock the thickness of the crushed zone is of the same size as, or slightly less than the radius of the hole. The system with radial cracks issuing from the center of the hole as a result of the tangential stresses, the so-called rose of cracks, extends considerably further. In the example mentioned it can extend from some decimeters to about a meter. Consequently, the first cracks have been completed (fig. 1:1) in fractions of a ms (1 ms = 1/1000 sec).

The lateral pressure in the shock wave is, to begin with, positive when the shock wave arrives, and then falls rapidly to negative values as shown by Selberg (fig. 1:2a). This implies a change from compression to tension. Near the hole ($r = r_0$ and $r = 2r_0$) the strain forces are even greater than the compression in the front of the shock wave. As the rock is less resistant to strain than to compression the primary cracks will chiefly occur under the influence of tensile forces with resulting pronounced radial cracks. It is interesting to observe calculations which show that there is also a tensile strain in a radial direction at great distances from the hole, at least in an ideally homogeneous, elastic medium. These, however, will not have any effect in practice.

18

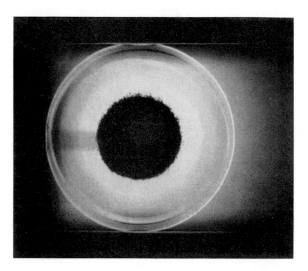

Fɪɢ. 1:1. The front of the shock wave (luminous circle) from a cylindrical charge propagates at a much greater velocity than the radial cracks (central dark section). Transparent lucite plate with a central hole and source of light from behind, exposed 15 μs after detonation (exposure 0.03 μs).

During this first stage of the cracking there is practically no breakage. If the drill hole with its charge goes straight into the rock without any adjacent surface parallel to the hole the shock wave fades out without any further effect. The remaining pressure of the gases in the drill hole slightly widens the cracks, but if the environment of the rock is studied after blasting, e.g. on a level some few feet below the top face of the rock, at right angles to the hole only the radial cracks will be found. The drill hole has been slightly widened to normally less than double the diameter by crushing and plastic deformation. As the cracks in this case have contracted again after the blasting it may even be difficult to discover them with the naked eye. They can easily be observed if some of the rock close to the hole is removed and sawn or treated in some other way. In model scale experiments in plexiglass (fig. 1:3), which is transparent, the entire picture of the cracks can be studied.

In blasting rock we usually have a free face of the rock in front of and parallel to the drill hole. When the compression waves reflect against these surfaces they will give rise to tensile stresses which may cause scabbing of a part of the rock near the surface. The mechanism of such reflections has been studied thoroughly by Fischer, Broberg, Pettersson, Rhinehart, Duvall. The process is the same as when a row of billiard balls is struck at one end and the blow passes from ball to ball until the last one in the row sets off

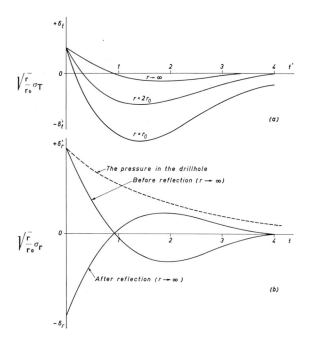

$V\sqrt{\dfrac{r}{r_0}}\,\sigma_T$

$V\sqrt{\dfrac{r}{r_0}}\,\sigma_r$

FIG. 1:2. The stress in a shock wave from a cylindrical hole with radius r_0 of the drillhole. (a) the tangential stress of the wave before reflection (b) the radial stress of the wave.

with full force. This would also happen if the balls were slightly cemented together. In rock blasting the so-called scabbing is generally of secondary importance. In granite with a relatively high tensile strength it is only of some importance when the charge considerably exceeds 1 kg per m³. In ordinary bench blasting the consumption of explosive is considerably lower, 0.15–0.60 kg/m³. In looser rock the exposed surface may, however, become worn. Even then it is of less importance in relation to the total volume of rock excavated. If the surface becomes at all torn it will happen within about a ms from the moment of detonation. If the movement of the rock is estimated with the aid of photographing or filming it must be realized that what can be seen there is only the superficial movement of the rock. If scabbing occurs the effect of it will dominate the picture for the first few milliseconds.

With an increasing size of charge the thickness, or the layer which is torn off, will increase. There may actually be a flaking of several layers. With a sufficiently large charge this effect can result in a crater which can reach up to, or close to the charge. In blasting so-called crater cuts, which have been developed and described by Hino, and where the charge per m³ is of

20

the order of magnitude 5 kg/m³ it is the reflected shock wave that causes the loosening of the rock.

Those two first stages in the process of loosening the rock, the radial cracks and the scabbing effect are, insofar as they occur, effected by the shock wave. When blasting with gun powder the shock wave is not intense enough to give radial cracks. Fractures or indications close to the drill hole are in such a case widened and the picture of the breaking will get a rather arbitrary character as the primary fractures usually have not the direction favorable for complete loosening of the burden. If in such a case two visually insignificant indications for cracks are chiselled to give directions for cracks which can result in an approximately 90° breakage, then the breaking force of the powder in relation to the energy and gas volume appears to be the same as for high explosives. This is both surprising and illustrative. It shows among other things that the main part of the energy required for complete loosening of the rock has no connection with the shock wave, the energy of which in the case of a high explosive is many times greater than for gun powder. For a high explosive the shock wave energy most probably amounts only to 5–15 % of the total energy of the explosive at a theoretical estimate. Fogelson and others have measured the intensity of the shock waves in the vicinity of the drill hole and have from these experiments concluded that the energy in the shock wave was about 9 % of the whole energy for a high explosive. As the shock wave is distributed all around a charge, at least $\frac{2}{3}$ of its energy will disappear without affecting the breakage for a single hole with an angle of breakage less than 120°. This would imply that only 3 % of the total energy of the explosive is distributed by the shock wave within the angle of breakage, and confirms what has been said above that the shock wave is not responsible for the actual breakage of the rock, but only for providing the basic conditions for this process.

The third and last stage of the breakage is, as pointed out by Johansson, a slower process. Under the influence of the pressure of the gases from the explosive, the primary radial cracks expand, the free rock surface in front of the drill hole yields and is moved forward. This may be described as a semistationary process in which the strain picture at any moment decides the continuation of the cracks just as in the case of a static load. When the frontal surface moves forward, the pressure is unloaded and the tension increases in the primary cracks which incline obliquely forward. If the burden is not too great, several of these cracks expand to the exposed surface and complete loosening of the rock takes place. The burden is consequently torn off and the maximum effect per drill hole and quantity of charge is attained if it is possible for the burden in front of a hole to move forward

21

a.

b.

Fig. 1:3a and b. The influence of the amount of explosive on the formation of cracks. Lower picture charge four times as large as in the upper one. The burden in both cases being the same, $V = 1$ cm (0.4 inch). Thickness of plate, 2 cm (0.8 inch).

freely when the charge detonates. It is of the utmost importance to make allowance for this fact when constructing drilling and ignition patterns. Experience shows that this is either not sufficiently known, or is disregarded. What may appear to be an insignificant change, for example in an ignition sequence in a tunnel round, can immediately and without other changes considerably increase the advance per round by improving the chances for free breakage of every individual hole.

The development of the cracks from within and out towards the free

22

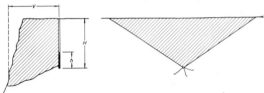

Fig. 1:4. Blasting in homogeneous rock

$d = 21.5$ mm $h = 0.33$ m
$V = 1$ m $Q = 212$ g LFB
$H = 1$ m   Throw: 4 m

surface can be studied when blasting in plexiglass (figs. 1:3a and b). With an insufficient charge (fig. 1:3a) the cracks are not fully developed, but it can be seen how some of them, at an angle of between 90 and 120°, have qualifications for giving full breakage if the static pressure is increased. It is also evident that it may be difficult to forecast an accurate angle of breakage. Local conditions near the drill hole may favor one or two neighboring cracks. This must all the more apply when blasting in more unhomogeneous rock. Complete breakage has been obtained in plexiglass in fig. 1:3b at an angle of 110°.

Some examples of blasting in homogeneous granite on a full scale show an angle of breakage at 115° (fig. 1:4), and at slightly over 90° (fig. 1:5), two cases with otherwise very similar conditions.

The natural angles of breakage that might be expected as an average in the general case are shown in fig. 1:6a and b with the two main cases to be taken into account, first with free bottom, secondly with fixed bottom.

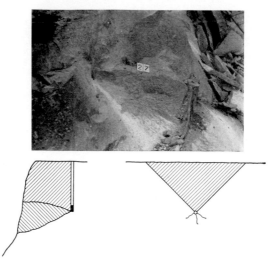

Fig. 1:5. Blasting in homogeneous rock

$d = 39$ mm	$h = 0.115$ m
$V = 1.02$ m	$Q = 200$ g
$H = 1.0$ m	Throw: 2–4 m

In both cases the angle of breakage α_u will be about 90° or more. With a free bottom, the bottom angle α_b will be about 135°. If the bottom is fixed, as is the case in ordinary benching and very often in stoping, the break at the bottom must follow another surface. The angle can then be between 90 and 135 degrees dependent on charging and depth of drill hole. If the calculation of the charge is correct, the bottom angle will be 90°. Smaller angles can only be expected under special circumstances, and then only for a considerably smaller angle of breakage, as is indicated by the shaded sections in fig. 1:6 b. If the drilling pattern is drawn up on the basis of 90° angle of breakage tearing at the bottom may be expected to take place perpendicularly to the drill holes, and not, as is often stated, with a bottom angle of less than 90°. The rules which apply for bench blasting can thus be adopted even for cuts and openings in tunnelling, and as long as the 90°-principle can be used, one has the most favorable case for the loosening of the rock.

The formation of cracks in blasting one row or more of drill holes will largely depend on the relation between the burden and the distance between the holes, and also if the ignition occurs simultaneously or with a certain delay. In instantaneous blasting the gas pressure in several holes close to each other cooperates and pushes the rock forward as the free face yields. If the spacing of the holes is comparatively close, the free face moves for-

24

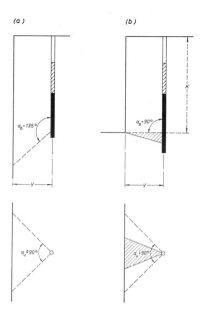

Fig. 1:6. Natural angles of breakage with free burden; (a) free bottom, (b) fixed bottom, K height of bench V burden.

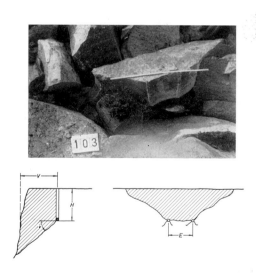

Fig. 1:7. Blasting in homogeneous rock

$d = 22$ mm $\qquad\qquad Q = 2 \times 24$ g LFB
$V = 0.5$ m
$H = 0.5$ m $\qquad\qquad v = 38°$
$E = 0.35$ m

FIG. 1:8a.

ward without much deformation and rather slight shearing, bending and tensile stresses inside the rock may be considered possible—of course with exception for the line joining the holes, where very high tensional stresses are obtained. In homogeneous rock one can also blast rounds where the common burden is torn off as a complete and single lump of rock (fig. 1:7). This cannot be done if the time for detonating the charges varies slightly. In such a case every individual hole must tear off a part of the burden. This retards the throw, but increases the tearing in the burden. It is one of the main reasons for the improved fragmentation obtained in short-delay blasting.

It is also possible to achieve increased shearing and strain in the section of the rock to be excavated, by increasing the distance between the holes in relation to the burden so that the exposed rock face cannot swing forward with a plane front but with a curved one, which will thus have a more pronounced bend in front of each individual drill hole. Model blastings with a ratio between burden (V) and hole spacing (E) of $V/E = 2.0$ (fig. 1:8a) and of $V/E = 0.5$ (fig. 1:8b), but with the same value of VE per hole, i.e. the same charge and footage of drilling per volume of rock, show the

26

FIG. 1:8 b.

FIG. 1:8. Influence of the spacing on the fragmentation. The left picture
(a) $EV = 2$, $E = V/2 = 1$ cm; picture above $EV = 2$, $E = 2V = 2$ cm. The same
specific charge (kg/m³) in both cases.

enormous difference in tearing in the material which can be achieved with
such simple means as a modified arrangement of holes, but with no change
of the simple instantaneous ignition.

The above applies to the homogeneous model material. The effect may
possibly not be just as spectacular in normally uneven, fissured rock. It
may be mentioned, however, that the improved fragmentation in the case
of short-delay blasting had a direct equivalent in results obtained in model
blasting. The fragmentation shown in fig. 1:8 b is definitely superior to
the effect of such short-delay blasting on a model scale. We know from
experiments that apparently insignificant irregularities or breaks in the
rock or model material can have a marked effect on the strain picture as
a whole, and conclusions for the practical application should, as always, be
drawn first after control tests have been made on a full scale. Figs. 1:8 a
and b are included here primarily in order to illustrate an important part
in the process of loosening the rock and also that there may still be room
for a fascinating development towards more effective blasting methods.

27

2. CALCULATION OF THE CHARGE
GENERAL PART

Determination of the arrangement of the holes, quantity of the charge and sequence of breakage can be indicated as the main problems of practical rock blasting. In simple cases, with an arrangement of the holes which can be repeated round after round, the quantity of the charge can easily be increased if there has not been sufficient breakage in previous attempts. In underground blasting stronger charges can often be employed, when special attention has not to be paid to the throw. The same applies to underwater blasting where the throw is eliminated by the water covering the surface of the rock. Even if the desired breakage can be obtained under such conditions there is no basis for judging if this corresponds to a minimum cost, or for calculating how this can be attained if the charge cannot be correctly calculated. In open air blastings it is necessary for the safety of the work to avoid too great a throw, while at the same time the most suitable fragmentation for the digging is desired.

This chapter deals with general connections. For *practical application* the reader is referred to the *Chapter 3*, which reproduces tables and simplified connections in treating the material. That chapter also deals with different factors occurring at the working-place and how they may affect the design of the drilling patterns.

Calculation of the charge primarily covers calculation of the minimum charge required for loosening the rock. In all circumstances this condition must be fulfilled. In practical application however this is far from sufficient and a final calculation must therefore also take into account the influence of faulty drilling, throw and swell, fragmentation and the effect obtained on surrounding rock or in neighboring buildings.

The calculations given here are based on systematic blasting experiments carried out according to theoretical deductions. In this way it has been possible to indicate the results in a general form.

During the ten years which have passed since the calculations first were presented it has been possible to make a careful check of them when blasting different kinds of rocks in practically every part of the world. The calculations have been successfully applied in the development of new methods in the technic of rock blasting.

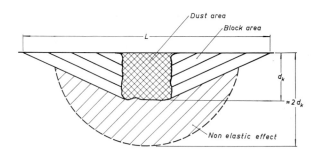

FIG. 2:1. Effect of external charges (according to Broberg).

2.1. The law of conformity

When a charge (Q) placed on the top of a homogeneous rock detonates, it crushes a part below the charge and pieces of the surrounding rock so that a crater is formed with the depth d_k and the diameter L (fig. 2:1). If such a test is repeated with a uniformly enlarged charge Q the crater will, according to Broberg, also be enlarged in all directions in the same proportion as the linear extension of the charge so that

$$\sqrt[3]{\bar{Q}/Q} = \bar{L}/L \qquad (2:1$$

This means that if the length, width and height of the charge are enlarged a certain factor b (that is, the volume of the charges is increased by a factor b^3), the diameter and depth of the crater will be increased by the same factor b. A simple consequence of this uniformity is that the charge per m³ of crushed rock is equally large if the trials are made on a large or on a small scale.

In the same way the crushed zone round a charge is enlarged when the charge is inserted in the rock. This enlargement is in direct proportion to the radius of the charge, whether it is a question of a cylindrical or a spherical charge.

In purely elastic processes the so-called "law of conformity" applies with great accuracy. It indicates that theoretically such physical magnitudes as pressure, velocity, compression and tearing remain unchanged in uniform enlargement. In the case of dynamic processes the time-scale shall be increased in the same proportion as the lengths.

This fact renders it possible to simplify essentially the relations for shock-waves from explosive charges in air, water and solid bodies, and implies that the pressure p of the shock-wave determined on the distance r from a charge of 1 kg is equally large at the distance $R = r \cdot Q^{1/3}$ from another charge Q. When for instance in Chapter 11 the pressure (p) from an under-water charge is indicated as a function $f(r)$ of the distance (r) from a 1 kg charge this connection can be made general by introducing $r = R/Q^{1/3}$ and the law of conformity gives

$$p = f(R/Q^{1/3}) \qquad (2:2)$$

This is the reason why the law of conformity can be applied also to the crushing zone in the two cases mentioned above. The pressure in this "fluid" causes the lateral loosening which is called "block area" in blasting according to fig. 2:1.

In ordinary rock-blasting the conditions are different and the law of conformity cannot be expected to apply other than as an approximation. A special case of importance may, however, be mentioned and that is in blasting crater-cuts (see 8.2c) where the loosening of the rock mainly takes place through the reflection of the shock-wave at the free face of the rock, so-called scabbing. In this case the law of conformity can probably be applied.

It is interesting to note that the simplest and probably also the oldest formula for calculating the charge, the formula of Vauban for land mine blasting, $Q_e = kV^3$, gives the same simple type of connection as (2:1) above, between the charge Q and a length, for example the burden V. Also according to the formula of Vauban the charge per volume of excavated rock is a constant representing the rock. It is said for instance that the consumption of explosive is 0.4 kg/m³ in a certain rock. In bench blasting and stoping this is, however, only an approximation. The consumption of explosive per m³ has been found here to vary considerably with geometrical conditions.

2.2. Older formulas

Belidor stated in 1725 that one part of the charge in blasting could be expected to be proportional to the volume excavated and one part proportional to the surface obtained. This gives a relationship of the type

$$Q = k_2 V^2 + k_3 V^3 \qquad (2:3)$$

where k_2 and k_3 are constants.

Höfer has given a connection which only includes the first term in (2:3) but in addition introduces a factor b_1 for the burst (Brisanz) of the explosive and b_2 for the strength of the rock

$$Q_H = b_1 b_2 \cdot V^2 \qquad (2:4)$$

30

Fraenkel has given a relation between the "blastability" (S) which is used as a measure of the resistance of the rock to blasting, maximum burden (V_{max}), depth of hole (H), height of the charge (h) and diameter (d) of the hole.

This relation gives

$$hd^2 = 50^{3.3} \, V_{max}^{3.3}/(S^{3.3} \, Hd^{2/3}) \qquad (2:5)$$

for the magnitude hd^2. Here $hd^2 = Q$ for an ordinary degree of packing $P = 1.27$ g/cm^3. In a uniform enlargement of V, H and d the charge Q increases in proportion to $V^{1.6}$. The exponent 1.6 is however found to be quite too low.

Over and above this a great number of other formulas are presented, but of all of these it may be said that they are today quite insufficient as a basis for an accurate performance of rock blasting. This was evident when introducing open air multiple-row rounds with short delays during the latter half of the 1940's. Such rounds must be fully calculated in advance. The bench height as well as the burden, spacing and diameter of the holes, vary within wide limits in each individual case. A calculation which applied to one round might give an incomplete breakage in another or a violent and unexpected throw. To get a fairly complete grasp of the technical problems of rock blasting the effect of all the different factors must be accounted for.

2.3. Basic principles regarding loosening of the rock

The problem of calculating the charge of explosive in rock blasting as a function of all the variables that influence the result can be given in the following form

$$Q = f_1(V, K, E, h, d, s, \varrho, u, c_i) \qquad (2:6)$$

Q is the amount of charge required for tearing the burden. V, K, E, h, d, are geometrical magnitudes; s, ϱ, u, are factors depending on the explosive;[1] c_i factors depending on the rock, on the degree of fixation and so on.

To determine experimentally the function f_1 it is necessary to reduce the number of variables. This can be done if the experiments are made with one and the same explosive with a definite density and with the same conditions in the rock. Furthermore if the blasting is restricted to single holes there remains a function f_2 of four variables.

$$Q = f_2(V, K, h, d) \qquad s, \varrho, u, c_i \text{ constant} \qquad (2:7)$$

As Q, on the other hand, is determined by d and h, we are actually only concerned with three independent variables so that there is a connection

[1] s = weight strength, ϱ = density, u = detonation velocity.

$$Q = f_3(V, K, h) \qquad\qquad (2:8)$$

V, K/V and h/V can also be taken as independent variables.

If this relation is to be determined in a series of test blasts, the variables should not be changed independently of each other but so that the geometrical proportions remain constant. Equation (2:8) for constant values of K/V and h/V can then be given in the following form

$$Q = f_4(V, K/V, h/V) = f(V) \qquad\qquad K/V, h/V \text{ constant} \qquad (2:9)$$

This means that Q is given as a function of a single variable, the linear dimension V. f is a positive function of V and can be developed in a power series

$$f(V) = k_0 + k_1 V + k_2 V^2 + k_3 V^3 + k_4 V^4 + \ldots \qquad\qquad (2:10)$$

so that the properties can be studied in the region where the series applies. As $f(0)=0$ we have $k_0=0$. If we further study a part of a long column charge with the extension V and vary the linear dimension, we have also a geometrically uniform case to which (2:9) and (2:10) apply. The charge per meter $Q/V = k_1 + k_2 V + \ldots$ must approach 0 when $V=0$. This means that $k_1=0$.

The problem is thus given in a general form and is so arranged that it remains for the experimental part of the investigation to determine the values of k_i $(i \geqslant 2)$. It has been found that $k_i=0$ for $i=4$ if there is no need to take into consideration what is required for the swelling. In general, however, k_4 must also be taken into account and we get

$$Q = k_2 V^2 + k_3 V^3 + k_4 V^4 \qquad\qquad (2:11)$$

The coefficients k_2 and k_3 depend on the elasto-plastic properties of the rock, k_4 on the weight of the rock to be excavated. In downward stoping the coefficient k_4 is negative, in vertical bench-blasting the third term in (2:11) may be ignored when single holes or a single row of holes of minor dimensions are considered. In ordinary Swedish bedrock and with geometrical conditions as in fig. 2:3a Q is given by

$$Q = 0.07\ V^2 + 0.35\ V^3 + 0.004\ V^4 \qquad\qquad (2:12)$$

The last figure $k_4 = 4\ gr/m^4$ is merely an estimate based on calculations, but gives the size for comparison with the two other terms. It can be called the "throwing" or "swelling" component.

The formula (2:11) is fundamental in rock mechanics and has been proved to apply in the entire range investigated, extending from $V=0.01$ m to $V=10$ m with charges which vary in proportions of 1 to 50 millions. It

TABLE 2:1. *Connection between burden* (V) *and a factor* (Q/V^3) *proportional to the specific charge* (g/m^3)

V m	$Q/V^3 = 70/V + 350 + 4\ V$ g/m³
0.01	$7000 + 350 + 0.04 = 7350.04$
0.10	$700 + 350 + 0.4 = 1050.4$
0.30	$233 + 350 + 1.2 = 584.2$
1.0	$70 + 350 + 4 = 424$
10	$7 + 350 + 40 = 397$
100	$0.7 + 350 + 400 = 750.7$
1000	$0.07 + 350 + 4000 = 4350.07$

seems reasonable to expect that the formula shall apply also to smaller linear dimensions and, for example, in such cases as crushing, where the so-called "Law of Rittinger" can be regarded as a special case, and in drilling when small pieces of stone are knocked off at every blow. The formula can be expected to be applicable also to very large dimensions, and consequently in atomic rock blasting, which could be used to determine the coefficient k_4.

The different terms in (2:11) and (2:12) have a clear physical meaning. The middle one represents the part corresponding to the law of conformity (2:1) or (2:2), the first is associated with losses of energy that arise in internal surface layers, for instance in flow and plastic deformation. The last term is the part required to heave the mass of rock sufficiently to give total breakage. How the charge per m³ will hereby vary with the size of the burden in dealing with an entirely uniform enlargement will be seen from table 2:1.

With a burden between 1.0 and 10 m the specific charge is approximately constant $q = 0.4$ kg/m³. This is due to the fact that the curve $a_4 = 4$ in fig. 2:2 in this region has a minimum.

The formula (2:11) gives the influence of the dimension on the size of the charge and can be generally applied. This, however, is only one part of the problem. In a uniform enlargement the coefficients k_2, k_3 and k_4 are constant, but in a complete calculation a knowledge is required of how they vary with the geometry, for instance in benching and stoping with the dimensionless magnitudes K/V and h/V. When they are changed the coefficients assume new values which means that for k we have

$$k_i = k_i(K/V, h/V) \qquad i = 2, 3, 4$$

The notation indicates that the coefficients are functions of two variables, bench height (K) and height of the charge (h) given in proportion to the

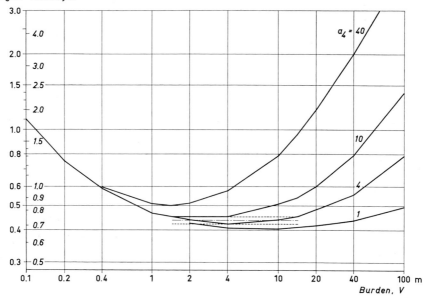

Specific charge, q
kg/m³ lbs/cu.yd.

FIG. 2:2. Charge in kg/m³ (lb/cu.ft) in conformal scaling with a swell component of $a_4 = 40$, 10, 4 and 1. The values for $a_4 = 4$ are given in table 2:1. For the corresponding curve the specific charge $q = 0.45 \pm 0.01$ kg/m³ can in the region $V = 1.4$–14 m (4.5–45 ft) with sufficient accuracy be regarded as a constant. Here the law of conformity applies.

burden (V). This applies in blasting in one and the same rock, with one and the same explosive with the same density and rate of detonation for different cases under comparison.

A noticeable simplification can be made in the analysis of a case according to fig. 2:3 with $V = K$ and with a height of the charge that is small compared with the burden, $h \ll V$.

Test blastings which have been carried out have shown that a certain quantity of charge has a breaking power which is independent of the height of the charge when it is concentrated ($h \ll V$). In tests in homogeneous granite according to fig. 2:3b with $V = K = 1$ m, approximately the same minimum charge for breakage ($Q = 0.200$ kg) was thus obtained with a hole diameter of $d = 39$ mm and a height of charge $h = 0.11$ m and with $d = 22$ mm and $h = 0.33$ m. In spite of a height of charge, which in this latter case was 3 times as large, the breaking power within the limits of test errors ($\pm 5\%$) has been the same. The reduction in the effect of the charge due to its greater extent may have been between 5 and 10 %. (The charge was 6 % larger in the latter case). For smaller h-values, $h \leqslant 0.3V$, the reduction in

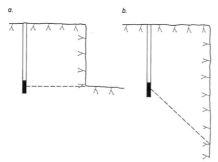

FIG. 2:3. Bench blasting with a) fixation at the bottom, b) free burden at the bottom.

breaking power at the bottom may be completely ignored. It has hereby been possible to determine with relatively simple test blastings the coefficients k_2 and k_3 for a case with $V = K$ which at the same time can represent the bottom part of a bench which may in other respects have an arbitrary height greater than V, as the additional part, that of the column, can be regarded as a different though also conformal case.

It will be shown here how a knowledge of those two typical cases make it possible to build up a complete calculation of charges for all types of stoping, that is to say where an extended charge in a drill hole tears up a burden, the size of which is indicated by the rectangular distance between the drill hole and the free rock face in front. Practically all civil rock blasting is carried out in principle as stoping: bench and tunnel blasting including blasting of plow, fan and cylinder cuts, and also shrinkage stoping, Janol driving and pillar blasting etc. in mining. Some important exceptions such as instantaneous cuts and crater cuts will be discussed later in another connection. Mine blasting, which may nowadays be considered of interest for its application in atomic rock blasting, is already included in formula (2:11) as it is a question here of a simple, conformal case. So-called chamber blastings, on the other hand, will not be described here as they have already been dealt with in a detailed and clear manner by Kochanowsky.

2.4. Formulas for a single drill hole

The calculation is first made for a schematic distribution as will be seen from fig. 2:5.

a. Bottom charge, concentrated

For calculation of the bottom charge Q_0 we may proceed from a case according to the above-mentioned fig. 2:3a, with $K/V = 1$, a fixed bottom and

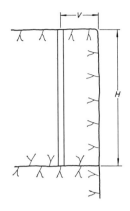

Fig. 2:4. Slab blasting.

the charge concentrated at the bottom of the drill hole. The charge is stemmed with dry sand to avoid loss of energy from exhaust. When converting into burdens of different sizes (2:11) applies

$$Q_0 = a_2 V^2 + a_3 V^3 + a_4 V^4 \qquad (2:13)$$

The coefficients a_i are special cases of k_i $(K/V, h/V)$ for $K/V = 1$ and $h/V = 0$ and we have

$$a_i = k_i (1,0).$$

Those a-values can be determined by test blasts. To determine a_2 and a_3 it is sufficient for instance to test with $V = 0.5$ and $V = 1.0$ m. It will be seen from (2:12) and tab. 2:1 that a_4, which is less than 1% in these tests, does not affect the result.

b. Column charge

Observe a part of a bench that is high in relation to the burden or, still better, a so-called slab blasting (fig. 2:4) where a crevice at the bottom renders an extra charge unnecessary. It may be assumed as evident that the size of the charge is in proportion to H in so far as losses at the ends of the hole can be ignored. To obtain a conformal case a part of the column with a length equal to the burden can be considered. The size of the charge in this part can be calculated for varying burdens according to the relation

$$Q_1 = b_2 V^2 + b_3 V^3 + b_4 V^4 \qquad (2:14)$$

in which b_i represents another special case of k_i. The charge per m (l_p) is indicated by Q_1/V and we get

$$l_p = b_2 V + b_3 V^2 + b_4 V^3 \qquad (2:15)$$

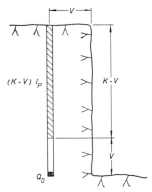

FIG. 2:5. Shot-hole with concentrated bottom charge and uniformly distributed column charge.

c. Total charge

The smallest charge that can be supposed to give full breakage is a function of bench height (K) and burden (V) and is

$$Q_t(K, V) = Q_0 + l_p(K - V) \tag{2:16}$$

as this holds good for a concentrated bottom charge, and a column charge for which no losses have been assumed. The formula (2:16) contains 6 constants a_i and b_i which depend on the rock and which must be determined by test blasting. In such tests it has been found that there is a relation between a_i and b_i which gives $b_2 = 0.4 \, a_2$ and $b_3 = 0.4 \, a_3$, and it has been assumed that this applies with satisfactory accuracy independent of the kind of rock. It is, of course, a condition that there is no crevice crossing the hole close to the intended bottom which in extreme cases can even give $a = b$. Strictly speaking in such a case it would be incorrect to speak of a-values as, in keeping with fig. 2:4 it is a column charge all along the hole.

The general formula for the total quantity of the charge in a hole will be

$$Q_t(K, V) = 0.4 a_2(K/V + 1.5) \, V^2 + 0.4 a_3(K/V + 1.5) \, V^3 + a_4 V^4$$
$$+ b_4(K/V - 1) \, V^4 \tag{2:17}$$

A comparison with (2:11) shows that for the case considered

$$k_2 = 0.4 a_2(K/V + 1.5)$$
$$k_3 = 0.4 a_3(K/V + 1.5) \tag{2:18}$$
$$k_4 = a_4 + (K/V - 1) b_4$$

which give the complete solution for the case with a concentrated bottom charge and an evenly distributed column charge without energy losses at the end of the hole.

37

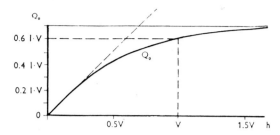

FIG. 2:6. Breaking capacity at the bottom of a charge with the depth h and the concentration of charge l. Q_0 is a charge concentrated at the bottom with the same bottom effect as the extended charge lh.

d. Distribution of the charge

The bottom charge has been calculated for a concentrated charge at the bottom. In applying this in practice it must normally be taken into account that as much explosive as possible is desired at the bottom of the hole so that a greater burden and bulk of rock can be blasted away with a given volume of the hole. The question of how the distribution of the charge affects the breaking power then arises, as an elongated bottom charge must have less effect at the bottom than if the entire charge is concentrated there. The relation between loading height (h) and effect will be seen from fig. 2:6, where Q_0 at the vertical axis indicates the concentrated bottom charge which gives the same breakage at the bottom. The diagram refers to a charge in a hole with a given diameter where consideration has not had to be taken to the swelling term (V^4). For small heights of the charge the effect is doubled when the height is doubled, that is to say the effect Q_b is in direct proportion to the size of the charge $Q = l \cdot h$. It can also be said that the charge is utilized here to 100 %. For heights of charge exceeding 0.25–0.30 times the burden the effect decreases. For $h = 0.96 \ V$ the total charge is $0.96 \ lV$, but the effect at the bottom is $Q_0 = 0.6 \ lV$. The total effect here is consequently only 63 %. Greater heights of the charge give practically no increased yield for loosening at the bottom—especially as normal cracks and crevices screen off the effect as the loading height rises.

If it is not worth while to allow the bottom charge to rise above $h \geqslant V$ the breaking power at the bottom can, on the other hand, be increased by drilling the hole below the intended bottom. The charge which is placed here gives practically a full yield down to a depth of 0.3 V below the bottom. The breaking power is then increased 50 %, from $0.6 \ lV$ to $0.9 \ lV$.

For such a hole charged in the usual manner according to fig. 2:7, the maximum burden is determined by the quantity of the charge (l_b) per m at the bottom of the hole. A bottom charge is defined as the charge which

38

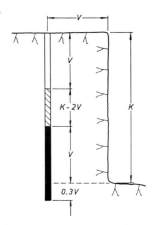

Fɪɢ. 2:7. Bottom charge and column charge.

fills the hole from $-0.3\ V$ to $+0.96\ V$. The concentration of the charge required is obtained by the relation

$$0.9\ l_b\ V = a_2\ V^2 + a_3\ V^3 + \text{terms with } a_4\ V^4 \qquad (2\!:\!19)$$

from which

$$l_b = 1.1\ a_2\ V + 1.1\ a_3\ V^2 + \text{terms with } a_4\ V^3 \qquad (2\!:\!20)$$

The total bottom charge is

$$Q_b = 1.26\ V l_b = 1.4\ a_2\ V^2 + 1.4\ a_3\ V^3 + \text{terms with } V^4 \qquad (2\!:\!21)$$

In test blasts which have been carried out the terms with V^4 have been too small to determine accurately. According to the formula (2:12) it can be expected to be about 1 % of the charge when $V = 1$ m. A concentrated bottom charge according to fig. 2:3 suffices for tearing the bottom face plus the side faces up to the height V above the bottom level. A bottom charge distributed from $-0.3\ V$ to $0.96\ V$ may in the same way be assumed to suffice for loosening up to the height 1.96 V above the bottom level, and the effect for loosening the rock at the bottom corresponds to 0.9 V of this charge concentrated at the bottom.

The charge indicated Q_b shall consequently have the same effect as $Q_t(K, V)$ according to (2:16) at a bench height of $K = 1.96\ V$. As Q_t represents a charge with the theoretically most favorable distribution and Q_b a practical case with the part of the charge corresponding to $l_p(K - V)$ in (2:16) displaced a distance V down into the drill hole, Q_b can be assumed to be at least as large as Q_t for $K = 1.96\ V$, which is indicated $Q_b \geqslant Q_t(1.96\ V,\ V)$. The comparison between (2:17) and the relation (2:21) which has been verified by test blastings shows that $Q_b - Q_t(2V, V)$ with satisfactory accuracy. It implies that the charge required for the loosening of the column part is approximately independent of the charge being pushed slightly farther down into the hole. To simplify the calculations it is assumed that this displacement can amount to $0.5 V - V$. This part is left unloaded at the top of the hole, or for example is stemmed with dry sand. This has been done in all test blastings.

As the bottom charge (Q_b) is sufficient for breakage at a bench height of $K = 2V$, the charge which is required in addition to this when the bench height exceeds $2V$ is defined as column charge. The size of the column charge is $Q_p = Q_t - Q_b$ and we get

$$Q_p = 0.4 (K/V - 2)(a_2 V^2 + a_3 V^3) + \text{ terms with } V^4 \qquad (2:22)$$

This charge is distributed over a length $K - 2V$ (fig. 2:7). The charge per meter l_p is already given by (2:15) and the coefficients b_i can be determined from (2:22).

If the bench height is less than $2V$ the column charge (2:22) will be negative. As the bottom charge cannot be diminished, the upper part of the rock will in such a case be too strongly loaded. With $V = K$, for example, the bottom charge would rise right up to the free surface and give an entirely unpermissible throw in open air blasting. The bench height should be $K - 2V$ when (2:19) and (2:20) are applied. If the bench height is smaller than $2V$ the bottom charge as well as the burden must be reduced. The connection between bench height and burden is given in diagram 2:2.

A comparison with (2:15) and (2:20) shows with reference to $a_2 = 2.5 b_2$ and $a_3 = 2.5 b_3$ that the relation between bottom charge and column charge per meter is 2.7. So much larger a charge is consequently required at the bottom than for the rest of the hole when blasting with the smallest possible charge. For sloping holes (fig. 2:8), where the fixation at the bottom is lower, the value is reduced, and is 2.0 in the event of entirely free fixation at the bottom (fig. 2:3 b).

If the entire hole is charged with the same explosive the upper part of the hole will either be overcharged or else only about 40 % of the volume of the hole will be utilized. In order to utilize the volume of the hole without overloading the column a more concentrated explosive can be employed at the bottom. If a factor (s) is introduced for the weight strength of the explosive (2:20) and (2:22) will give respectively

$$l_b = \frac{1.1}{s_b} (a_2 V + a_3 V^2 + 0.7 a_4 V^3 + 0.7 b_4 V^3) \qquad (2:23)$$

$$l_p = \frac{0.4}{s_p} (a_2 V + a_3 V^2 + 2.5 b_4 V^3) \qquad (2:24)$$

In the following the reference value $s = 1.27$ for blasting gelatine is used to define the weight strength s. This gives the value $s = 0.87$ for AN-oil and $s = 1.0$ for a dynamite with 35 % NG.

2.5. Formula for straight bench with several drill holes

For rounds of several holes there are two additional variables, the spacing of the holes, E, and the number of holes n. If all the charges are ignited simultaneously the size of the charge per hole can be diminished when the number of holes is increased. In rock with normal cracks, charges which lie at greater distances than twice the burden in all probability do not noticeably cooperate. For rounds of four or more holes and $E = V$ the charge in every hole can be diminished to about 80 % of the charge according to (2:16) and (2:17). In the case of a constant burden the amount of rock blasted per drill hole is proportional to E/V. If the factor for the weight strength (s) of the explosive is introduced, as well as a factor for the fixation (f), where $f = 1$ when blasting in a straight bench with a fixed bottom and vertical holes, we obtain the formula

$$Q = f \cdot \frac{1}{s} \cdot \frac{E}{V} 0.8 \, Q_t \qquad (2:25)$$

2.6. Degree of fixation and slope of the holes

Compared to the bottom charge in vertical benching only 75% is required in blasting with a free bottom as in fig. 2:3 b. The corresponding figures are 85% and 90%, or $f = 0.85$ and $f = 0.90$ for a bench with a slope of 2:1 and 3:1. In these cases, as shown in fig. 2:8, one has greater angles of breakage at the bottom, which makes it easier to tear and loosen the rock.

With inclined holes one can thus have more rock per drilled meter. This being proportional to $V \cdot E$, the relation $VE \cdot f = \text{const.}$ holds good for a given diameter. With $f = 0.90$ the VE-value can be increased 11 % as compared with a vertical bench.

A further advantage with inclined holes is that there is less risk of back break at the top of the hole. Most essential, however, is that with inclined holes, say an inclination of 3:1, the bottom excavated can at a slope of 1:3 descend till it reaches the surface of breakage originally intended, if this for one reason or another should not be reached for one previous row in the round.

41

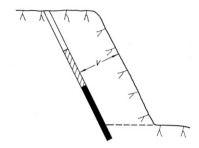

Fɪɢ. 2:8. Inclined holes give a favorable angle between the holes and the intended bottom.

In the table below the values for the fixation factor f are given for some actual cases. (The influence of the fixation in stoping towards a parallel hole cut, where the hole in its length has a possible angle of breakage of less than 90° is given in chapter 7, table 7:2.)

For the relief holes that have to be ignited one by one, the fixation is greater than in a bench row. The essential difference is that in bench blasting the line of loosening must pass the holes and thus give a defined face for the subsequent row to tear against, whereas for such relief holes in a tunnel that are fired one by one, the angle of breakage is not strictly defined. The subsequent holes can then have a higher resistance to the blasting than in a pattern where an angle of breakage of 90° is assumed.

Just as a considerably greater loading density is required at the bottom of a hole than in the upper part, the more restrained corner holes in a row call for more explosive than the central holes. In simultaneous ignition it is a matter of an increase of 50% for the side holes. This arrangement is, however, no longer required. In short delay blasting extra charging of the edge holes can be avoided if they and the sequence of firing can be arranged so that they are no longer wedged at the moment of detonation.

In tunnel blasting with parallel holes and when the holes are blasted one at a time towards a central opening the potential angle of breakage is less than 90° for the first holes, and one has to be prepared for a wedging at the sides which substantially increases the charge required (see 6.2).

TABLE 2:2. *Fixation factor (f) at the bottom.*

Incline	∞:1	3:1	2:1	Free bottom
Bench, one row of holes	1	0.9	0.85	0.75
Relievers, one row of holes	1	(0.9)	(0.85)	0.75
Downward stoping one row of holes	0.8	0.7	0.65	0.6
Relievers, single holes	1.45	1.3	1.25	1.1

2.7. The rock constant c

The rock is characterized in blasting by 3 constants a_2, a_3 and a_4, and, in addition, by b_2, b_3 and b_4 which can be determined if the a-constants are known. It will be seen from table 2:1 that for these the middle term (a_3) for burdens between 1.0 and 20 m (3 ft and 60 ft) dominates. It is indicated in the following as the factor to represent the influence of the rock and is defined c_0 when it refers to a limit charge. By c is indicated the value which includes a technical margin for a satisfactory breakage, for example $c = 1.2\,c_0$. This factor has been found in trial blastings to be $c_0 = 0.17$ kg/m³ in blasting in brittle crystalline granite, and $c_0 = 0.28 - 0.35$ kg/m³ in practically all other rock species hitherto examined, without a pronounced direction of cracks and fissures. In blasting single holes at right angles to the stratification in slaty rock species, for example, one can find values corresponding to $c = 1.0$ kg/m³ and similar results for single holes in other kinds of rock if the cleaving direction obstructs the loosening of the rock.

In blasting a single row of holes and for burdens 1.0–20 m, and in otherwise normal rock the two other constants, a_2 and a_4 are responsible for only about 20% of the share. To simplify the practical use of the calculations the values will be fixed in the following at $a_2 = 0.07$ kg/m² and $a_4 = 0.004$ kg/m⁴. An error in these values only slightly affects the final figure for the charge.

This means that it can nowadays be considered unnecessary in most blasting operations to make a trial blasting. The job can be planned directly with $c = 0.4$ kg/m³. With this value as a starting point calculations have been made for various types of blasting jobs and applied in all parts of the world with satisfactory results. In some of the instances, where considerably higher values have been reported, it is possible that the reason is the fact that the local tendency for cracking had rendered breakage for single holes difficult. When this has been the case it may be regarded as a geometric or fixation effect. It varies with the direction of the blast in one and the same rock and is consequently not a constant representative of the rock.

The calculations can be applied for any c-value and there is no objection in principle to accepting also considerably higher values. But in practical application they obviously don't occur very frequently.

An important case in practice is blasting in sedimentary rock with a stratification more or less perpendicular to the direction of the holes. Here too $c = 0.4$ is an ordinary value. However, as the stratification facilitates loosening at the bottom a full bottom charge is not required but one with a concentration between what is demanded for a column charge (l_p) and

43

a bottom charge ($2.7 l_p$), though closer to the last-mentioned value as there is no question here of open crevices. For such eventualities the burden and distance of holes can be increased by 10–15 % each, compared with normal rock.

2.8. Diameter of drill holes, degree of packing

The diameter of the hole does not in itself exercise any influence on the quantity of the charge required for breakage. In the case of a larger or smaller diameter the same breaking force is obtained with the same quantity if the height of the charge does not exceed $0.3 V$ from the bottom level.

In the same way it holds good for the hole in its entirety that with a moderate variation in the diameter of the hole, but with unchanged concentration of the charge per meter the breaking force remains unaltered. This appears to apply as long as the density of the charge does not fall much below 1.0 kg/dm³. (2.0 kg/m or 1.5 lb/ft in a 2″ hole). The size of the charge per m at the bottom of the holes consequently alone decides how large a burden can be loosened.

When we speak of the drill hole diameter we usually mean the diameter (d) of the drill bit. This is not entirely correct, but we will adopt the same notation, because, among other things, the actual hole diameter is difficult to determine in each individual case.

In consequence, the degree of packing is defined as the quantity of charge in kg per dm³ of the nominal volume of the hole. As this may be 5–15% less than the true volume, the degree of packing is somewhat larger than the real density of the explosive in the hole. We have the relation

$$l = P(d/36)^2 \qquad\qquad (2:26)$$

in which l indicates the quantity of the charge per m of the hole. In loading with a tamping pole $P = 1.0$–1.4, with the pneumatic cartridge loader $P = 1.3$–1.6 kg/dm³. For a diameter of the drill bit $d = 36$ mm the concentration of the charge in kg/m equals the degree of packing, $l = P$.

As the charge quantity per m at the bottom of the drill hole must be 2.0–2.7 times as large as in the column part it is often difficult to utilize the volume of the hole in the column which in bench blasting often is overcharged. To make better use of the volume of the hole the degree of packing can be increased at the bottom, for example by employing a loader and by using a bottom explosive of greater bulk strength. Sloping holes can be employed to increase the effect of the holes. In a slope of 2:1 the volume of rock excavated per hole can be increased by 10–15 %. The overcharge in the column is hereby reduced in the same proportion.

2.9. Hole spacing

With a given diameter of the drill at the bottom the product of burden and spacing VE is also given, thus $VE=$ constant. Ordinary practice has been to let the spacing be of the same size as the burden, or somewhat above, $E \leqslant 1.3\ V$. The spacing has a great influence on the fragmentation and on the character of the face left behind in the remaining rock. Small distances between the holes as compared to the burden will tend to create large blocs and an even face behind. With "wide spacing", which is a new concept in benching where $2\ V < E < 8\ V$ an essential increase in the desintegration can be achieved without increasing the specific charge.

2.10. Maximum burden

The maximum burden depends on the charge per meter at the bottom of the hole and its height, the weight strength of the explosive, the rock constant, the degree of fixation and the spacing of the hole. The height of the bottom charge is for ordinary benches assumed to be $1.3\ V$, its concentration is determined by the diameter of the drill bit d in the lower part of the hole. The charge per m for $d=50$ mm (2 in) is 1.8 kg/m (1.2 lb/ft) in loading with ammonium nitrate in the form of prills; 2.2 kg/m (1.5 lb/ft) with crystalline AN and oil or other powder explosives in bulk; 2.4 kg/m (1.6 lb/ft) with dynamite containing 25–50 % nitroglycerin and 2.8 kg/m (1.9 lb/ft) with AN–TNT slurries. The charge per meter consequently varies considerably with the explosive. In addition it varies with the temperature at the loading, which for example in tunneling, can effect the rate of advance (see 7.6).

The concentration of the charge required for loosening at the bottom is given in (2:20). If we include, as in (2:25), the degree of fixation f, weight strength of the explosive at the bottom s_b and spacing of the holes E we get

$$l_b = 0.88 \frac{f}{s_b} \cdot \frac{E}{V} (a_2\ V + a_3\ V^2 + \text{terms with } V^3).\qquad (2:27)$$

The bottom of the hole can accommodate per m according to (2:26)

$$l_b = P(d_b/36)^2 \qquad (2:28)$$

in which d_o gives the diameter of the drill bit at the bottom of the hole. These two l_b-values are equal when the burden is the greatest possible. From this we get the relation deciding the maximum burden

$$(0.070/V + c + \text{terms with } V) = P \frac{(d_b/36)^2 \cdot s_b}{V^2 \cdot E/V \cdot f \cdot 0.88} \qquad (2:29)$$

45

Here a_3 is denoted c and $a_2 = 0.070$ according to (2:12). The three terms on the left hand side indicate the charge per m³ and shall together be denoted $\bar{c}$. Their magnitude and mutual relations are seen in tab. 2:1. With greater burdens than 20 m (66 ft) experiments are required to determine the magnitude of the third "terms". For normal burdens the value $a_4 V = 0.004 V \text{ kg/m}^3$ should be satisfactory.

The middle term corresponds to the law of conformity. It would alone give the relation $d/V = \text{const.}$, that is to say that the burden is in direct proportion to the diameter if we have the same explosive and degree of packing and a constant relation between E and V, for example $E = 1.25 V$. As is seen in table 2:1 the sum of two variable terms has a minimum 0.0335 kg/m³ for $V = 4$ m (13 ft) in the expression Q/V^3. As it is a question of a continuous function it is in the vicinity of the extreme value approximately constant for small variations of V. This is evident in fig. 2:2. In the region $V = 1.4$–15 m (5–50 ft) it can with an accuracy of ± 0.016 kg/m³ be indicated that the left hand side of (2:29) is constant

$$\bar{c} = 0.070/V + c + 0.004\,V \approx c + 0.05 \quad V = 1.4\text{–}15 \text{ m} \qquad (2:30\,\text{a})$$

For smaller burdens than 1.4 m (5 ft) the term $0.004\,V$ can be disregarded, which gives a maximum error of 0.005 kg/m³. The first term must here be included at its full value. With a maximum deviation of 0.005 kg/m³ we can put

$$\bar{c} = 0.070/V + c + 0.004\,V \approx 0.070/V + c \quad V \leqslant 1.4 \text{ m} \qquad (2:30\,\text{b})$$

$$V = d_b/33 \sqrt{\frac{P \cdot s}{\bar{c} \cdot f \cdot (E/V)}} \qquad (2:31)$$

in which the value $\bar{c}$ for $V > 1.4$ m is therefore approximately constant. For example with $s_b = 1$, $f = 1$, $E = 1.25\,V$, $P = 1.25$ and $c = 0.40$ ($\bar{c} = 0.45$), which are rather ordinary values, we get $V = d_b/22$, that is to say for $d_b = 33$ mm we obtain $V = 1.5$ m.

In mining it is not unusual to blast between two free surfaces so that there will be no need for any special bottom charge. The maximum is then given by (2:24) which shows that the requirement of the concentration of the charge (Q) is reduced in proportion $0.4/1.1 = 1/2.7$ compared to an ordinary bottom charge l_b with a fixed burden. In eq. (2:29) the value for the diameter d_b can be substituted by

$$d_p = d_b/\sqrt{2.7} \qquad (2:32)$$

for loosening the same burden. The maximum burden is

$$V = d_p/20 \sqrt{\frac{P \cdot s_p}{\bar{c} \cdot f \cdot (E/V)}} \qquad (2:33)$$

when the charge is not expected to give any loosening at the bottom.

46

$t=0$

$t=0.5\,s$

$t=1.0\,s$

FIG. 2:9. Throw in blasting a row of holes with an over-charge of 0.21 kg/m³ (0.013 lb/cu. ft), 0, 0.5 and 1 sec. after ignition. Instantaneous ignition. 4 holes, $V = 1.0$ m (3.3 ft). Slope of the holes 2:1 (acc. to Forsberg and Gustavsson).

TABLE 2:3. *Length of throw of the center of gravity as a functon of the excess charge.*

Excess charge in	kg/m³	0	0.10	0.20	0.30	0.40
	lb/cu yd	0	0.17	0.33	0.5	0.67
The rock is moved forward a distance,	m	0	6	12	18	24
	ft	0	20	40	60	80

2.11. Throw

The throw, which affects the main part of the rock, increases with an increased charge. In addition to this stones or minor parts of the rock may be expected to be thrown 5–10 times as far and in unfavorable circumstances still farther. This occurrence will be called *scattering.* The throw can nowadays be fully controlled and estimated in advance, but not, on the other hand, the scattering. It has other affinities and does not for example necessarily increase with the charge.

First some simple rough relations will be given to serve as a basis in estimating the results of a round.

Fundamental investigations of the throw in blasting have been carried out by Forsberg and Gustavsson. They have shown that the total throwing

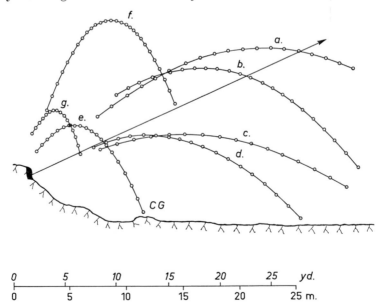

FIG. 2:10. Study of the movement of individual pieces of the rock in fig. 2:8. The position of a piece is indicated every 0.1 sec (acc. to Forsberg and Gustavsson). The arrow indicates the direction perpendicular to the holes.

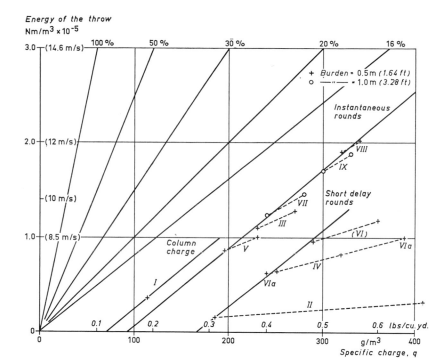

Fig. 2:11. Energy of the throw (acc. to Forsberg and Gustavsson) as a function of the specific charge in instantaneous and short delay blasting.

energy (e_{throw}) is in direct proportion to the excess charge ($q - q_1$) where q_1 indicates the charge in kg/m³ for a limit charge that does not give any throw

$$e_{\text{throw}} = \text{const.} \ (q - q_1) \tag{2:34}$$

The calculation of the throw must thus be based on that of the limit charge. Fig. 2:9 shows the throw of an instantaneous round with 0.21 kg/m³ excess charge. In fig. 2:10 throw paths have been indicated for individual stones and marking for every 0.1 second. It is seen that the observation of the hits obtained on the ground, or from where the weight of the rock falls, cannot be taken as a direct basis for calculation of the throwing energy as this is divided up among different equations. The center of gravity of the loosened rock has been moved forward 13 m. ⅔ of the rock has been thrown less than 15 m.

For a series of such rounds $I - IX$ with different ways of ignition and excess charge the total throwing energy has been indicated as a function of the charge per m³ (fig. 2:11). The energy is given in Newton meters (Nm) which is approximately 0.1 kpm. In excavating the rock the main part has been thrown out in the usual manner, while some neighboring parts which were also torn out were only moved some meters away. This has meant that the rock excavated in the test rounds had a volume which could not be exactly foreseen. This is only to be expected when blasting small rounds, as variations in the cracks and frequency of crevices in the rock then have a comparatively large influence. The pieces of the rock which have been loosened in this way have obviously acquired the energy for their movement from the other part of the round. For every round there has been set out in fig. 2:11 the charge per m³ for the expected volume (the point to the right) as well as the point corresponding to the actual breakage obtained. For these last-mentioned values there is a distinct, clearly defined linear connection between the specific charge (kg/m³) and the energy of the throw, which is evident from the straight lines that have been inserted for the two groups "Instantaneous rounds" and "Short delay rounds".

The slope of the lines can be compared with the curves departing from origo and which show the throw energy if 100–16 % of the total power of the explosive should be converted into throw. The comparison shows that about 16 % of the energy of the excess charge is transmitted to throw energy irrespective of the size of the burden and for instantaneous as well as short delay rounds, that is to say as far as can be judged also regardless of the manner of ignition. This may be regarded as an unexpectedly high "efficiency".

The limit charge, on the other hand, is affected by the way the round is ignited—it is strongly emphasized when the cracking tendency of the rock is parallel to the row of holes. Then the loosening is facilitated in instantaneous rounds, but not in short delay ones where the loosening occurs hole by hole.

These facts give us a very instructive illustration of the difference in the mechanism of breakage for instantaneous and short-delay rounds. In short-delay blasting loosening of the rock entails more work because of increased tearing in the bulk of rock to be excavated. This gives better fragmentation and less throw. With $q = 0.20$ kg/m³ in fig. 2:11, for example, a throw energy of $0.86 \cdot 10^5$ Nm/m³ has been obtained in instantaneous blasting. In short-delay blasting the corresponding value is only $0.16 \cdot 10^5$ Nm/m³. The difference, $0.70 \cdot 10^5$ Nm/m³ has been consumed in tearing the burden into smaller pieces.

It is interesting to note that when the loosening has actually taken place an increase in charge gives an equally large increase of the throw energy in short-delay as well as in instantaneous blasting. In multiple row blasting with short-delays those stones in the rear lines which have had the greatest speed will catch up with the slower parts of the front rows and be held up.

Such a collision reduces the total throw energy insofar as it concerns a non-elastic shock, or if part of the translation energy at the shock causes rotation. Trials which have been carried out with so-called collision blasting, where different sections of the rocks are blasted out against each other, have shown that this offers a possibility to take advantage of the throw energy in order to get better fragmentation. The collisions which occur in a multiple row blast contribute in the same way to improve the fragmentation. Much more important than this effect is the fact that the rock as a whole acquires considerably less dispersion in the throw velocity. As it is the greatest length of throw that in many cases limits the specific charge the charge can therefore in multiple-row blasting be increased without extending the front limit of the round.

The "scattering" of small stones seems mainly to be due to their having been torn off by the gases which pass out through the holes or cracks at a high velocity if the rock only cracks in some few large blocks. The flow of the gases can be concentrated more readily in a single direction if the charge is not capable of loosening and breaking up the burden. This may be due to too great fixation of the rock, to faulty drilling of the holes so that the burden proves to be too large, or to failure in ignition in these holes. It is even claimed that too small column charges may for this reason be a cause of scattering.

2.12. Swelling

In multiple-row rounds, or in rounds where the burden is loaded with previously blasted rock a larger charge than the limit charge is required. There must be a certain amount of swelling to complete the loosening. With large over-charges, which sometimes are needed in underground blasting, a volume of 20 % for the swelling may be sufficient. In open-air blastings, or where there can be greater freedom of movement for the rock on the whole and with specific charges of less than 0.6 kg/m³ (1.0 lbs/cu.yd) the swelling is 40–50 %. Ordinarily there is reason to count on the higher value.

In blasting according to fig. 2:12a the bottom charge must be large enough also to remove the excavated rock which hampers the free face of the front of the round. For a multiple-row round with a vertical arrangement of holes (fig. 2:12b) the overcharge at the bottom of a row must be so large that the center of gravity of the rock has moved a distance of about 20 % of the distance (V) between the rows when the subsequent row detonates. At $V=1.25$ m and delay of $\tau=20$ ms the bottom part of the bench shall consequently have had a mean velocity of $v=0.25$ m/0.020 s $= 12.5$ m/s. According to fig. 2:11 the corresponding overcharge is 0.25 kg/m³ for a

a.

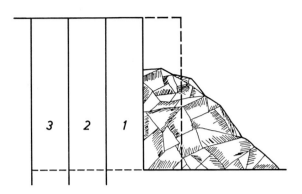

3 | 2 | 1

b.

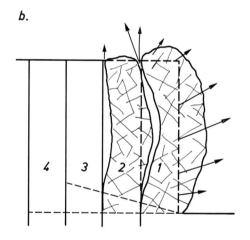

4 | 3 | 2 | 1

c.

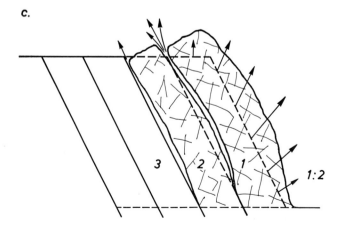

3 | 2 | 1 | 1:2

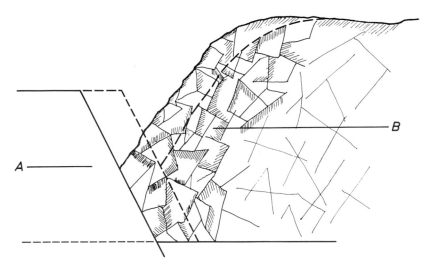

FIG. 2:13. In continuous blasting without mucking the center of gravity of the rock must be heaved from A to B.

free throw if all sections of the rock move forward at the same speed. As this is not the case a larger charge is needed, so that the center of gravity shall acquire the speed of 12.5 m/s. In addition to this the rock masses in the bottom part of the bench are soon retarded by friction against the bottom as they gradually fall downwards under the influence of the force of gravity.

For a sloping bench, for example, where the initial direction of the throw is 1:2 at the bottom (fig. 2:12 c), conditions differ entirely from those at the beginning of the throw and for about half a second there is an upward movement so that the friction of the rock masses with the ground is insignificant.

As the blasting proceeds row after row the pile of rock in front of the round will gradually increase and lie closer to the burden which is about to be blasted. Also when blasting bottom benches in a tunnel, a canal, in road blastings or such like where the swelling cannot take place laterally, we almost arrive at a state of continuation as in fig. 2:13. The center of gravity of the excavated rock must be raised c. 45 %, which calls for extra

FIG. 2:12. In blasting of multiple-row rounds the rock in front of the back row is obstructed in its forward movement. This calls for an overcharge to attain sufficient loosening at the bottom. a) Vertical bench with delay greater than 0.5 sec, after blasting of the first row. b) and c) short delay interval of 25 ms between the rows.

energy (e_{sw}) per m³. If given in tonm/m³ for a rock with a density of $\varrho =$ 2.8 ton/m³ (g/cm³) we have

$$e_{sw} = 0.64\,K \qquad (2:35)$$

For an explosive with a specific energy of 500 tonm/kg (5.10⁶ Nm/kg) and which has an efficiency of 16 % for the throw, 80 tonm/kg will be available for moving the rock. With a direction of throw of 1:2 20 % of the energy of the throw is associated with the lifting component. The additional charge which is needed to provide sufficient swelling is of the order of

$$q_{sw} = 0.04\,K \qquad (2:36)$$

The charge per m³ for the bottom part of a bench is $q_0 = 0.36$ and the total specific charge which is required when allowing for the swelling is

$$q_b = 0.36 + 0.04\,K \qquad (2:37)$$

This can be attained by increasing the diameter of the holes from d to d_1

$$d_1/d = \sqrt{q_b/q_0} \qquad (2:38)$$

$$d_1 \approx d(1 + 0.05\,K) \qquad (2:39)$$

Instead of increasing the diameter of the holes the burden will usually be diminished in proportion to d/d_1, and we get

$$V_1 = V/(1 + 0.05\,K) \qquad (2:40)$$

With a direction of throw of 1:3 only 10% of the energy is used in the lifting component and twice as large an overcharge is needed. The swelling can thus under common conditions demand a larger quantity of explosive than just for loosening the rock and in the case of multiple-row rounds it is in all circumstances considerable. If we observe a case with uniform enlargement the share that corresponds to the required swelling is increased in proportion to V^4. The law of conformity cannot, of course, be applied.

Example. Blasting of a bottom bench in a tunnel with $K = 8$ m, $V = 2$ m. A limit charge for a single row of holes is $q \simeq 0.20$ kg/m³. If we assume for the sake of simplicity that the throw takes place at right angles to the drill holes, the swelling with holes sloping 2:1 needs according to (2:36) a charge of 0.32 kg/m³ and demands in all $q = 0.52$ kg/m³.

With a slope of the holes of 3:1 the value is $q = 0.84$ kg/m³ with the same assumptions in other respects. Blastings which have been made have confirmed that the value $q = 0.52$ is required for a slope of the holes of 2:1. It was established with an inclination of holes of 3:1 that $q = 0.65 - 0.75$ kg/m³ was not sufficient to give complete loosening at the bottom in continuous blasting without unloading the broken material after each round.

In the case of high benches the motion of the burden close to the bottom probably takes place at right angles to the holes. The gases from the explosive, which are partly forced through the loosened rock, partly upwards along the holes and adjacent cracks, give the upper parts of the rock an upward motion. They consequently acquire a higher elevation dependent on the placing of the charges; this is suggested in fig. 2:12 b and c by the inserted arrows. The same applies near the bottom at lower benches where the rock is not debarred from moving upwards. In such cases there may be a direction of throw of 1:2 and the relation (2:36) can thus be applied already at an inclination of 3:1 or may be 4:1.

2.13. Fragmentation

When blasting a single hole the required bottom charge per m^3 is $q = Q/V^3$, when blasting a row of holes with a spacing of $E = 1.25\,V$, $q = Q/1.25\,V^3$. The discussion will here be limited to conditions applying to normal rock with $c_0 = 0.350$ kg/m^3 referring to limit charge and $c = 0.400$ kg/m^3 when an ordinary (and minimum) technical margin is included. With q_0 in g/m^3

$$q_0 = 0.8\,Q_0/V^3 = 320 + 56/V + 3.2\,V \qquad (2:41)$$

The rock excavated by a single drill hole with a bench height equal to the burden $(K = V)$ has an approximate volume of V^3. In blasting a 0.5 m burden with a limit charge, the greatest part of the volume may be in a single boulder. With increasing dimensions it is more difficult to get the increasing volume to hold together in a single boulder. If we assume that the share of the largest boulder of the total body of rock decreases in inverse proportion to the burden, when this exceeds $V = 0.5$ m, this should provide a comparatively satisfactory approximation for practical application. If the average length of the side of the largest boulders is indicated by L, and the final volume consequently by L^3 we obtain

$$\begin{aligned} L^3 &= V^2/2 & V &\geqslant 0.5\ \text{m} \\ L &= V & V &\leqslant 0.5\ \text{m} \end{aligned} \qquad (2:42)$$

If this is inserted in (2:41) we get the relation between q_0 and L

$$\begin{aligned} q_0 &= 320 + 40/L^{3/2} + 4.5\,L^{3/2} & V &\geqslant 0.5\ \text{m} \\ q_0 &= 320 + 56/L + 3.2\,L & V &\leqslant 0.5\ \text{m} \end{aligned} \qquad (2:43)$$

In bench blasting and stoping the bench height is usually 3–5 times as large as the burden. The specific charge required for breakage at $K = V$ is indicated by (2:41). For $K = 3\,V$, which may here represent a normal case, the quantity of charge per m³ for the bottom part is q, for the column part $0.4q$. This gives the average value $q_1 = \frac{1}{3}(q + 2 \cdot 0.4q) = 0.6q$.

We obtain

$$q_1 = 190 + 24/L^{3/2} + 2.7\,L^{3/2} \qquad V \geqslant 0.5$$
$$q_1 = 190 + 34/L + 1.9L \qquad\quad\ V \leqslant 0.5 \qquad\qquad (2{:}44)$$

where it has been assumed that with such a bench $K = 3V$ we obtain the same average size of boulder as in the preceding instances. The said relation is given in Diagram 2:5 (p. 67) by the lower unbroken curve.

In blasting multiple row rounds with short delays and a burden of $V = 1.1 \pm 0.2$ m the following relation between the specific charge and the largest boulder has been obtained.

TABLE 2:4. *Fragmentation in short-delay rounds with* $V \approx 1.1$ *m.*

Spec. charge, q kg/m³	0.20	0.24	0.28	0.33	0.40
Largest boulder, L³ in m³	1	1/2	1/4	1/8	1/16

If these values are inserted in Diagram 2:5 it is seen that with $q = 0.20$ we will hardly get a sufficient breaking force as we are below the unbroken curve. The fragmentation curve (dashes) should start from the point on the unbroken limit curve, where $V = 1.1$ m, according to (2:44), corresponds to $L = 0.84$ m. In the same way it can be indicated where the fragmentation curves begin when blasting with other burdens $V = 0.25$–10 m.

The fragmentation is obviously affected in a high degree by the nature of the rock. In sedimentary species of rock a smaller fragmentation is obtained than that given here. In granite and other primitive rock formations occasional boulders can be larger if there are large crevices or cracks which shut off parts of the rock from strains acting in the rest of the rock.

Fragmentation in rock blasting is one of the most important of the remaining problems in this sphere of technics. It is not merely a question of the relation between charge and fragmentation, but also the influence of fragmentation on the digging capacity and the economy, which plays an important part in rock blasting operations. The subject matter which has been discussed in this section is intended to serve as a basis for further investigations, but also and above all to give in a condensed form a record of the connections which have been determined in experiments and are required in practical application.

2.14. Rock blasting with nuclear explosives

When blasting with nuclear explosives the throwing or swelling component becomes responsible for a great part of the energy. The size of k_4 in formula

(2:11) depends on geometrical conditions and on how much heaving of the rock is necessary to get the desired result, whatever this may be, whether forming a crater or just loosening the rock.

In a test series with equal results and under equal geometrical and rock conditions the k_4 is a constant. The influence of k_2 is so small that it can be disregarded in this connection. The resulting formula for the charge as a function of the depth (V) is

$$Q_n = k_3 V^3 + k_4 V^4 \qquad (2:45)$$

and can be satisfactorily described in a certain region of V by the approximation

$$Q_a = k_a \cdot V^{3+\gamma} \qquad (2:46)$$

where $0 < \gamma < 1$ and γ for small V-values is close to 0 and at extremely great depths approaches 1. The volume of broken rock may be denoted kV^3, where k is a constant in uniform scaling. The ratio $\varkappa = k_4/k_3$, is of the order of 0.01 per meter in bench blasting in hard rock. The figure is higher if the bench has to move upwards when being loosened. It is lower in blasting land mines than in benching. These two factors compensate each other more or less and it is reasonable to assume that $\varkappa$ is of the order of 0.01. The specific charge is

$$q_n = k_3/k[1 + \varkappa V] \qquad (2:47)$$

and the approximation

$$q_a = k_a/k \, V^\gamma \qquad (2:48)$$

The approximation gives an error ε in the specific charge

$$\varepsilon = q_n - q_a \qquad (2:49)$$

In a region close to a value V_0 the best approximation is obtained if

$$\begin{cases} \varepsilon_0 = 0 \\ (d\varepsilon/dV)_0 = 0 \end{cases} \qquad (2:50)$$

from which a simple relation between γ and V_0 is derived

$$\gamma = \frac{\varkappa}{\varkappa + 1/V_0} \qquad (2:51)$$

When $\gamma = 0.5$ this gives $\varkappa = 1/V_0$

Example. If a series of tests at depths of the order $V_0 = 100$ m has given $\gamma = 0.5$ we have $\varkappa = 0.01$ m^{-1}. With $\gamma = 0.33$ the corresponding value is $\varkappa = 0.005$ m^{-1} and for $\gamma = 0.67$ we get $\varkappa = 0.02$ m^{-1}.

When tests are carried out in a large range, $V_1 < V < V_2$, a good approximation is obtained with a γ=value denoted γ_0 between γ_1 and γ_2, corresponding to (2:51) with V_0 replaced by V_1 or V_2. In this case the constant k_a/k can be increased above the value that makes $\varepsilon_0 = 0$ so as to make $\varepsilon = 0$ at two other V-values in the interval $V_1 - V_2$. This will give lower values for the maximum error, $\varepsilon_{\max}$.

Example. Tests with depths between 33 m and 150 m will give $\gamma_1 = 0.25$ and $\gamma_2 = 0.6$ at the ends of the interval if $\varkappa = 0.01$ m^{-1}. From the investigation it will be found that the exponent in the approximate formula is in the vicinity of 3.4 ($\gamma = 0.4$).

However, the use of the approximation is not to be recommended. It is to prefer to determine the actual coefficients k_3 and k_4 or the $\varkappa$-value. Then the exact formula is given by (2:45) and can be used for any depth.

2.15. Summary of formulas for bench blasting and stoping

Single hole. $s = 1.0$, $f = 0.9$ (slope of the holes 3:1–2:1)

Notation $\bar{c} = (0.07/V + c + 2.5 b_4 V)$

$$K' = 1.05 \, K \qquad \text{for a slope } 3:1$$
$$K' = 1.12 \, K \qquad \text{for a slope } 2:1$$

Concentration of bottom charge:

$$l_b = \bar{c} \, V^2$$

Concentration of column charge:

$$l_p = 0.4 \, \bar{c} \, V^2$$

Bottom charge when $h \leqslant 0.3 \, V$ (concentrated charge)

$$Q_0 = 0.9 \, \bar{c} \, V^3$$

Total charge:

$$Q_t = Q_0 + (K' - V) l_p$$
$$= (0.5 + 0.4 \, K'/V) \bar{c} \, V^3$$

A row of several holes. $f = 0.9$

Total charge per hole:

$$Q = \frac{f}{s} \cdot E/V \cdot 0.8 \cdot Q$$

Specific charge: $q = Q/K'EV$

Maximum burden (in m) for a drill diameter d_b mm at bottom

$$V = d_b/33 \sqrt{\frac{P \cdot s}{\bar{c}f(E/V)}}$$

$$\bar{c} = c + 0.05 \qquad \text{for} \qquad V = 1.4 - 15 \text{ m}$$

$$\bar{c} = c + 0.07/V \quad \text{for} \qquad V < 1.4 \text{ m}$$

Height of bottom charge: $\qquad h_b = 1.3 \ V$

Bottom charge, equally distributed:

$$Q_b = \frac{f}{s} \cdot E/V \cdot 0.8 \cdot (1.3 \, \bar{c}V^3)$$

2.16. The formulas expressed in diagrams

Maximum Burden

The maximum burden (V) depends on the diameter (d) of the drill hole at the bottom, on the degree of packing (P), weight strength of the explosive in the bottom charge (s), degree of fixation (f) and rock factor (c). The inclination of the holes should ordinarily be $3:1-2:1$. The charge extending from $-0.3\,V$ to $+V$ above the bottom level contributes to the breakage at the bottom. A length of V m (ft) above the charge in the hole can be unloaded and, if possible, stemmed with dry sand.

If the length of the hole is less than $2.3\,V$ (e.g. $K \leqslant 1.8\,V$ for a slope of $2:1$) the height of the bottom charge must be diminished and consequently also the burden dependent on the value of K. There are two principal cases: ordinary benches $K \geqslant 1.8\,V$, and low benches $K < 1.8\,V$.

In multiple-row blasting the burden is the distance between consecutive rows, the spacing (E) is the distance between the holes in a row. The spacing in the diagrams is generally assumed to be $E = 1.0-1.25\,V$.

The values for maximum burden and spacing thus obtained should not be exceeded at the intended bottom where satisfactory breakage is desired. As there is always a certain drill-hole deviation this requires the values used in the drilling pattern for V and E to be less than those given in the diagrams. In the following Chapter 3 such questions as arise in practical application of the calculations are discussed.

Ordinary benches, $K > 1.8\,V$ (Diagram 2:1)

To avoid too many variables the diagram is made for $f = 1$ (vertical holes) and $s = 1.0$ (as for a dynamite with 35 % NG). With the slope generally recommended, $3:1-2:1$, the burden and spacing can each be 5 % larger.

Example 2:1a. Assumptions: Bench height $K = 2.5$ m (8 ft), diameter of the drill bit at the bottom $d = 32$ mm (1.28 in). Degree of packing $P = 1.25$. Ordinary value for the rock factor $c = 0.4$.

Calculation: Start from the left-hand horizontal scale $d = 32$ mm and proceed vertically upwards to the line $P = 1.25$, turn right to the vertical scale "Concentration of the charge at bottom" which gives $l_b = 0.97$ kg/m (0.64 lb/ft).

Continue to the line for rock factor $c = 0.4$ and turn down to the right-hand horizontal scale for the maximum burden. The value $V = 1.42$ m (4.7 ft) is obtained.

Example 2:1b. Assumptions: Bench height $K = 10$ m (33 ft), drill hole diameter $d = 100$ mm (4 in). Concentration of the charge at the bottom is found to be $l_b = 6$ kg/m (4 lbs/ft) in practical application.

Calculation: Start from the value $l_b = 6$ kg/m on the central vertical scale and proceed horizontally to the line for rock factor $c = 0.4$. Turn down to the right-hand scale for the maximum burden, which gives $V = 3.55$ m (10.2 ft).

To get the degree of packing, start from the value $d = 100$ mm (4 in) and proceed upwards to a horizontal line passing through the value $l_b = 6$ kg (4 lb/ft). The two lines intersect at a point a on the line $P = 0.8$.

Low Benches, $K < 1.8 V$ (Diagram 2:2)

For a slope 2:1, a rock factor $c = 0.4$ (normal value) and various concentrations of the charge the Diagram 2:2 shows how the maximum burden increases with the bench height up to $K = 1.8 V$. For greater heights the burden no longer increases with K and is then given by Diagram 2:1.

Example 2:2a. Assumptions: Bench height $K = 0.7$ m (2.3 ft), drill hole diameter $d = 33$ mm (1.32 in), concentration of charge $l_b = 1.1$ kg/m (0.75 lb/ft).

Calculation: Start from $K = 0.7$ m (2.3 ft) to the right, turn down from a point between the curves $l_b = 1.0$ (0.7) and 1.6 kg/m (1.0 lb/ft). The line meets the horizontal scale at $V = 0.8$ m (2.7 ft).

Example 2:2b. With a bench height $K = 2$ m (6.7 ft) and $d = 40$ mm (1.6 in), and the charge concentration $l_b = 1.6$ kg/m (1.05 lb/ft) the burden $V = 1.55$ m (5 ft) is obtained.

Minimum bottom charge and total charge per hole (Diagram 2:3)

For burdens between 0.8 and 8 m (3 and 27 ft) and various rock factors the Diagram 2:3 gives a minimum bottom charge (Q_b), e.g. the total charge up to $1.3 V$ from the bottom of the hole. The reading of the diagram is then taken from the actual value for the burden (horizontal scale) up to the rock factor c in question and then turning *left* to the left hand vertical scale for Q_b. When, for instance, the value $Q_b = 1.65$ kg (3.1 lbs) is obtained for a burden $V = 1.4$ m (4.7 ft) it means that with this burden at the bottom the charge Q_b is required with the further condition that this charge shall not extend beyond $1.3 V$ m (ft) from the bottom of the hole.

The bottom charge Q_b can also be obtained from Diagram 2:1 read in inverse direction from the burden (right-hand horizontal scale) to the concentration l_b of the charge. The extension of the bottom charge being 1.3 V gives by the simple relation $Q_b = 1.3 l_b V$ the amount of the bottom charge.

For the total charge Q_t the reading is made proceeding from the burden on the horizontal scale, up to the value for the rock factor c and then turning *right* to the dotted line for $K'/V = 2$. If a higher value for this relation is concerned, e.g. $K'/V = 4$, one has to deviate upwards from the dotted line to the line for $K'/4 = 4$ before again turning to the right and to the right-hand vertical scale that gives the total charge.

Example 2:3a. Assumptions: The burden is $V = 5.0$ m (16.5 ft), the bench height $K' = 20.0$ m (66 ft) the rock factor $c = 0.4$.

Calculation $K'/V = 20/5 = 4$ (66/16.5 = 4). Start from $V = 5.0$ m (16.5 ft) on the horizontal scale up to the line $c = 0.4$ (a), turn right to the dotted line (b), and proceed from the intersection point (b) vertically up to the line $K'/V = 4$ at (c) from which one has to continue to the right where the right hand scale gives the value $Q_t = 110$ kg (240 lb).

Compare the value Q_b received if one continues to the right, passing the line $K'/V = 2$ (and the point (b)) without deviation. $Q_b = 70$ kg (155 lbs).

Example 2:3b. Assumptions: The burden is $V = 1.4$ m (4.6 ft) the bench height $K = 1.3$ m (4.6 ft), the rock factor $c = 0.4$. $K' = 1.4$ m.

Calculation: $K'/V = 1$. Start from $V = 1.4$ m (4.6 ft) and follow as in the previous example a path vertically upwards to $c = 0.4$, and then to the right to the dotted line. In this case, however, the K'/V-value being less than 2, one has to deviate vertically downwards to the line $K'/V = 1$ and then to continue to the right. The value $Q_t = 1.1$ kg (2.45 lb) is obtained. Note the condition that the charge shall not reach higher in the hole than to a distance V from the upper free face. To satisfy this condition the diameter of the hole must be big enough for the Q_b to be contained in a length $(K' - 0.7 \, V)$ of the hole, H being the depth of the hole $H = 1.1 \cdot K + 0.3 \, V = 2.0$ m (6.6 ft). The height of the charge must then be 0.6 m (2.0 ft) and the concentration $l_b = 1.1/0.6 = 1.8$ kg/m (2.45/2.0 = 1.2 lb/ft). (From Diagram 2:1 it is found that this corresponds to $d = 42$ mm with an ordinary value for the degree of packing, $P = 1.25$.)

Limit charge concentrated at the bottom for $K = V$ (Diagram 2:4)

This diagram is mainly used for trial blasting. Concentrated charge means that the height of the charge is small compared with the burden which can be given quantitatively as the condition $h \leqslant 0.3 \, V$. A "limit charge" is defined as a charge that gives a complete loosening of the rock without throwing it more than 0–2 m (0–6 ft). A greater throw indicates that the rock factor c used in the calculation has been too high. As some information is obtained from the throw (2 m for 10 % excess charge) it is better to begin with too high a value c than one that is too small. The diagram deals with two main types, the one with "fixed bottom", where the surface that has to be broken at the bottom is perpendicular to the hole, the other with an angle for free breakage at bottom as the lower figure in the diagram shows.

The diagram is read starting from the actual value for the burden V at the bottom of the hole and passing vertically up to the curve for the rock factor c, e.g. $c = 0.4$, and then turning to the left in the case of "fixed bottom". The "Limit charge" is read off on the left-hand scale. For free burden at the bottom the reading off is done on the right vertical scale after a vertical deflection from "fixed bottom" down to "free bottom".

Example 2:4. To decide the magnitude of the rock factor c_0 a part of the rock is taken where no evident cracks and crevices influence an intended trial blast with one hole of a burden $V = 0.5$ m (1.65 ft) and a free bottom.

Calculation: Assume a value $c_0 = 0.4$, start from $V = 0.5$ m (1.65 ft), proceed up to the line $c = 0.4$ (a), turn right to the dotted line (b) and deflect vertically downwards to the line "free bottom" (c) and continue to the right-hand vertical scale (d) that gives $Q_0' = 0.05$ kg (0.11 lbs).

Result: With this charge the main mass of rock is thrown some 4 m forward. It corresponds to 20 % excess charge, that is $Q_0' = 0.042$ kg (0.09 lbs). The c_0-value is obtained in starting from this Q_0'-value and reading in reverse direction from (d') to (c'), (b') and (a'), all these points now being some mm:s lower in the diagram than the a, b, c, and d before them. The intersection of this path with the line starting from the burden $V = 0.5$ m (1.65 ft) takes place at the line $c_0 = 0.3$, and this gives the value sought for application: Owing to small variations in the rock and the general need for some technical margin the c_0-value corresponding to limit charge should not be used in practical application, but has to be increased to $c = 1.2\ c_0 = 0.36$. This value has to be used in calculating the maximum burden and necessary charge in a bench row.

Note: As the c-value thus obtained mainly covers ordinary variations in the rock, additional margins required to compensate for deviation in the drilling and for the swell in multiple-row blasting have to be added according to the conditions in every special case (see chapter 3).

Fragmentation (Diagram 2:5)

The correlation between size of the boulders and the specific charge is given for multiple-row short delay blasting and various burdens 10–0.25 m (33–0.8 ft) with a bench height of about $K' = 3V$. The size of the boulders is decreased with a higher specific charge, but also with smaller burdens. With a burden $V = 1.1$ m (3.6 ft) and specific charge $q = 0.28$ kg/m³ (0.47 lb/cu.yd) an average side length of $L = 0.6$ m (2 ft), whereas for $V = 2.5$ m and the same charge the value $L = 1.0$ m (3.3 ft) is obtained. This figure means a volume $L^3 = 1.0$ m³ (1.3 cu.yds). To get the same L-value as with the smaller burden the specific charge has to be increased to $q = 0.40$ kg/m³ (0.66 lb/cu.yd).

The greater charge means greater throw of the broken material and also that a greater part of the rock is crushed to very small pieces.

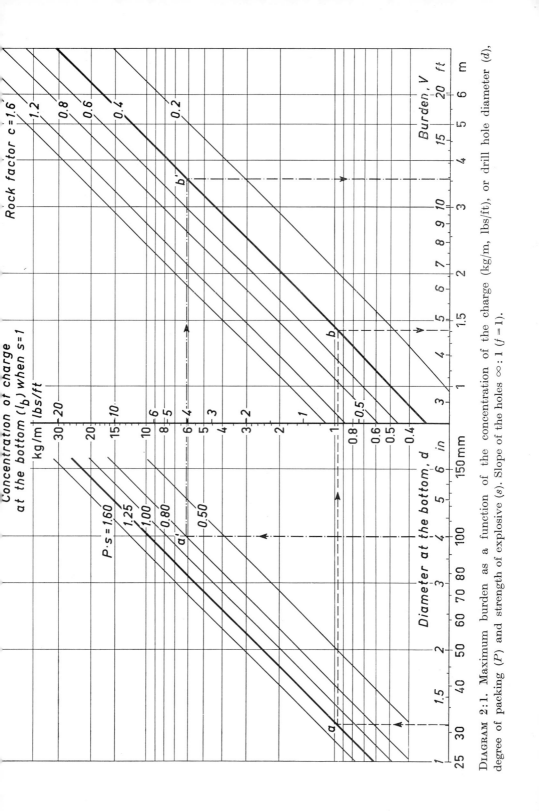

DIAGRAM 2:1. Maximum burden as a function of the concentration of the charge (kg/m, lbs/ft), or drill hole diameter (d), degree of packing (P) and strength of explosive (s). Slope of the holes ∞ : 1 ($f = 1$).

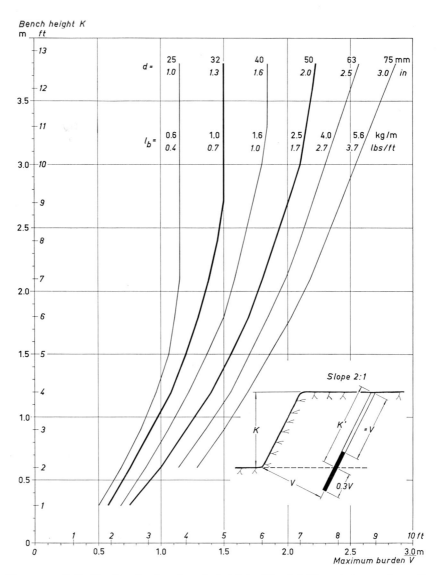

DIAGRAM 2:2. Burden for low bench heights ($K < 1.8\ V$) and for various diameters of the drill hole (normal rock $c = 0.4$).

2.17. Problems

1. Test shot with $V = K = 0.60$ m (20 ft) and free burden according to fig. 2:3. Calculate the charge corresponding to a rock factor $c = 0.4$.

2. If this charge gives an ordinary breakage not (too much) influenced by cracks and crevices and the center of gravity of the broken material is moved

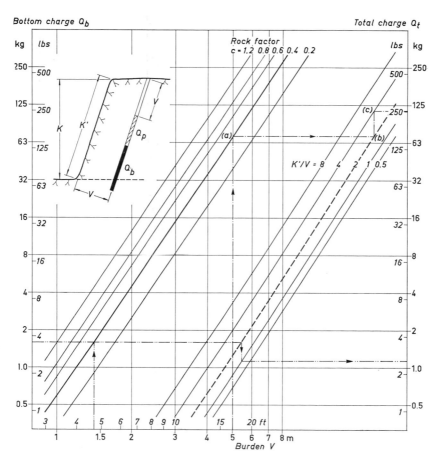

Bottom charge Q_b

Total charge Q_t

Rock factor c = 1.2 0.8 0.6 0.4 0.2

K'/V = 8 4 2 1 0.5

Burden V

DIAGRAM 2:3. Minimum bottom charge (Q_b) and total charge (Q_t) in a drill hole as a function of the burden at various values of the rock factor (c) and of the relation between bench height (K') and burden (V).

6 m (20 ft) forward, estimate the excess charge. Calculate the limit charge (Q_1) and the corresponding value for the rock factor c_0.

3. a) Bench blasting. Diameter of the hole at bottom $d = 40$ mm (1.6 in), degree of packing $P = 1.10$. Weight strength $s = 1.0$, rock factor $c = 0.4$. Calculate the concentration of the charge (l_b) in kg/m (lbs/ft) and the maximum burden (V).

    b) The same problem with $P = 1.60$.

    c) Calculate for a bench where $V = E$ and $K = 3$ V the m (ft) of drill hole per m³ (cu.yd) rock.

4. Bench blasting. Diameter of the hole $d = 75$ mm (3 in), degree of packing $P = 1.45$. $c = 0.4$. Calculate the maximum burden (V) if the rock factor is

    a) $c = 0.3$.

    b) $c = 0.6$.

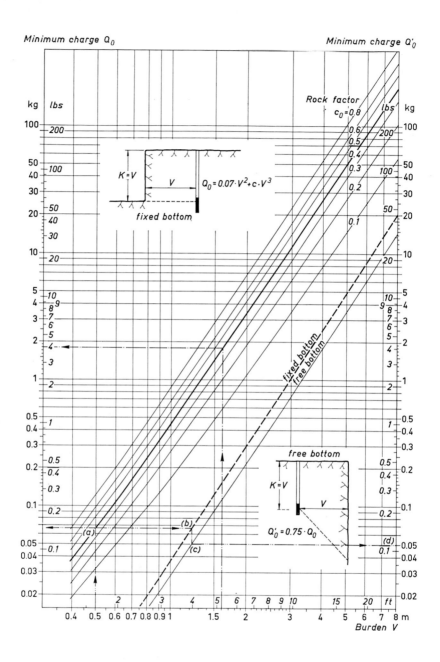

DIAGRAM 2:4. Minimum bottom charge for loosening the rock when the burden equals the vertical bench height $(K = V)$ and when the charge is concentrated at the bottom of the hole $(h \leqslant 0.3\ V)$.

Size of the boulders L

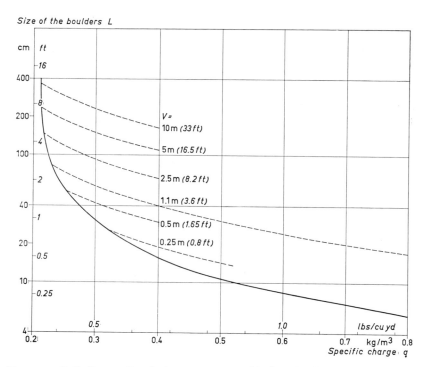

DIAGRAM 2:5. Connection between average side length (L) and specific charge (kg/m³, lb/cu. yd) for various burdens $V = 0.25$–10 m (0.85–33 ft) in multiple-row blasting with short-delays (preliminary diagram). Sloping bench height $K' = 3 V$. The unbroken curve represents for every V-value the specific charge required just for loosening the rock.

5. Rock factor $c = 0.4$ is assumed for one case to be dealt with. A maximum burden of $V = 3.9$ m (13 ft) is desired.

a) Find the amount of explosive required per m for the bottom charge, when $s = 1.0$.

b) The same problem if ammonium nitrate (AN) in prills and oil is used, $s = 0.9$.

c) What diameter of the hole should be used if the hole is loaded with AN-oil in prills, for which the degree of packing is $P = 0.9$ and the strength $s = 0.9$.

d) What diameter of the hole should be used if the hole is loaded with a gelatinized dynamite ($s = 1$) in cartridges by the aid of the pneumatic loader, $P = 1.5$.

6. A bench height $K = 5.4$ m (18 ft) is given. Discuss with regard to the drilling equipment what diameters of drills you can use in this case. What are the corresponding values for the maximum burden?

7. Discuss for the case of problem 6 the relationship between size of the burden and of the boulders obtained. Discuss also this question regarding the digger equipment available and the size of the shovel.

8. Tunnel round. In a drilling pattern for a tunnel round with an advance of $A = 4.2$ m (14 ft) the diameter of the drill is $d = 50$ mm (2 in) for the relievers. Rock factor $c = 0.6$, degree of packing $P = 1.25$; $s = 1.0$. Calculate the maximum burden (V) and the total charge (Q_b) in such a hole.

9. Low bench-blasting $K = 1.2$ m (4 ft) $c = 0.4$, slope of the holes 2:1. $P = 1.25$. Determine d, V and H.

10. Calculate the total charge Q in a hole according to problem 9.

11. Calculate the specific charge in kg/m³ (lb/cu.yd) in the case above.

12. Please learn and adopt the metric system.

3. CALCULATION OF THE CHARGE
PRACTICAL APPLICATION

In this chapter, calculation of the charge will be given for practical applications in *surface and underground* operations. Since the material is mainly provided in tables, the work is reduced to selecting the proper table and figures for the actual case to be considered. It should be kept in mind, that all the material given here is based on ordinary rock conditions $(c=0.4)$.

For other c-values (usually greater than 0.4), the diagrams of Chapter 2 can be used. It may be easier in many cases to use the tables provided, changing the values for the concentration of the charge, l, in direct proportion to $\bar{c}$, or approximately to c. If $c=0.6$ instead of $c=0.4$, the l-values needed in a certain case are roughly 50 % larger than that of the table. This means that the diameter of the hole should be increased by a factor $\sqrt{0.6/0.4}=1.23$.

Blasting in sedimentary rock with a stratification more or less parallel to the intended bottom of breakage represents an important special case; directions for the calculations are given in 3.3d.

3.1. Some simplified connections

Attention may first be drawn to some simple, useful connections between the diameter and the concentration of the charge and the burden.

With a degree of packing $P=1.27$ the amount of explosive (l) in kg/m is given by the simple relation

$$l=0.001\,d^2 \tag{3:1 M}$$

where d is given in mm. M in (3:1 M) indicates metric system. For example with $d=31.5$ mm, $d^2=1000$ we have $l=1.0$ kg/m. This is one reason why $P=1.27$ is used as a standard reference. Another is that it happens to be close to the value obtained in careful loading with a tamping pole.

In English units an even simpler formula is obtained if $P=1.36$. This however presupposes charging with the pneumatic loader (see 4.2). Then for $d=1$ inch the value $l=1.0$ lb/ft is obtained and

$$l=d^2 \tag{3:1 E}$$

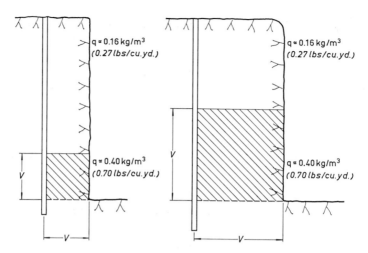

FIG. 3:1. The specific charge q in kg/m³ (lb/cu yd) required for loosening the rock is in the bottom part 2.5 times as much as the corresponding mean value for the rock above the $+V$ level. Thus the total average value for the consumption of explosive increases with the burden for one and the same bench height.

For the region 1.4–14 m (4.6–46 ft), the maximum burden V in m can with sufficient accuracy be considered to be directly proportional to the diameter d in mm

$$V = 0.046\,d \qquad\qquad (3:2\ \text{M})$$

In English units this relation with V in feet and d in inches is

$$V = 3.8\,d \qquad\qquad (3:2\ \text{E})$$

The value (V_1) to be used in the construction of a drilling pattern must be less than V due to the deviation in the drilling (see below 3.3a). This may amount to a reduction of some 10 to 20%. Thus we have a rule of thumb that is extremely easy to remember, if the practical burden (V_1) is given in meters and the diameter of the holes in inches

$$V_1 = d \qquad\qquad (3:3\ \text{M–E})$$

This means that for $d = 2$ inches the practical burden approximates $V_1 = 2$ m, for 3 inches 3 m and so on, assuming that $P = 1.27$, that one simple row of holes is used, the burden is not loaded with previously broken rock, slope of the holes 3:1–2:1 and ordinary rock constant $c = 0.4$.

These relations may serve as a guide for preliminary estimates, prior to carrying out the calculations in greater detail.

70

3.2. Maximum and minimum charge in single-row bench blasting

a. Ordinary benches. $K \geqslant 1.8\ V$. *Table 3:1*

Practically everything required for the calculation is given in the table 3:1 which is valid for ordinary rock conditions. It should be kept in mind that the value for the burden V applies at the bottom of the hole, not at the free upper face of the rock in the vicinity of the collar.

In vertical drilling the burden V and also the spacing E should be 5 % less, with the same bottom diameter and bottom charge. This means that the rock volume per drill hole proportional to $V \times E$ is reduced 10 % due to a higher fixation at the bottom. The minimum column charge, not being influenced by the conditions at the bottom, is to be reduced 10 %.

Drilling below the bottom level should be $-0.3\ V$. The bottom charge extends to $+ V$ so that the height of the bottom charge is $1.3\ V$. Especially with greater burdens, 4–14 m (13–130 ft), as in many quarries, a drilling of only $0.15\ V$ is often used. In such cases the breaking force of the charge is reduced 10 %. Then the concentration of the charge in kg/m (lbs/ft) should be increased, or the VE-value reduced 10 % if a higher frequency of stumps at the bottom is to be avoided.

If a lower or higher degree of packing is adopted one should start by reading the second column of table 3:1 "Concentration of the bottom charge" as this is the decisive factor, not the diameter in itself.

The figures given represent the minimum amount of explosive required to give a complete loosening of the rock and specifies how the charge should be distributed in this case. This can be regarded as a necessary condition to be fulfilled under all circumstances. It naturally does not mean that the minimum charge is recommended—in ordinary practice it is almost always exceeded because of the need of better fragmentation, corrections due to deviation in the drilling, need for swell in multiple-row blasting and provision for an extra margin for greater variations than expected in some of the conditions.

All these factors change from place to place and they should be included in the calculations only to the extent they are justified (see 3.3) and, in any case, each one must be known and observed separately.

b. Low benches. $K \leqslant 1.8\ V$. *Tables 3:2, 3:3*

In the calculations it is generally assumed that a distance V is left uncharged in the upper part of the hole. This means that, for a vertical bench with heights K less than $2\ V$ $(H < 2.3\ V)$ the height of the charge must be less than $1.3\ V$. There is no room for a column charge. The bottom charge—and thus

TABLE 3:1. *Bench blasting and stoping. One row of holes. Bottom and column charges and maximum burdens. Weight strength of explosive s = 1.0. Slope 2:1–3:1. Degree of packing P = 1.27 (tamping pole). Spacing E = 1.25 V. Drilling below the bottom level 0.3 V. The charge extends up to a distance V from the top of the hole.*

Diameter of the hole d		Concentration of the bottom charge l_b		Concentration of the column charge l_p		Max. Burden V		Total bottom charge Q_b	
mm	in	kg/m	lbs/ft	kg/m	lbs/ft	m	ft	kg	lbs
14		0.18	0.12	0.09	0.06	0.6	2.0	0.15	0.33
16		0.25	0.17	0.10	0.07	0.7	2.3	0.25	0.55
19	0.75	0.36	0.24	0.14	0.09	0.85	2.8	0.43	0.95
22		0.48	0.32	0.19	0.13	1.00	3.3	0.65	1.4
25	1	0.63	0.42	0.25	0.17	1.15	3.8	0.95	2.1
29		0.84	0.56	0.34	0.22	1.3	4.5	1.5	3.3
32		1.0	0.68	0.40	0.27	1.5	5.0	2.0	4.4
38	1.5	1.4	1.0	0.60	0.40	1.8	5.9	3.3	7.3
44		1.9	1.3	0.80	0.50	2.0	6.8	5.2	11
50	2	2.5	1.7	1.0	0.7	2.3	7.8	7.6	17
63		4.0	2.7	1.6	1.1	3.0	10	15	33
75	3	5.6	3.8	2.3	1.5	3.5	11.5	26	57
88		7.7	5.2	3.1	2.1	4.1	14.5	41	90
100	4	10.0	6.7	4	2.7	4.7	15.5	60	130
125	5	15.5	10.5	6	4	5.9	19.5	120	260
150	6	22.5	15	9	6	7.0	23	200	440
175	7	30	20	12	8	8.2	27	320	700
200	8	40	27	16	11	9.4	31	500	1100
225	9	50	34	20	13	10.6	35	700	1500
250	10	62	41	25	17	11.8	39	950	2100
300	12	90	60	36	24	14.0	46	1600	3500

also the burden—is smaller than in ordinary benching where $h = 1.3 V$ (table 3:1). With inclined holes, 2:1, which is recommended in the case of low benches, the same applies when the bench height K is lower than 1.8 V.

The calculation for "low benches" is more complicated as the burden is not only a function of the diameter of the hole (d) and the degree of packing (P), but also of the bench height (K). The minimum charges for low benches are given in table 3:2.

The maximum burden is given in table 3:3 for various diameters between 25 and 100 mm (1–4 in) and for bench heights below 6 m (20 ft). As seen from the table the burden increases with the bench height. For example, with $d = 32$ mm (1.3 in) the burden increases from $V = 0.58$ m (1.9 ft) up to $V = 1.5$ m (5 ft), which is the value given in table 3:1 for ordinary benches.

TABLE 3:2. *Minimum charge in low bench blasting. One single row of holes. The slope of the holes 2:1. Degree of packing P = 1.27, weight strength for the explosive s = 1.0. E = 1.25 V. Drilling below the bottom 0.3 V, but not less than 10 d. (d = diameter of the hole).*

Bench height K		Minimum charge per hole, kg.																			
m	ft	V=0.5	0.6	0.7	0.8	0.9	1.05	1.2	1.35	1.5	1.65	1.8	2.1	2.4	2.7	3.0	3.3	3.6	3.9	4.2	4.5 m
		1	2		3		4		5		6		8		10		12		14		ft
0.3	1	0.06	0.09	0.12	0.17	0.23															
0.6	2	0.09	0.11	0.16	0.22	0.29	0.41	0.58													
0.9	3	0.11	0.14	0.19	0.26	0.34	0.48	0.67	0.9	1.2	1.5	1.9	2.9	4.1	5.7	7.6	10	13	16	20	24
1.2	4	0.14	0.17	0.22	0.30	0.39	0.55	0.76	1.0	1.3	1.7	2.1	3.2	4.5	6.1	8.1	11	14	17	21	25
1.8	6	0.19	0.23	0.29	0.39	0.50	0.70	0.94	1.2	1.6	2.1	2.5	3.7	5.2	7.0	9.2	12	15	18	23	27
2.4	8	0.24	0.29	0.35	0.48	0.60	0.85	1.1	1.4	1.9	2.4	2.9	4.3	5.9	7.9	10.5	13	17	20	26	30
3.0	10	0.30	0.34	0.42	0.56	0.71	1.0	1.3	1.7	2.1	2.7	3.2	4.8	6.5	8.8	12	15	18	22	28	32
3.6	12	0.35	0.40	0.49	0.65	0.82	1.15	1.5	1.9	2.4	3.0	3.6	5.4	7.2	9.7	13	16	20	24	30	35
4.2	14	0.40	0.46	0.55	0.73	0.92	1.3	1.7	2.1	2.6	3.3	4.0	5.9	7.9	10.5	14	17	21	26	32	37
4.8	16	0.45	0.50	0.62	0.82	1.0	1.4	1.8	2.3	2.9	3.7	4.4	6.4	8.6	11.5	15	18	23	28	34	40
5.4	18	0.50	0.56	0.69	0.90	1.1	1.6	2.0	2.5	3.2	4.1	4.8	6.9	9.3	12	16	19	24	29	36	42
6.0	20	0.55	0.62	0.75	1.0	1.2	1.7	2.2	2.8	3.5	4.4	5.2	7.5	10	13	17	21	26	31	38	45

To get the values in lbs multiply by 2 and add 10%.

c. *Various rock factors*

The tables are made for ordinary rock conditions with a rock factor $c = 0.4$. The specific charge for a bench height equal to the burden is for ordinary burdens ($V > 1.0$ m) close to a constant. The constant, denoted $\bar{c}$, is $\bar{c} = 0.45$ kg/m³ (0.75 lb/cu.yd) for a single hole, and $\bar{c} = 0.36$ kg/m³ (0.6 lb/cu.yd) for a row of holes. The same value applies to the bottom part of a bench with a height greater than the burden. The part above V m (ft) from the intended bottom of breakage requires only 0.15 kg/m³ (0.25 lb/cu.yd). In fig. 3:1 figures including 10 % greater margin are given.

The rock factors may vary from $c = 0.1$–1.4, and the corresponding figures for the consumption of explosive are $\bar{c} = 0.15$–1.45 kg/m³ (0.25–2.5 lb/cu.yd) for the bottom part.

How should the tables intended for $c = 0.4$ ($\bar{c} = 0.45$) be used if one finds that another rock factor applies? Actually test blasting could be employed according to Diagram 2:4. The diameter can also be increased in full scale tests in benching (one row) till the required value is obtained. If an investigation shows that the diameter should be increased 1.20 (20%), the tables 3:1–6 can then be used with that correction for all other cases. The l-values should be increased $1.2^2 = 1.44$ (44 %).

If the job has to be carried out with the same diameter of the drill holes as before, even if the rock is found to require more explosive than expected, the actual value of the hole diameter should be divided by 1.20 before reading the tables. In the case mentioned $\sqrt{\bar{c}}$ is 1.20 larger than for the tables in their present form, and for the rock investigated $\bar{c} = 0.65$ kg/m³ (1.1 lb/cu.yd) and $c = 0.6$.

3.3. Practical application

a. *Correction for deviations in the drilling*

The "difference" between theory and practice in rock blasting is mainly due to the fact that the drill holes are not at the exact spot and do not run in the exact direction indicated by the drilling pattern. Therefore, this must be taken into account when constructing the drilling pattern. Instead of taking the maximum burden (V), a lower value for the practical burden (V_1) must be applied.

The corrections to be made depend on faults in the collaring and in the alignment of the holes (fig. 3:2). The first mentioned can be 0.1 m (0.3 ft) and should not be permitted to exceed that value for holes with a diameter less than 100 mm (4 in). The alignment faults can, and should be kept below the value 0.03 m/m (1.2 in/ft). A deviation for a depth of 1 m should have

74

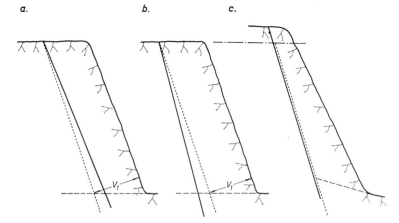

a.   b.   c.

FIG. 3:2. Influence of erroneous drilling. The dotted lines indicate intended position of the holes according to the drilling pattern with a burden V_1. The broken line indicates the bottom obtained. The real burden V at the bottom may be greater, due to faults in a) collaring and b) alignment. The influence of the latter increases with the depth. In tunnel blasting it often occurs that the bottom of the holes deviates from the intended position as in c, due to irregularities of the previously broken round if the length of the drill is not changed.

a lower value than 0.03 m, for 2 m less than 0.06 m and so on, increasing with the depth as shown in fig. 3:2 b. The practical burden is then

$$V_1 = V - 0.1 - 0.03\,K \qquad\qquad (3\!:\!4\ \mathrm{M})$$

$$V_1 = V - 0.3 - 0.03\,K \qquad\qquad (3\!:\!4\ \mathrm{E})$$

This relationship can be approximated to

$$V_1 = V - 0.05\,K \qquad\qquad (3\!:\!5)$$

equal in form for both metric and English units and for $K = 5$ m (15 ft) in exact agreement with (3:4). For lower bench heights the formula (3:5) calls for greater accuracy in the collaring, that e.g. for $K = 1.0$ m (3 ft) should not be made with a greater error than 2 cm (1 in). For a greater bench height than $K = 5$ m (15 ft) the formula (3:5) gives a greater reduction in the practical burden than (3:4), but, on the other hand another factor comes into the picture: deviations inside the rock, the influence of which increases with the depth. An analysis of the influence of that factor is made in the Chapters 7 and 8. For ordinary tunnel driving and bench blasting the formula (3:5) is simple and accurate enough.

The formula presupposes alignment faults less than 3 %. In hand-held drilling, it is not unusual to find values as high as 10 % (0.10 m/m, 1.2 in/ft). It is necessary to avoid such large deflections.

The accuracy obtained in drilling with mechanical equipment or fixed drilling machines is almost always over-estimated. Only with very special arrangements should it be assumed that the accuracy is better than 3 %.

b. Need for swell in multiple-row blasting

When blasting one single row of holes where the rock is not loaded with previously broken material the burden at the bottom (V) given in table 3:1 can be used. In multiple-row blasting the charge shall not only loosen the rock but also move and heave the masses in front of the row in question as shown in fig. 2:12. The requirements, of course, change from a single-row round to multiple-row rounds in channels and bottom-benches in tunnels, continuously blasted round after round (fig. 2:13). Here with a throw component of 1:2 even without deviations in drilling the burden, has to be reduced according to (2:40) that gives approximately

TABLE 3:3. *Maximum burden in low bench blasting. One single row of holes. Slope of the holes 2:1. Ordinary degree of packing ($P = 1.27$) as obtained when loading carefully with tamping pole. $E = 1.25\ V$. Drilling below the bottom level 0.3 V, but not less than 10 d.*

Bench height K		Maximum Burden V, m											
		$d =$ 25	29	32	36	40	45	50	63	75	100	150	mm
m	ft	$l =$ 0.62	0.82	1.0	1.3	1.6	2.0	2.5	4	5.6	10	12	kg/m
0.3	1	0.50	0.55	0.58	0.62	0.67	0.70	0.73					
0.45		0.60	0.62	0.66	0.70	0.78	0.84	0.90					
0.6	2	0.68	0.72	0.76	0.80	0.87	0.94	1.0	1.15	1.3			
0.9	3	0.85	0.87	0.90	1.00	1.05	1.10	1.20	1.35	1.5	1.8		
1.2	4	0.95	1.00	1.10	1.20	1.25	1.30	1.40	1.55	1.7	2.0	2.6	
1.5	5	1.05	1.10	1.20	1.30	1.35	1.45	1.50	1.7	1.8	2.2	2.8	
1.8	6	1.10	1.20	1.30	1.40	1.50	1.60	1.70	1.8	2.0	2.3	3.0	
2.1	7	1.15	1.30	1.35	1.45	1.55	1.70	1.80	2.0	2.2	2.5	3.2	
2.4	8		1.35	1.45	1.55	1.65	1.80	1.90	2.1	2.3	2.7	3.4	
2.7	9			1.5	1.60	1.75	1.90	2.00	2.2	2.4	2.9	3.5	
3.0	10				1.65	1.80	2.00	2.10	2.3	2.5	3.0	3.6	
3.3	11					1.85	2.00	2.10	2.4	2.6	3.2	3.8	
3.6	12						2.05	2.20	2.5	2.8	3.3	4.0	
3.9	13						2.10	2.25	2.6	2.9	3.5	4.2	
4.2	14							2.30	2.7	3.0	3.6	4.4	
4.8	16								2.8	3.2	3.8	4.6	
5.4	18								2.9	3.4	4.0	4.8	
6.0	20	1.15	1.35	1.5	1.65	1.85	2.1	2.3	2.9	3.4	4.2	5.0	

TABLE 3:4. *Multiple-row bench blasting. Diameter of the drill bits 34–29 mm (1.35–1.16 in) P = 1.27. Length of the rod 0.8–4.8 m (2.6–16 ft). Slope of the holes 3:1–2:1. E = 1.25 V.*

Bench height K		Depth of the hole H	Diameter at the bottom d	Max. burden V	Practical burden V_1		Minimum charge/hole Q	
m	ft	m	mm	m	m	ft	kg	lb
0.3	1	0.67	34	0.60	0.55	1.8	0.09	0.2
0.45		0.80	34	0.70	0.65	2.1	0.14	0.3
0.6	2	1.00	33	0.77	0.7	2.3	0.20	0.45
0.9	3	1.35	33	0.92	0.85	2.8	0.34	0.75
1.0		1.45	33	1.00	0.95	3.1	0.45	1.0
1.2	4	1.65	33	1.10	1.00	3.3	0.60	1.3
1.5	5	2.0	32	1.20	1.10	3.6	0.85	1.9
1.8	6	2.4	32	1.30	1.20	3.9	1.10	2.4
2.1	7	2.7	31	1.35	1.25	4.1	1.30	3.0
2.4	8	3.1	31	1.40	1.25	4.1	1.60	3.5
2.7	9	3.4	30	1.40	1.25	4.1	1.70	3.7
3.0	10	3.8	30	1.40	1.25	4.1	1.80	4.0
3.3	11	4.2	30	1.40	1.20	4.1	1.90	4.2
3.6	12	4.5	29	1.35	1.15	3.8	1.90	4.2
4.0		4.9	29	1.35	1.15	3.8	2.00	4.4

$$V_1 = V(1 - 0.03\,K) \qquad (3:6\ \text{M})$$

$$V_1 = V(1 - 0.1\,K) \qquad (3:6\ \text{E})$$

For smaller values of the maximum burden than $V = 1.7$ m (5 ft), corresponding to diameters of the hole $d = 37$ mm (1.5 in), the V_1-values according to (3:6) are greater than those given by (3:5). This means that the practical burden is more restricted by the deviations in the drilling than by the swell.

For larger diameters of the holes than $d = 37$ mm ($1\frac{1}{2}$ in) the practical burden in multiple-row blasting is given by (3:6), e.g. it depends on the conditions for the swell.

One consequence of this is that with an increasing diameter one can permit greater deviation in the drilling—in fact proportional to the diameter. With $d = 75$ mm (3 inches) the formula (3:6 M) is $V_1 \approx V - 0.09\,K$, and at the bottom of a $K = 7.5$ m (25 ft) bench a deviation of 0.67 m (2.2 ft) can be permitted.

Some other important consequences are that the consumption of explosives increases with the bench height, and that for higher benches it is much more important to incline the holes. If the holes have such a slope that the throw component is 1:3 the extra charge for sufficient swell must be doubled as compared with a throw of elevation 1:2 (see 2.12).

Bench height K		Depth of the hole H		Diameter at the bottom d	Max. burden V	Practical burden V₁		Minimum charge/hole Q
m	ft	m		mm	m	m	ft	kg
		Slope						
		3:1	2:1					
0.3	1		0.73	40	0.67	0.6	2.0	0.11
0.45			0.90	40	0.78	0.7	2.3	0.19
0.6	2		1.05	39	0.85	0.80	2.5	0.26
0.9	3		1.4	39	1.02	0.95	3	0.40
1.0			1.5	39	1.13	1.05	3.5	0.60
1.2	4		1.7	38	1.20	1.10	3.6	0.76
1.5	5		2.1	38	1.30	1.20	4	1.0
1.8	6		2.5	37	1.4	1.30	4.3	1.3
2.1	7		2.8	37	1.5	1.40	4.6	1.8
2.4	8		3.2	37	1.6	1.45	4.8	2.3
2.7	9		3.5	36	1.6	1.45	4.8	2.4
3.0	10	3.7	3.9	36	1.65	1.5	5	2.7
3.6	12	4.3	4.5	35	1.60	1.4	4.6	2.8
4.2	14	5.0	5.2	34	1.55	1.3	4.3	2.8
4.8	16	5.5	5.8	33	1.50	1.2	4	2.9
5.4	18	6.1	6.5	33	1.50	1.2	4	3.2
6.0	20	6.8	7.2	32	1.50	1.2	3.6	3.5
6.6	22	7.5	7.9	31	1.45	1.1	3.6	3.5
7.2	24	8.1	8.5	30	1.40	1.0	3.3	3.6
7.8	26	8.7	9.1	29	1.35	0.9	3	3.6
8.4	28	9 3	9.8	29	1.35	0.9	3	3.6

Note: for d = 42 − 31 mm the values V_1 for the "Practical burden" should be increased 0.1 m for every bench height equal to, or above $K = 1.5$ m.

c. Some tables for practical application. Tables 3:4, 3:5, 3:6

Complete calculations are carried out for some of the most frequent series of monobloc steels. These series begin with a bit gage between 42 and 34 mm (1.7 and 1.35 in). The lengths of the drills in a series are 0.8 m, 1.6 m, 2.4 m and so on. For every new length the bit gage is 1 mm less than for the previous one. At a length of 4.8 m (16 ft) and more the bit gage is 37 mm (1.5 in) or less, which means that in practice, when using monobloc steels the deviation in the drilling must be taken into account but not the requirements for the swell.

TABLE 3:6. *Practical burden* (V_1)* *in multiple-row bench blasting. Diameter of the drill bit d = 44, 50, 57, 63, 75, 100 mm. Slope of the holes 3:1 − 2:1. E = 1.25 V.*

| Bench height K | | Depth of the hole for d = 63 m H | | Practical burden V_1 | | | | |
m	ft	3:1	2:1	d = 44 l = 1.9	50 2.5	63 4.0	75 5.6	100 mm 10 kg/m
		Slope						
0.3	1			0.65	0.70			
0.45				0.80	0.85			
0.6	2	1.25		0.90	0.95	1.1		
0.9	3	1.6		1.05	1.15	1.3	1.4	1.7
1.2	4	1.9		1.2	1.3	1.4	1.5	1.9
1.5	5	2.3		1.3	1.4	1.6	1.7	2.1
1.8	6	2.6		1.45	1.5	1.6	1.8	2.2
2.1	7	2.9		1.6	1.6	1.8	2.0–2.1	2.3–2.4
2.4	8	3.1		1.6	1.7	1.9	2.1–2.2	2.5–2.6
2.7	9	3.5		1.7	1.8	2.0	2.2–2.3	2.6–2.8
3.0	10	3.8	3.9	1.8	1.9	2.1	2.3–2.4	2.7–2.8
4.0	13	5.0	5.2	1.8	2.0	2.3	2.5–2.7	3.1–3.3
5.0	17	6.0	6.3	1.8	2.0	2.4–2.5	2.8–3.1	3.3–3.6
6.0	20	7.3	7.6	1.7	1.9–2.0	2.4–2.6	2.8–3.1	3.4–3.9
7.0	23	8.3	8.7	1.7	1.8–2.0	2.3–2.5	2.7–3.1	3.6–4.0
8.0	27	9.4	10	1.6	1.7–1.9	2.2–2.5	2.6–3.0	3.5–4.1
9.0	30	10.5	11	1.5	1.7–1.9	2.1–2.4	2.5–3.0	3.4–4.1
12.0	40	13.5	14	1.4	1.5–1.7	1.8–2.3	2.2–2.8	3.0–4.0
15.0	50	17.0	18	1.2	1.3–1.6	1.6–2.1	1.9–2.7	2.5–3.8
18.0	60	20.0	21	1.0–1.1	1.1–1.4	1.3–2.0	1.6–2.5	2.1–3.7

* The lowest values in notations, e.g. 1.7–1.9, indicate the burden when the swell can only take place vertically. For shorter rounds the higher value (1.9) gives the practical limit.

For extension drill steels the diameter of the bits is more than 37 mm (1.5 in) and the swell condition (3:6) must be regarded when it is a question of multiple-row rounds. For those drill steels the maximum burdens are found in table 3:1 for ordinary benches and can be calculated from table 3:2 for small benches. The practical burden is given in table 3:6 for diameters 44–100 mm ($1\frac{3}{4}$–4 in).

The values given in the tables can be used directly when marking the holes on the rock surface and when ordinary rock is encountered. The figures for the minimum charges Q, however, are only given for information. It is recommended to exceed these values, when throw conditions permit.

79

Fig. 3:3. Sedimentary rock, with bedding planes almost horizontal. (Rio Furnas, Brazil.)

d. Blasting in sedimentary rock

When the rock is geologically formed through sedimentation in a sea where the bottom has subsequently heaved into high plateaus or mountains, the rock often has a horizontal stratification. For bench blasting where the intended bottom ordinarily is horizontal, the same breaking force at the bottom as in ordinary rock is not required. This is in fact a geometrical effect that also appears in other kinds of rock if crevices at the bottom of a bench render a full bottom charge unnecessary. If it is a question of a large crevice, only a pipe charge may be required. The rock factor in itself is about the same as for ordinary rock and can be taken as $c = 0.4$.

In sedimentary rock it is more a question of weak faces than crevices, and the concentration of the charge at the bottom of a hole is something between a bottom charge and a pipe charge—the latter requires only 0.4 as much charge per m (ft).

Figure 3:4 shows a large bench blast for a canal through sedimentary rock (Brazil, South-America). Fig. 3:3 gives a close-up of the stratification of the rock. Here a charge of only 80% of an ordinary bottom charge is required. For the column charge the same concentration as for ordinary rock

Fig. 3:4. Excavation in sedimentary rock. Approach canal, Rio Furnas, Brazil. (Courtesy of Georg Wimpey and Co, Ltd., London.)

TABLE 3:7. *Blasting in rock with horizontal stratification (allowing 25% more rock per drill-hole). Bottom charge of dynamite (s = 1.0) with a degree of packing P = 1.0, or powder explosive (s = 0.9) with P = 1.1. V_1 and E_1 practical burden and spacing*

a) With height of bottom charge 1.3 $\sqrt{VE}$ and 0.3 $\sqrt{VE}$ sublevel drilling.

Diameter d		(VE)	($V_1 E_1$)			Bottom charge	
			K = 6	10	15 m		
mm	in	m²		m²		kg	m
36	$1\frac{1}{2}$	3.2	2.2	1.7	—	2.4	2.3
50	2	6.5	5.0	4.2	3.2	6.6	3.3
75	3	14.5	12	11	9.3	23	5.0
100	4	26.5	23	21	19	54	6.7
125	5	41	37	35	32	105	8.3
150	6	60	55	52	49	180	10
162	$6\frac{1}{2}$	70	65	62	58	230	11

b) With height of bottom charge 0.8 $\sqrt{VE}$ and no sublevel drilling.

Diameter d		(VE)	($V_1 E_1$)			Bottom charge	
			K = 6	10	15 m		
mm	in	m²		m²		kg	m
36	$1\frac{1}{2}$	2.2	1.4	—	—	1.2	1.2
50	2	4.5	3.3	2.6	—	3.4	1.7
75	3	10	8.2	7.0	5.8	11.5	2.5
100	4	18	15	14	12	27	3.4
125	5	28	25	23	21	54	4.3
150	6	41	37	35	32	91	5.1
162	$6\frac{1}{2}$	48	44	41	38	115	5.5

can be used. These figures can be regarded as rather typical and used when the same geometrical conditions apply. In tables 3:7a and b some calculations are given for different diameters of holes, and a slope of 2:1.

The degree of packing $P = 1.0$ is rather low and for diameters up to 100 mm (4 in) can be increased up to 50% if the pneumatic loader is used with dynamite cartridges. The VE value must also be increased in the same proportion.

The values for the maximum burden in table 3:1 apply to table 3:7a), as the lower degree of packing is compensated by the easier breakage at the bottom. The burden given in table 3:1 can be used when $E = 1.25\,V$.

If a drilling pattern with equal values for the burden and spacing is preferred, the burden can be increased, but in such a way that the area (and thus also the volume) per drill hole remain the same. In a square pattern the burden and spacing will then have the same value $\sqrt{VE}$.

3.4. Distribution of the charge

At the bottom of the holes, where more explosive is needed than in the upper part, the explosive should be well tamped up to V above the bottom level and $0.3\,V$ below if the maximum effect of a hole is desired. The deck or column charge in the rest of the hole extends up to V m (ft) from the upper face of the rock and has no influence on the break at the bottom. The concentration of the charge in the column need not be more than 0.4 of that at the bottom. The distribution of the charge in a hole should then be as shown in fig. 3:5, when the smallest amount of explosive is desired. The calculations given in most of the tables and diagrams presuppose conditions as in this figure. Note that V_1 is the burden in the drilling pattern, V is the value to be permitted at the bottom, (maximum burden) with the bottom charge in question. The height of the bottom charge is given in relation to this maximum burden, and so is the length of the sublevel drilling.

To get better fragmentation, or to move the masses excavated, one has to put more explosive in the rock. This can be done either by drilling the holes closer or giving every hole more explosive. There are really great possibilities for increasing the charge per m³ or cu.yd without increasing the drilling, if the extra space in the column is used, and further if the stemming (or otherwise unloaded part) is reduced. This can often be done down to 0.6–$0.8\,V_1$ instead of V.

The excess charge put into the holes gives a throw, the energy of which is proportional to the excess charge per m³, as shown in Chapter 2.11. The table 2:3 indicates that the center of gravity of the masses is moved forward a distance of 6 m (20 ft) per 0.1 kg/m³ excess charge. The main movement of the rock can thus be calculated or estimated in advance.

More difficult and more hazardous is the throw of small pieces of rock that cannot be predicted. This throw increases if a high concentration of the charge happens to be close to a free face of the rock. On the other hand, it has been found that a heavy charge placed at the bottom of the hole with an unloaded upper part can give a great throw of stones if the upper part is not broken into pieces but only loosened in large blocks. When the expanding gases from the charge below are released through small cracks behind the blocks they acquire a very high velocity that accelerates small pieces of rock before the blocks have moved. If the upper part of the rock

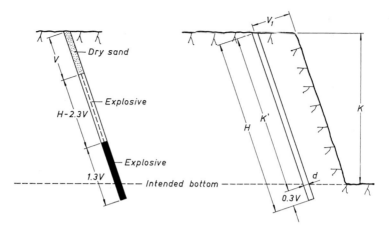

FIG. 3:5. Distribution of explosive in a hole when a minimum charge is desired.

is well broken up, the expanding gases are spread into and affect a larger volume of rock and single pieces are not subjected to extreme velocities. If some cartridges are placed above the main charge, this promotes fragmentation of the upper part of the rock. It is claimed—but not yet proved—that the extreme throw can be reduced in this way. However this may be, one must pay attention to these problems in the distribution of the explosive in the hole. For the relation between specific charge and fragmentation reference is made to Chapter 2.13, table 2:4 and Diagram 2:5.

Example 3:1. A bench with a height $K = 4.8$ m (16 ft) is to be excavated in ordinary rock and with a slope of 3:1.

Table 3:5 gives: $H = 5.5$ m (18 ft), $d = 33$ mm (1.3 in), $V = 1.5$ m (5 ft), $V_1 = 1.2$ m (4 ft), bottom charge 2.9 kg. Calculate the magnitudes V, E, l etc. mentioned below:

Burden	V	$= 1.5$ m (5 ft)	V_1	$= 1.2$ m (4 ft)
Spacing	E	$= 1.9$ m (6.3 ft)	E_1	$= 1.5$ m (5 ft)
	VE	$= 2.8$ m² (32 sq.ft)	$V_1 E_1$	$= 1.8$ m² (20 sq.ft)
Concentration of the charge	l_b	$= 1.1$ kg/m (0.73 lb/ft)		—
	l_p	$= 0.44$ kg/m (0.73 lb/ft)		—
Height of the bottom charge	h_b	$= 2.0$ m (6.7 ft)		—
	h_p	$= 2.0$ m (6.7 ft)		—
Total bottom charge	Q_b	$= 2.2$ kg (4.8 lb)		—
Total column charge	Q_p	$= 0.9$ kg (2 lb)		—
Total charge	Q_t	$= 3.1$ kg (6.8 lb)		—
Minimum specific charge	q	$= 0.22$ kg/m³ (0.37 lb/cu.yd)		0.34 (0.57)

Possible excess charge	1.8 kg (4 lb)	—
corresponding spec.	0.35 kg/m³	
charge	(0.6 lb/cu.yd)	0.54 (0.9)
Excess charge with	2.8 kg (6.1 lb)	—
stemming 0.8 m		
corresponding spec.	0.43 kg/m³	
charge	(0.7 lb/cu.yd)	0.65 (1.1)

Discussion: For blasting of a single row it is possible to perform the blasting with a VE-value near the maximum 2.8 m² given above if one has good drilling precision. Even in such a case a specific charge of 0.35 kg/m³ (0.6 lb/cu.yd) can be obtained with 1.5 m (5 ft) uncharged at the top of the hole. As indicated in table 2:4 this corresponds to a size of boulders of less than 1/8 m³ (1/6 cu.yd), which in most cases should be sufficient. Some bigger ones will of course be obtained from the top of the bench which is not charged. If a round gives unsatisfactory fragmentation it is suggested that some cartridges be placed in the upper part of the hole, e.g. 0.3–0.5 kg (0.7–1.1 lbs) distributed from 1.5–0.8 m (5–2.7 ft) from the top. (The value 1.0 kg calculated as above is given as an upper limit for the charge and should be used only if it is proved to give a throw that is permissible under the circumstances in question.)

With ordinary deviation in the drilling the V_1E_1-value is only 1.8 m². One has to drill c. 60 % more holes and have a *minimum* specific charge of 0.34 kg/m³ (0.57 lb/cu.yd). This is the same charge per m³ as with maximum VE-value and the hole filled with explosive up to 1.5 (5 ft) from the top, a case in which the charge is better distributed and gives a better fragmentation of the upper part of the rock.

When drilling deviation is considerable one has not only to drill more holes, but a more unfavorable fragmentation will be obtained with the same amount of explosives. The same effect is even more pronounced with greater bench heights. Compare e.g. with $K=8.4$ m (28 ft) where the $VE=2.3$ m², the $V_1E_1=1.0$ m² and the proportion $VE/V_1E_1=2.3$. More than twice as many holes have to be drilled in that case and too much explosive has to be placed in the bottom part of the bench. The example shows how important it is in single row blasting for the economy of the job to have good drilling precision, to arrange for it and check the drilling on the site.

In multiple-row blasting it is not as important to cut the deviations in the drilling to much lower values than 5 cm/m (alignment faults of 3 cm/m), as the need for swell requires an excess charge also at the bottom of the hole. This is obtained by reducing the burden 0.03 VK according to (3:6). With conditions as in Ex 3:1 this means that the burden is reduced 0.045 K or 4.5 cm/m and consequently the number of holes cannot be reduced with a smaller deviation in the drilling.

Example 3:2. The same conditions as in Example 3:1, but the geometrical magnitudes K and d are three times as large.

For one row bench blasting all simple geometrical factors are to be multiplied by a factor 3, the VE-values by a factor $3 \cdot 3 = 9$. The *specific charge is unchanged*. This represents a big bench blast with $K = 14.4$ m (48 ft), $d = 100$ mm (4 in) and a maximum burden $V = 4.5$ m (15 ft). The relative influence of faulty drilling is the same if $K = 4.8$ m in Example 3:1 is compared with $K = 14.4$ m in Example 3:2.

In multiple-row benching the simple scaling cannot be applied as the relative influence of the swell factor increases with the bench height. For $K = 14.4$ m, $V_1 = 0.57 V = 2.6$ m instead of three times the V_1-value of Ex. 3:1, which would have been 3.6 m. The practical burden for a round with only a few rows is between these two values.

In the practical case the bench height is often given. Then the relative influence of errors in the drilling is reduced by the bigger diameter.

3.5. Problems

1. Calculate the maximum burden and the spacing in a bench where $d = 75$ mm, and the degree of packing is $P = 1.27$.
2. The same problem as above but a) with a lower degree of packing $P = 1.0$, b) with $P = 1.5$ (pneumatic loader). Compare the VE-values.
3. Decide the practical burden to be used in the three cases above for a bench height $K = 10$ m (33 ft) and an average deviation in the drilling of 5 % of the bench height (3 % alignment faults). Calculate the $V_1 E_1$-values.
4. Calculate the minimum bottom and column charges and specific charge for the holes in a row of relievers in a tunnel round. $d = 50$ mm (2 in), advance per round 4.8 m (16 ft). (See table 3:1.) Maximum burden and spacing? Practical $V_1 E_1$-value with deviation 5 % of the advance?
5. Discuss for the previous problem a) the consequences for the bottom charge of a possible error in the drilling as in fig. 3:2c. b) the extension of the pipe charge in the outer part of the hole.
6. If it is a question of downward stoping and all the holes in the row are fired in the same interval the degree of fixation f is less than in vertical benching (see table 2:2). In what proportion can the VE-value be changed.
7. Relievers with $d = 50$ mm round a square opening of dimensions 3×3 m² $(10 \times 10$ ft²) and a depth of 4.8 m (16 ft). Decide maximum burden and practical burden if the drilling deviation is less than 0.06 m/m.
8. Low bench blasting. Bench height $K = 0.7$ m (3 ft) $d = 39$ mm (1.5 in), slope of the holes 2:1. Find the practical burden V_1 and spacing E_1 and the charge to be used.
9. Low bench blasting. $K = 0.6$ m, $d = 50$ mm, slope 2:1. Find the practical burden V_1, height and concentration of the charge.
10. Calculate the column in table 3:3 for $d = 75$ mm with the aid of the law of conformity (see 2.1), starting from the values given for $d = 50$ mm.
11. Discuss the fragmentation in the bottom, pipe and upper part of the burden in example 3:1 according to diagram 2:5.

4. LOADING OF DRILL HOLES

While there has been a rapid and intense development in the tools and technics for drilling the holes there has been a definite time lag in adopting new loading methods. Hence on many highly mechanized jobs the amazing result may be that it takes more time to load the holes than to drill them. But it is often not made quite clear that the loading operation is a highly qualified one, calling not only for attention to the technical efficiency but also to safety considerations and a correct distribution of the charge which is finally to do the job.

The delay in the introduction of new loading equipment is due to the demand for extensive safety testing of every detail and for every situation that might possibly occur, and it may also be a question of control in practice for several years. Then in many countries there are still old-time regulations demanding a tamping rod of wood or such like, and these will have to be changed. In some cases there must also be a change in manufacturing and packing the explosive.

Specialized cartridges for specific purposes, pneumatic and pressure-water loaders for dynamites and other explosives in cartridges, and powder loaders for AN-oil or various prefabricated types of powder explosives are on the way. They will in the coming years change the whole procedure of loading the drill holes.

4.1. Simple loading methods

a. Rod tamping

For explosives in cartridges rod tamping is employed as a rule. It is a primitive method and experience shows that the average degree of packing is comparatively low. To improve this (the degree of packing) cartridges are used with a perforated case, which rips up when compressed. Another method is to cut the case with two diametrical incisions lengthways. This is not appropriate, however, in fissured rock where the cartridges are apt to stick in the drill hole. There is less need to perforate or cut the case if the cartridge case is made of thin paper. In Sweden the case for dynamite cartridges consists of thin paraffined paper two times round, weighing only 65 g/m² with 20–25 g/m² paraffin. If the degree of packing is to get anywhere near the density of the explosive in the cartridges only one cartridge at a time should be inserted and compressed in the drill hole. In deep holes this is a very slow job.

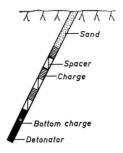

Fig. 4:1. Stick loading for trench-blasting ($d = 32$ mm).

b. Long rigid cartridges

To facilitate the loading of drill holes long, rigid cartridges which are pushed into the drill hole without being deformed are sometimes used. The maximum degree of packing will not be particularly high, partly because of the thickness of the rigid case, partly because of the difference in diameter required between drill hole and cartridge. If this is made too small there will be the risk of the cartridges sticking when they are being pushed into the hole.

In holes drilled with chisel bits the cartridge diameter should be 5–6 mm less than the diameter of the bit. In diamond-drilled holes the difference may be smaller. According to recommendations in Canada $1\frac{1}{8}''$ cartridges can be used in $1\frac{1}{4}''$ hole, $1\frac{1}{4}''$ cartridges in $1\frac{1}{2}''$ holes and $1\frac{3}{4}''$ cartridges in $2''$ holes. This corresponds to a difference in diameter of 3 mm in $1\frac{1}{4}''$ and 6 mm in $1\frac{1}{2}''$ and $2''$ holes. In the LKAB iron ore mines at Malmberget (Sweden) for many years 30–40 m (100–130 ft) long 36 mm (1.4'') diamond-drilled holes were loaded with 32 mm ($1\frac{1}{4}''$) cartridges. In this work cartridges were sometimes getting stuck so that holes had to be left unloaded.

c. Blowing in the cartridges with compressed air

In Germany and Austria a method has been developed for blowing in the cartridges with compressed air through a metal tube which is inserted in the drill hole. The cartridges are given a great velocity so that they are crushed when striking the bottom. The method was originally employed for loading the chambers of sprung drill holes. In recent years it has become adopted to a certain extent for loading deep drill holes, especially in fissured rock. To increase the degree of packing the speed has been increased so that the safety margin is rather small. In Germany it would seem to be permitted to use an apparatus which gives 30 mm ($1\frac{3}{16}''$) cartridges of a gelatinized AN-dynamite with 22 % NGl (nitroglycerin–ethyleneglycol–dinitrate) a velocity of 50 m/s (170 ft/s) and 25 mm (1″) cartridges 60 m/s (200 ft/s). According to Berthmann and Christmann 25 mm (1″) cartridges of this explosive detonate at 88 m/s (300 ft/s) and 30 mm (1.2″) cartridges at 103 m/s (340 ft/s).

In spite of the high striking speed the degree of packing is moderate. According to Hintze is an approximate degree of packing of $P = 1.0$ kg/dm³ obtained in 10–30 m (33–100 ft) drill holes with a diameter of 60–95 mm (2.4–3.8″). In two 90 mm (3.6″) drill holes which sloped down 3.5° and contained some water although they had been blown with compressed air, the degree of packing was only $P = 0.83$.

Plastic hose is now employed instead of the metal tubes. With the pressures and velocities used in practical loading the paper is sometimes peeled off so that the inside of the hose becomes sticky with dynamite. When this is the case the pops from the cartridges striking at the bottom of the hole are no longer heard, but only a rustle. The lower striking speed implies that the degree of packing will be poor. It has therefore been suggested to blow a pad of cotton waste through the hose after loading each drill hole. On account of the risk of an electrostatic charge from the great velocities of the cartridges employed here it is forbidden to use electric detonators or to have such in the neighborhood when loading.

d. Loading with sticks

In such cases as trench blasting, etc., where only a concentration of 0.25 kg/m (0.17 lb/ft) is desired, even cartridges of ⅞″ diameter put into the hole without tamping give far too much explosive (0.5 kg/m, 0.33 lb/ft). For the moment there is no special cartridge for the desired concentration, and the lower concentration of the charge has to be attained with so-called stick loading. With a gelatinized AN-dynamite with 35% NGl (LFB-dynamite) wooden pegs of about the same length as the cartridge can be placed between every cartridge, thus halving the concentration, with ⅞″ cartridges to 0.25 kg/m (0.17 lb/ft) and with 1″ cartridges to 0.35 kg/m (0.23 lb/ft) (fig. 4:1). A more even distribution is obtained with 0.1 m (4″) sticks and 1″ cartridges with half the length. This should be used for explosives with a low gap test distance.

e. Charges for smooth blasting

Special charges for perimeter blasting have been developed by Nitroglycerin AB, Sweden. They were introduced on the market in 1955. They consist of pipes about 0.5 m (1.6 ft) long with a special explosive of gurit (fig. 4:2). The charges can be joined together by sleeves to obtain the desired lengths

Fig. 4:2. Charges for smooth blasting.

and the external diameter at the joints would then be 18.5 mm ($\frac{3}{4}''$). The gurit charges constitute a complete tool for all kinds of smoothing roofs and walls in tunnels, and in other underground premises. By using them a gentle treatment of the finished rock face is ensured. The concentration is equivalent to 0.18 kg/m (0.12 lb/ft) of LFB-dynamite.

Charges for smooth blasting can also be made by cutting up ordinary dynamite cartridges lengthways and putting them on a wooden rod. By using such halves of 25 mm cartridges placed 5 or 10 cm apart a concentration of less than 0.25 kg/m (0.17 lb/ft) is obtained. Another charge which gives comparatively good results can be made with 10 cm wooden sticks alternating with halved cartridges of LFB-dynamite, as mentioned in the preceding paragraph.

90

FIG. 4:3. Loading a round with 22,500 kg (50,000 lb) of dynamite in 1300 holes, 50 mm (2 in) in diameter, 15 m (50 feet) deep with the pneumatic cartridge loader.

4.2. The pneumatic cartridge loader

a. Introduction

The pneumatic cartridge loader has been developed according to a principle suggested by C. H. Johansson. A high degree of packing is obtained with gelatinized AN-dynamite even in very deep drill holes. By using the pneumatic cartridge loader the drilled volume is fully exploited. Thus the drilling costs are reduced to a minimum. The loader as well as its improvements are patented.

By comprehensive experiments at the laboratories of the Nitroglycerin Company and the experience gained from practical use over a period of 10 years the pneumatic cartridge loaders have been greatly improved and they satisfy very high requirements of safety and efficiency. They have become quite indispensable for underwater blasting, for blasting without removing the overburden and for long-hole blasting. They are also used increasingly for tunneling and other sorts of rock blasting. Fig. 4:3 shows the loading work for blasting without removing the overburden. The round comprises 1300 drill holes 2 inches in diameter and 50 feet deep, 8500 m³ rock, 50,000 lbs of dynamite and 4280 blasting caps. In some cases of underwater blasting the charging can be done from the surface, and the diver has only to put the loading tube into the drill holes. If divers carry out the loading from the bottom they need not insert the cartridges into the holes as these will be put into the tube at the surface. In cracked rock there is the advantage of noticing if the hole is blocked before the loading is started.

In the beginning loading-tubes of brass and an Al-alloy were used. Early experiments with flexible loading tubes gave negative results due to the difficulty of getting suitable tubes from materials then available and also due to the demand that the loader should be usable for drill holes with diameters down to 30 mm. In connection with the method introduced by AB Skånska Cementgjuteriet for blasting rock without having to remove the overburden, experiments were started with flexible tubes for drill holes of 40 mm diameter on the initiative of C. Anger. These experiments led to the use of flexible tubes S. Ljungberg has suggested and togther with L. Sternhoff developed a semiautomatic breech-piece for feeding in the cartridges continuously and a robot loader for the movement of the tube in the drill hole. These improvements have multiplied the loading speed, simplified the handling and reduced the necessary working power. Nowadays it is possible to load dynamite cartridges to a density $\varrho = 1.3$–1.5 kg/dm³ (80–95 lb/cu.ft) at a gross capacity of 300–500 kg/hour (650–1100 lb/hour).

Before the pneumatic cartridge loader with the metallic loading tube was permitted in Sweden a comprehensive investigation had to be carried out in order to ensure a satisfactory degree of safety. A considerable difficulty was that the knowledge of the mechanism of initiating and the conditions for initiating by impact or friction were and still are very incomplete. Test methods were therefore worked out reproducing as far as possible the strain which the explosive would be exposed to in practical loading work. On the basis of the results obtained, the metallic loading tubes are permitted in dry bore holes for loading a gelatinized ammonium nitrate dynamite with 35% NGl. In wet drill holes the metallic tubes can be used for all sorts of gelatinized dynamites including blasting gelatine with 95% NGl.

Loaders with polythene tubes are permitted for all types of civil explosives also in dry drill holes.

b. The design and function of the loader

The hand-operated loader with polythene tube (fig. 4:4) has a reducing valve (1), giving 3 Atm (overpressure) when connected to the system for compressed air with a maximum of 10 Atm and a foot-operated three-way

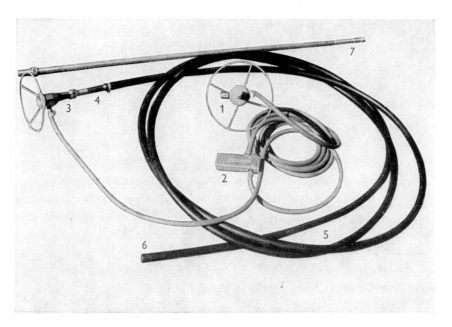

FIG. 4:4. The pneumatic loader with polythene tube.

air valve (2), which lets out the overpressure in the loading tube when the valve is closed. The breech-piece (3) for inserting the cartridges has a lid which closes automatically when the compressed air is turned on (fig. 4:5). A throttle flange with 5 mm diameter restricts the air supply so that the maximum speed of the cartridges in the loading tube will be 10 m/s. The wheel-like protection ring holds the breechpiece above the ground and keeps out gravel, mud and so on. The loading tube (5) is made of antistatic

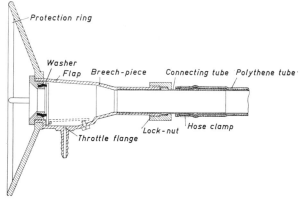

FIG. 4:5. The breech piece.

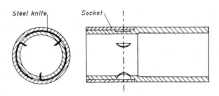

Steel knife   Socket

FIG. 4:6. The jet.

polythene with inner longitudinal grooves. The jet has three internal knives
(fig. 4:6) and is welded together with the loading tube by means of a
special welder. The intermediate tube (7) is in some cases placed between
the breech-piece and the connecting tube to facilitate the inserting of
cartridges.

While charging the loading tube is passed down to the bottom of the drill
hole. The compressed air flows through the tube and pushes the cartridges
forward, the diameter of the latter being approximately 1 mm less than
the inner diameter of the tube. As soon as a cartridge reaches the jet, which
has about the same diameter as the cartridge, the air exhaust will be
checked, and the pressure in the tube increases. The cartridge is then pressed
out at a very low velocity and the paper is cut up by three knives in the
jet. If the loading tube is moved slowly outwards with repeated light counter-
movements the ejected cartridges will be packed to a high density.

There are apparatus for 22, 25 and 29 mm ($\frac{7}{8}$, 1 and $1\frac{1}{8}''$) cartridges and
experiments are being made with 40 mm ($1\frac{1}{2}''$) cartridges. Loaders for
25 mm (1″) cartridges can be used for drill holes with 40–80 mm (1.6–3.2 in)
diameter, and they give a density of 1.5–1.4 kg/dm³. With these apparatus
one man loads approximately 600 pounds per 8-hour shift, and two men
load 1100–1300 pounds per shift in drill holes of 50 feet depth.

By using the semiautomatic breech-piece and the robot loader (fig. 4:7)
the loading will be done more conveniently and very much quicker. The
apparatus is operated by one man, who only has to put the cartridges into
the breech-piece. He will be spared the rather strenuous work of packing
the explosive in the drill hole, and he will not be troubled by the returning
air with nitroglycerin vapours, or the water that is thrown out from wet
drill holes.

The semiautomatic breech-piece contributes on its own merits an important
increase in the loading speed. It acts as a sluice which passes the cartridges
into the loading tube while retaining pressure in the tube. Thus the loading
can proceed without intervals for inserting the cartridges. The way it
functions is evident from fig. 4:8 which shows four stages of the passage
of the cartridges through the sluice. An automatic valve gives the volume

Fig. 4:7. Loading with the semiautomatic breech-piece and the robot loader.

of the sluice alternatively 3 Atm and 0 Atm (over-pressure) with a frequency of 1 per second. This corresponds to a gross capacity of 800 lb/h for 1 in and 1100 lb/h for $1\frac{1}{8}$ in cartridges.

In loading with semiautomatic breech-pieces two men loaded three bottom charges of 55–75 lbs each per hour with 1.3–1.4 kg/dm³ density in

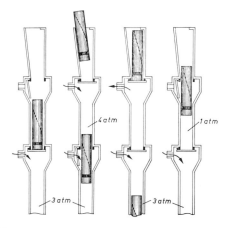

Fig. 4:8. The semiautomatic breech-piece.

95

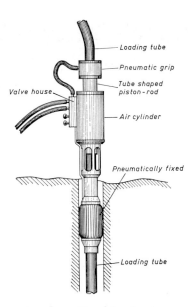

FIG. 4:9. The robot loader.

115 feet deep drill holes with 3 in diameter. The work included opening the dynamite packages, inserting the blasting caps and removal of the loading tubes between the holes. With good planning and some practice it may be possible for two men to load 2000–3000 lbs in an 8-hour shift.

The robot loader (fig. 4:9) consists of a pneumatic cylinder which runs a tubular piston-rod through which the loading tube is inserted in the drill hole. To the piston-rod is connected a pneumatic grip arrangement, a "hand" which holds the tube by friction so that it makes an inward and outward movement with a 10–15 cm amplitude. If the inward movement of the tube is held up by extruding cartridges the tube will slide through "the hand" and compress the cartridges with a force equal to the sliding friction. In the outgoing movement the tube follows the grip and new cartridges are fed through the jet. The degree of packing depends on the air pressure of the grip. Thus it is possible to have a maximum degree of

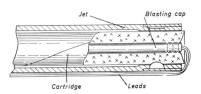

FIG. 4:10. The blasting cap inserted in a cartridge in the jet of the loading tube.

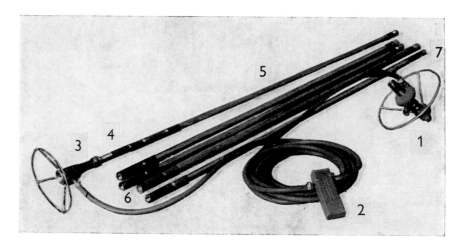

FIG. 4:11. The pneumatic cartridge loader with metal tubes.

packing at the bottom and then obtain a lower value in the column of the hole simply by lowering the air pressure of the grip.

The blasting cap is inserted into the drill hole with the loading tube. Fig. 4:10 shows the cap placed in a cartridge at the end of the tube well protected against mechanical damage. The dynamite cartridge is either forwarded to the jet through the tube, or a cartridge of somewhat smaller diameter is put in through the jet. When the cartridge with the blasting cap is pressed out of the loading tube the leads should be kept loose so that they do not cause the cap to tilt.

The pneumatic cartridge loader with metallic loading tubes (fig. 4:11) consists of a jet-tube of brass, extension tubes of an Al-alloy and jointing sleeves and lock sleeves of brass. The ends of the tubes are cone-shaped and the jointing sleeve forms a double cone. In loading vertical drill holes and in all cases where the free space is insufficient for the whole length of the tube the parts of the tube are put together gradually as the tube is inserted into the drill hole. Inversely it is taken to pieces successively when ejected during the loading. Experience shows that after a short period of practice this is quickly and easily done. It is important that all parts of the joints are kept clean.

The metallic loading tubes have the advantage that they can be used for drill holes of smaller diameter. Table 4:1 (p. 116) shows the dimensions necessary for loaders with polythene tubes.

With metallic tubes 25 mm cartridges can be loaded in drill holes 33 mm in diameter and 22 mm cartridges in 30 mm holes.

FIG. 4:12. Testing apparatus for the effect of friction on explosives.

c. Safety considerations

The pneumatic cartridge loaders which are used in Sweden are approved by the National Workers' Protection Board. The authorities have based their decision on investigations of the conditions for initiating the explosive by friction, impact and electrostatic discharge.

Friction. For loaders with metallic loading tubes the friction test is especially important. The jet-tube and the jointing sleeves that expose the explosive to friction are made of brass which has proved a suitable alloy. The extension tubes are protected by the protruding jointing sleeves, and in order to keep down the weight they are made of an Al-alloy. The friction test is carried out with a piece of a loading tube 25 mm in diameter which is pulled to and fro in a semicylindrical slot 30 mm in diameter on the surface of a block of granite. The explosive is smeared on the rough surface of the slot. The length of the tube is 8 cm (3 in), the amplitude 12 cm (4.8 in), the frequency 2 p/s and the maximum speed 3.5 m/s (10.5 ft/sec). The pressure against the granite surface can be adjusted between 10 and 200 kp, and at each test the tube is driven 3 times in 3 seconds. Fig. 4:12 shows the testing apparatus with pneumatic cylinders for the movement and the pressure against the granite surface.

The severest test is obtained when the cleaned granite surface is dried with a blow-torch and dynamite applied when the temperature of the surface is 5 to 10°C higher than the air temperature. With regard to the results obtained the use of loaders with metallic tubes in dry drill holes is limited to gelatinized AN-dynamite with 35 % NGl. In very wet ones or those full of water they may however be used for all gelatinized dynamite qualities. Loaders with plastic tubes are permitted for loading all civil explosives also in dry drill holes.

Impacts. Fig. 4:13 shows the frequency of detonations as a function of the velocity when 25 mm cartridges of gelatinized ammonium dynamite with 35 % and 50 % NGl respectively are thrown against a steel plate. The cartridges were accelerated with compressed air in a loading tube 7.5 m in length. Every point represents 15 tests. The investigation shows that none of the cartridges of dynamite with 35 % NGl was initiated until the velocity exceeded 106 m/s at 15°C, and 80 m/s at 30°C respectively. As to dynamite with 50 % NGl the corresponding results were 80 m/s at 15°C, and 57 m/s at 30°C and 40°C respectively. By exhausting through a tube 1.5 m in length a frequency of detonations of 30 % at 40°C was obtained at 53 m/s. No detonation was obtained at 30°C and 15°C respectively. The higher sensitiveness may be due to the high over-pressure of 6 atm which was necessary in this case.

The stationary velocity in the loading tubes is limited to 10 m/s by the throttle flange of the breech-piece. Considerably higher velocities, however, can be reached if the cartridges are stuck, so that the pressure in the tube is increased, and then suddenly released.

In order to prevent too high velocities the pressure is limited to 3 atm over-pressure. The foot valve which supplies the compressed air is of a type which blows out the over-pressure of the loading tube when shut off. The internal longitudinal grooves of the tube render the formation of tight plugs more difficult. Cartridges with a spiral paper cover must be inserted so that the compressed air does not penetrate the cover and peel it off. Then a plug of dynamite can stick in the tube. In Sweden nowadays the cartridges are marked with red arrows to indicate the insertion. Since these measures were taken the formation of plugs seems to have ceased completely. Should the tube be blocked by dynamite the plug can easily be removed by water pressure. A bucket of water and a simple water pump connected to the loading tube instead of the breech-piece are sufficient. *Attempts to blow out plugs with compressed air must on no account be made.*

Electrostatics. The loading tubes are manufactured of an antistatic polythene. This contains an admixture which gives rise to electric conductive layers on all free surfaces thus preventing the formation of electrostatic charges.

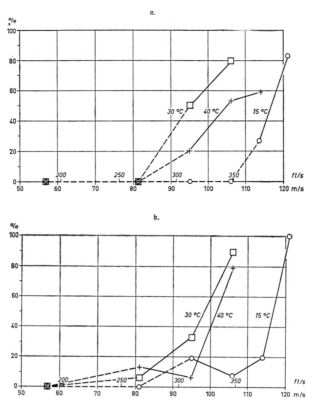

FIG. 4:13. Frequency of initiation of cartridges of gelatinized ammonium-dynamite hitting a steel plate. a) 35 % NGl. b) 50 % NGl.

d. Practical experience

The pneumatic cartridge loader has been in practice in Sweden for 10 years and it is nowadays used on a large scale for many kinds of rock blasting. Some examples are given.

Loaders with metal tubes were employed as early as 1950 by the Swedish State Power Board at Harsprånget for loading 2500 drilling meters of submerged drill holes, partly with internal water pressure. The rock was, in addition, very fissured. The depth of hole was up to 18 m (60 ft), and the diameter of the bits 44–34 mm. The loading was carried out with 25 mm (1 in) cartridges of blasting gelatine and the degree of packing was $P = 1.0–1.2$.

Loaders with metal tubes have been employed since 1953 for underground blasting with long holes in an LKAB iron ore mine at Malmberget, in Lappland. The holes are drilled with jointed drill rods and 42 mm steels.

100

Holes up to 40 m (135 ft) long and an upward slope of 60° are loaded without difficulty. The explosive is gelatinized ammonium dynamite with 35 % nitroglycerin in 25 mm (1 in) cartridges. The average degree of packing is $P=1.1$, and the loading effect with one man per loader is 100–125 kg (220–275 lb) per shift. Before the loader was introduced 1.5 m (5 ft) long 32 mm ($1\frac{1}{4}$ in) cartridges in rigid casings were used in 36 mm (1.45 in) diamond-drilled holes. The degree of packing in this was $P=1.0$. It often happened, however, that cartridges stuck so that holes had to be abandoned incompletely loaded. The rod tamping previously used took considerably longer and gave an average degree of packing of only $P=0.8$.

In blasting without removing the overburden the pneumatic cartridge loader is a necessary tool. For the Lindö Canal at Norrköping the Skånska Cementgjuteriet used in 1957–1960 the loader for blasting 200,000 m³ (260,000 cubic yards) of rock. In the course of this work 30,000 drill holes with a depth of 16 m (53 ft) and a diameter of 40–50 mm were loaded. To begin with loaders with metal tubes were used, but most of the work has been done with polythene tubing for 25 mm (1 in) cartridges. The degree of packing has been $P=1.3$–1.5. (See table 4:1, p. 116.)

4.3. The pressure water loader

For loading under water good results have been obtained by using a loader with compressed water instead of compressed air. The upward force of air-filled loading tubes is apt to be troublesome, especially when the loading is carried out by divers. The tendency of the escaping air to stir up the bottom-mud and completely obstruct the divers' vision is also a great drawback. These difficulties are eliminated by loading with pressure water.

The potential energy of the pressure medium can be ignored because of the slight compressibility of the water. In this case the cartridges cannot be accelerated to high velocities by rapidly lowering the pressure. The loading pressure must, however, be moderated so that the detonating capacity of the explosive is not damaged. The explosive must of course be suitable for underwater blasting. Gelatinized AN-dynamites with 50 % NGl (LF50) and 60 % NGl (LF60) respectively have been used in Sweden, in certain cases also with 35 % NGl. In loading to a depth of 15 m (50 ft) it has been found that 4–5 Atm are a suitable working pressure if the operator stands above the surface of the water. When divers do the loading the pressure should not be more than c. 3 Atm, at a depth of 15 m (50 ft), and at greater depths c. 1 Atm more for every 10 m (30 ft). Considering the detonating capacity, pressures above 7 Atm should not be used for gelatinized ammonium dynamites.

101

These loaders are in principle constructed in the same way as the pneumatic ones, but all fittings are designed with larger flow areas. Between the breech-piece and the plastic tube there is an intermediate tube which is drained when cartridges are inserted. It is 1.8 m (8 ft) long and holds c. 12 cartridges. At its lower end there is an outlet cock and in front of the breech-piece there is a valve. Both are operated simultaneously with a foot-operated release valve which opens the one when the other is shut off, and inversely.

The reducing valve is made for a maximum of 13 Atm, but can be used for higher pressure after a change of springs. It is fitted with a manometer and connected to the valve on the breech-piece by a $\frac{3}{4}$ in pressure water hose with a length of 10 m (33 ft). The breech-piece is fitted with a manometer on which the process of loading can be followed. The pressure goes down when the loading hose is emptied.

In downward drill holes the water remains in the hose when the intermediate tube is drained for inserting a new set of cartridges. Thus the system can be readily filled up with water when continuing to load.

Care must be taken to see that the loading of the drill hole runs smoothly. If not, the operator may not feel when the cartridges are pressed out and waterfilled interspaces can be formed in the explosive column. If a cartridge gets stuck in the tube the water pressure can be increased until the plug is dislodged without removing the loading tube from the drill hole. The manometer of the breech-piece shows when all cartridges have been ejected.

4.4. Loading of AN-mixtures and other powder explosives

To make explosives by mixing ammonium-nitrate (AN) and oil is a rather old invention made by the Swedes Olsson and Norrbin only some few years after Alfred Nobel invented the dynamite. As the AN-crystals have a pronounced tendency to coalesce and form coherent cakes the AN-oil composition was however not suitable for loading in drill holes.

In the 1950's coated prills of AN which does not cake were manufactured in U.S.A. When mixed with carbon or oil and poured into downward holes of large diameters it could be initiated to detonate and used in practice in rock blasting as was shown by R. Akre with carbon black and AN. Blasting with AN-mixtures has since then become very widely adopted in large diameter bench blasting, especially in U.S.A.

The invention of treatments of crystalline and prilled AN so that it remains free-running has made it possible to produce bulk explosives which can easily be poured or blown into the holes. Ivansen, Johansen, Pousette and others found that by using such crystalline AN it was possible to get a

102

comparatively stable detonation also in holes of small diameter, down to 25 mm (1 in). This opened the way for the use of AN-mixtures also for underground operations in mining and tunnel driving. In recent years prilled AN has also been introduced for this purpose in Canada.

In underground operations special care must be taken to avoid the formation of dangerous fumes. Small changes in the composition, or incomplete mixing, can have a very noticeable effect on the gases. It has been found that they can vary considerably from one round to another in practical full scale application, especially in the case of the do-it-yourself mixtures. Attention must be paid to the priming so that the detonation is maintained through the entire column of explosives also in long holes. In loading powder it is important to avoid air inclusions, which may cause misfire or incomplete reaction. The risk of toxic gases being produced is definitely increased by incomplete detonation.

a. Different types of powder loaders

In downward holes of a larger diameter than 75 mm (3 in) free running mixtures can simply be poured down through a funnel. If it is question of big rounds and holes of $d = 100$–250 mm (4–10 in) the holes can be filled directly from a big container on a car, or from a tank car, etc., as has been used in many operations in U.S.A. and Canada. It can then be arranged for the AN and oil to be mixed automatically when pouring into the holes.

For horizontal and upward drill holes special loaders are used. They consist, as a rule, of an air pressure container from which the powder is blown into the drill hole through a hose. Some loaders are equipped for mixing AN and oil at the drill hole immediately before the loading. It is important that the container is designed and equipped so that no channels are formed in the powder through which the air blows into the hose without conveying powder. Another problem is to dimension the hose and adjust the quantity of air so that the powder is fed into the drill hole without forming plugs and gusts of air.

A very simple powder loader with a high air velocity is shown in fig 4:14, the so-called Penberthy Anoloader. It consists of an open feed tank and an ejector at the bottom. The AN-oil flows by gravity to the ejector intake and is emitted through a $\frac{5}{8}''$ opening with an annular ring of air jets. The apparatus has been widely used in Canada in underground work for charging holes with a depth of up to 4 m (14 ft). It is ordinarily used for prilled AN and gives a loading rate of 2–4 kg/min (4–8 lb/min). With crystalline AN the loading rate is higher, as is also, however the percentage of AN that returns with the outflowing air.

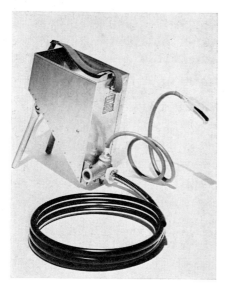

FIG. 4 : 14 a.   FIG. 4 : 14 b.

To keep the "return waste" low the powder loaders commonly operate with a low air velocity. The length of the hose should not exceed 9–12 m (30–40 ft) when the inner diameter is 15–25 mm (0.6–1.0 in), and 15–20 m (50–70 ft) when it is 30 mm (1.2 in) for crystalline AN if gusts of air are to be avoided.

In Canada there is the 'blasthole charger' manufactured by C. I. L. This is capable of loading 2″ holes of more than 60 feet in length at a rate of 20 kg/min (50 lb/min) through a 30 m (100 ft) $\frac{3}{4}$″ tube with an air pressure of 6.5 Atm (95 p.s.i.). The Swedish apparatus Ammon 70 and Ammon 30 for 70 and 30 liters respectively are similar in appearance and operation. They are manufactured by the Nitroglycerin AB. From the very beginning they have been designed for use as loaders of explosive. The safety points of view have been fundamental from its vere conception. One man can work both the loading tube in the drill hole and the diaphragm valve that regulates the flow of powder. The latter is managed by compressed air and a hand or foot-operated valve. The apparatus are used for AN-oil, as well as for prefabricated powder explosives.

Fig. 4:15. Charging blastholes underground with Pneumatisk Transport ANOL-500 litres mounted on the PT-40 Charging Truck.

105

The normal working pressure is 1.5–2 Atm, but it can be varied in keeping with the depth, direction and diameter of the holes. To prevent the formation of channels in the container of the loader this is furnished with a drain cock in the lid. The pressure air is fed in at the bottom of the container and the part of the air that flows out at the lid aerates and stirs the mixture on its way through. Another method which has given satisfactory results is to introduce the pressure air alternatingly through different openings. The air-mixed powder flows out at the bottom of the container through the diaphragm valve. Immediately ahead of this valve an air current is blown into the hose. This extra air is used to prevent the intrusion of water and stone dust when the hose is put into the drill hole, to facilitate the transportation of the powder through the hose and to clear out the explosive remaining in the hose when the valve is closed after the loading is finished. The hose is made of an "antistatic plastic" with an admixture which ensures that all free surfaces have good conductivity.

b. Directions for the operation of the powder loaders

The following instructions for the NAB powder loaders may give a concrete idea of the procedure. This is more or less the same for other kinds of powder loaders. The main difference is that these apparatus are from the very beginning designed for the use of explosives. In regard to safety the pressure is kept low and no cocks are used in which the explosive powder will be exposed to crude mechanical strains from mobile metallic pieces.

Instructions for the loader

Before the hose from the compressed air line is connected to the loader clean it from water and particles by exhausting a few seconds.

Before the explosive is put into the loader check that it is free from foreign matter. If necessary wipe the inside of the container.

After putting explosive in the container close the lid tight and turn the outlet cock to a position between half and fully open.

Turn the main air-cock fully open. Adjust the reducing valve with regard to the length and diameter of the hose between 1.5 and 2.5 Atm.

Instructions for the loading

Blow out water or drilling dust in the drill holes.

Put the primer at the bottom of the drill hole. It should consist of at least 0.2 kg dynamite in two or three cartridges. The cap is placed in the second cartridge from the bottom.

Insert the loading hose in the hole. The ignition wires (fuse) should then be kept well stretched so that they do not slide into the hole.

When the loading hose is in touch with the primer it is withdrawn some 3 cm and the loading is started by opening the diaphragm valve. It is kept open throughout the loading of the hole.

The loading hose is kept in its position until the powder is felt or heard to flow out from the mouth of the hose. When powder is packed round the primer the loading hose is drawn backwards while being moved gently to and fro. The powder is effectively packed in the drill hole by the forward movements.

The diaphragm valve for the powder flow should be closed a little before the intended length is loaded. With a loading hose of 22 mm ($\frac{7}{8}$ in) inner diameter in 34 mm (1.4 in) drill holes this should be done when about 1.2 m (4 ft) remain.

Moisture and drilling dust must be removed from the loading filter before the hose is inserted in a new drill hole. This can be done by tapping the mouth of the hose against the rock.

When the explosive is running low in the container the loading hose begins to vibrate strongly. Then release the pedal.

If any dirt or rubbish such as clips for closing the sacks, pieces of sacking, etc. have got into the container with the explosive, or if moisture should have collected in the lower part of the container, the powder can stick to the walls and a channel be formed in the middle. Compressed air will then flow into the loading hose without powder and give a strong puff. Close the diaphragm valve at once, shake the container lightly and load again.

Before filling up the container check again that it is free from obstructive matter and that the air filter of the outlet cock in the lid is not chocked up.

When loading work is finished for the day the container and hose must be emptied and the container cleaned and dried internally.

c. Practical experience

In *bench-blasting*, which is the main field for use of ANFO-explosives of do-it-yourself or prefabricated type, the conditions vary with the diameter. In large holes the highest capacity is obtained if the explosive is simply poured into the hole.

For holes with $d = 75$ mm (3 in) a capacity of 75 kg/h (165 lb/h) is then obtained for a hole depth of $H = 10$ m (33 ft) and 150 kg/h (330 lb/h) per man for $H = 15$ m (50 ft). These figures apply for Ammonit, a pre-fabricated, specially treated free-running AN-explosive.

In loading direct from tank cars with automatic mixing of the AN and oil a net capacity of about 90 kg/min (200 lb/min) in 200 mm (8 in) holes has been received. A round of 8000 kg (18,000 lbs) can be loaded by two men on one shift. This corresponds to some 500 kg/h (1100 lb/h) per man for hole depths of $H = 15$ m (50 ft).

In *horizontal holes* as in tunnel blasting the loader can give a practical capacity of 20–50 kg/h (40–110 lb/h) per man for diameters of 30–50 mm (1.2–2 in) and at a depth of 3–8 m (10–25 ft) holes. "Return waste" may be about 5 % for crystalline AN-oil, 2–1 % for prills and less than 1 % for crystalline AN-explosives loaded with the Ammon apparatus and filter.

In *upward loading*, as in over-hand stoping in top slicing, the problem is to avoid too much explosive returning from the drill hole with the escaping air or simply running downwards. In Canada good results have been obtained with AN prills and fuel oil.

The waste is reduced if an adhesive tape is wound round the jet of the tube so that a smaller inner diameter can be combined with a larger outer one of the loading tube. The waste value for an 80° slope has thus been reduced to 12 % (Hoberstorfer) for crystalline AN.

The return waste is a big problem in bulk powder-loading especially in upward holes, not so much because of the increased cost of the explosive, which may be negligible, but because of the corroding effect of ammonium nitrate which is spread round, and also because of the discomfort for the operators getting salt on their skin and their clothes.

The transportation of the powder upwards requires a pressure that must be increased with the height. This gives a limitation of 25 m (80 ft) to heights for pressures below 3 Atm. Higher pressures and greater heights entail a greater risk for jamming of the powder in the loading tube.

The AN powder explosives don't detonate in *holes with much water*. Trials have been made by lining the holes with thin sausage casings of plastic. They are easily broken and make the loading much more complicated.

In wet holes in bench blasting with diameters exceeding 50 mm (2 in) a mixture of AN, TNT, and water can be used, as shown by Cook. It can be mixed on the spot as has been done for example in the Babbit Mine in the Mesabi Range, U.S.A., where three bags of 50 lbs. AN, and one of 50 lbs TNT were emptied into the holes together with water. Compressed air tubes were passed to the bottom of the holes. The bubbles passing through the slurry gave a good mixture. However, it is now usually prefabricated and packed in plastic bags that can be lowered into the holes, the first of the bags with a primer (see fig. 4:17). If only the bottom of the holes is wet the top of the charge can be of an AN-oil mixture, or a similar AN-explosive.

Fig. 4:16. A bag of prefabricated slurry (HYDROMEX) slit to show the consistency of this blasting medium. (Courtesy of Canadian Industries Limited).

Fig. 4:17. Lowering Pento-mex-primer into Hydromex bag. (Courtesy of Canadian Industries Limited.)

4.5. Degree of packing in different loading procedures

a. Shape and volume of drill holes

The effective diameter (d_e), i.e. the diameter of a circular hole which has the same area as the drilled hole, is larger than the diameter of the drill bits, here called the nominal diameter (d). Fig. 4:20 shows the result of measurements of volume for horizontal holes drilled in granite with detachable four-wing bits. The unbroken lines indicate the diameter of the bits and the open rings the values observed for the effective diameter (d_e). Measurements have been carried out for granite with chisel-bits with a diameter of 33–38 mm and four-wing bits with a diameter of 40–65 mm ($1\frac{1}{2}$–$2\frac{1}{2}$ in). The effective volume of the holes is 6 % more than the nominal volume.

The degree of packing of the explosive has been defined as the loading weight per unit of nominal volume, which always is known. Its unit is kg/dm³. The degree of packing defined in this way is 6 % greater than the density of the explosive in the drill hole. In horizontal holes which had not been effectively flushed clean from drilling cuttings the effective volume was only 2 % greater than the nominal.

109

FIG. 4:18. Lowering primed bag of Hydromex into bore hole. (Courtesy of Canadian Industries Limited. The photographs of fig. 17 and 18 are from the CIL Explosives Bulletin. August 1960.)

When using cartridges in rigid casings, and when loading with the pneumatic loader it is important for the loading speed that the cartridges and the loading tube pass easily into the drill holes. The clearance required depends on the nature of the rock and the drilling equipment. Holes which have been drilled with chisel bits have a circular form which is triangularly deformed, and experience shows that the diameter of bit should be at least 5 to 6 mm larger than the diameter of the tube. In drilling with loose four-point bits the difference in diameter can be a couple of millimeters smaller. Experience shows that many interruptions are avoided and a greater speed in loading is attained if there is sufficient clearance.

b. Loading with tamping pole

Experience shows that the degree of packing obtained with a tamping pole is, on an average, comparatively low. According to Kallin and Janelid it is ordinarily about $P = 0.8$, and rarely rises to $P = 1.0$. This means that the volume of the drill holes will not be fully exploited and that the drilling cost will be high.

The marked influence of the temperature on the degree of packing is often disregarded. The effect has been studied in an apparatus where 25 mm (1 in) cartridges with and without a split in the paper are statically compressed in a 30 mm ($1\frac{1}{4}$ in) tube. The explosive was a gelatinized AN-

110

Fig. 4:19. De-Watering Truck and Mix Truck in large open pit mine (Courtesy CIL, Canada).

dynamite with 35 % NGl (nitroglycerin-glycoldinitrate). The paper was folded twice round the cartridge and weighed 65 g/m². It was of the type ordinarily used in Sweden and treated with 20–25 g/m² of paraffin. When the paper of the cartridges is split it is not so important to have a high pressure, as a favorable density (1.4) is obtained even with the comparatively low pressure 30 kp, and even at a low temperature of $-25°C$ ($-13°F$). With ordinary paper a higher pressure is required to get the same density as with split paper (fig. 4:21). At low temperatures it is definitely necessary to have good pressure on the tamping pole. An increase from 30 to 60 kp at $-25°C$ ($-13°F$) has increased the density almost 30 % from 1.08 to 1.38 g/cm³.

111

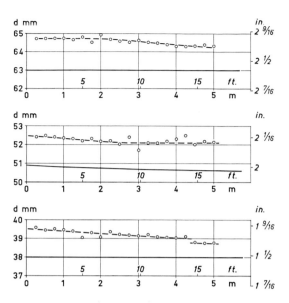

FIG. 4:20. The effective diameter (d_e) of the hole compared to the diameter of the drill-bit (d) after blowing or washing the pulverized rock from the drilling out of the holes.

With short drill holes a favorable degree of packing can be attained by carefully loading one cartridge at a time. The packing is facilitated if the cartridge casing is perforated or slit, or if the cartridge is broken before being inserted in the drill hole. As a rule, however, 2–4 cartridges are inserted at a time and the degree of packing will then be low in all circumstances. Fig. 4:22 shows the degree of packing for gelatinized AN-dynamite with 35 % NGl as a function of the number of cartridges at a time. The investigation has been carried out by the Swedish Highway and Waterway Department in collaboration with Nitroglycerin AB, and deals with 22 and 25 mm ($\frac{3}{4}$ and $\frac{7}{8}$ in) cartridges respectively with casings of the comparatively thin, paraffined paper which is used in Sweden. The holes were 6 m (20 ft) deep and drilled with monobloc chisel bits with a diameter of 35 to 29 mm (1.4–1.2 in). The same values were obtained with both types of cartridges.

c. Pre-arranged cartridges

Fig. 4:23 shows the degree of packing for an explosive with a density of 1.45 kg/dm³ in rigid cartridges as a function of the diameter of the hole. In this case the thickness of the cartridge casing has been assumed to be

112

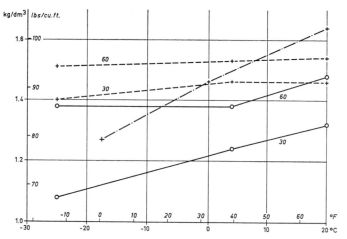

Fig. 4:21. Density of 25 mm cartridges of 35% ngl dynamite in a tube of 32 mm ($1\frac{1}{4}$ in) diameter --- the paper of the cartridges split before applying the force 60 or 30 kp. —— the paper intact before loading –·–·–· careful loading with the pneumatic loader in bench blasting.

1 mm, and the outer diameter of the cartridges (d_e) 5 mm less than the diameter of cut (d). The unbroken curve indicates the effective, and the dashes the nominal degree of packing on the assumption that the effective volume of the hole is 6% greater than its nominal. It is seen from the diagram that the degree of packing (P) will be comparatively low with 0.8 kg/dm³ for 25 mm (1 in), 1.0–1.1 kg/dm³ for 50 mm (2 in) and 1.2 kg/dm³ for 75 mm (3 in) holes.

d. The pneumatic cartridge loader

The pneumatic cartridge loader gives a high degree of packing also in deep drill holes, and this is the case even if the diameter of the drill hole is considerably larger than that of the cartridge. From experience with loaders for 25 mm (1 in) cartridges with hand-operated tubing the degree of packing in drill holes 15 m (50 ft) deep will be $P=1.4$–1.5 with drill hole diameters between 40 and 80 mm (1.6 and 3.2 in).

The value mentioned is not a limit value. In tests in shaft driving as well as in benching, values of up to $P=1.7$ have been attained. It is virtually contradictory to the fact that the proper density of the explosive is about 1.5 kg/dm³, which together with the greater effective volume should give only $P=1.6$. The reason for the higher degree of packing is that the pneumatic loader gives a compression of the air inclusions in the dynamite.

Even for the pneumatic loader the temperature plays an important part,

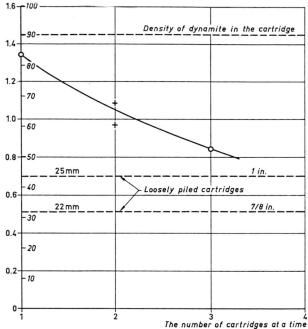

FIG. 4:22. Degree of packing obtained with tamping pole with 25 mm (1 in) and 22 mm (7/8 in) cartridges in a 38 mm (1½ in) hole.

as has been shown in practical tests in bench blasting (line-dotted curve in fig. 4:21). At a temperature of $-15°C$ ($0°F$) a density of 1.27 kg/dm³ ($P=1.35$) has been obtained, and at $20°C$ ($68°F$) about 30 % more, 1.64 kg/dm³ ($P=1.74$).

In ordinary practice a degree of packing of $P=1.35$–1.5 could be counted on.

When using the robot loader the degree of packing depends on the air pressure of the grip as is shown in fig. 4:24 for 35 m (110 feet) deep drill holes with 50 mm (2 in) diameter. Thus it is possible to have a maximum degree of packing at the bottom and then obtain a lower value in the column of the hole simply by lowering the air pressure of the grip.

4.6. Problems

1. Calculate the volume per m for a cylindrical hole with diameter a) $d = 35.5$ mm, b) $d = 50$ mm, c) $d = 75$ mm.

2. Calculate the weight per m for dynamite cartridges ($P = 1.5$) with a diameter a) $d = 22$ mm, b) $d = 25$ mm, c) $d = 29$ mm, d) $d = 40$ mm.

114

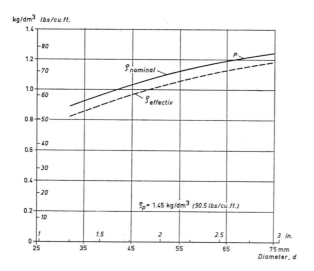

FIG. 4:23. Pre-arranged cartridges with a density of $\varrho = 1.45$ kg/dm³, 0.5 mm ($\frac{1}{50}$ in) thickness of the tube, the outer diameter of which is 5 mm ($\frac{1}{5}$ in) less than the drill-bit, give a higher degree of packing (P) in bigger holes.

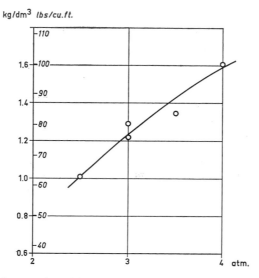

FIG. 4:24. The degree of packing of a 35% NG dynamite as a function of the air pressure of the robot loader grip.

115

Cartridge diameter		Hose diameter in mm		Drill hole diameter	
mm	*in* nominal	Inner	Outer	mm	*in*
22	¾	24	30	⩾ 36	⩾ 1 ⅜
25	⅞	26	33	⩾ 39	⩾ 1 ½
29	1 ⅛	30	37	⩾ 44	⩾ 1 ¾
40	1 ½	41	51	⩾ 63	⩾ 2 ½

With metallic tubes 25 mm cartridges can be loaded in drill holes 33 mm in diameter.

3. The same for cartridges with powder explosive with $P = 1.15$ in the cartridges.

4. Calculate the concentration of charge in a hole drilled with a 40 mm drill bit if the volume of the hole is 6 % greater than a cylinder with $d = 40$ mm and if the hole is loaded with tamping pole taking 2 cartridges at a time.

5. The same hole as in problem 4 is charged with the pneumatic cartridge loader to a concentration of $l = 2.0$ kg/m. Determine the degree of packing P. Compare it with the density in the hole.

6. A bench round has a drilling pattern for holes drilled with a drill bit of $d = 50$ mm and loaded with dynamite ($s = 1.0$) with the pneumatic cartridge loader to a degree of packing of $P = 1.4$ at the bottom of the holes. What diameter of the drill bit is required if the holes are loaded with prills, $P = 0.9$, and with a weight strength $s = 0.9$.

7. An AN–TNT-slurry is used in 150 mm holes, $P = 1.55$, $s = 0.78$. Calculate the diameter to be used with the same burden and spacing if the holes are charged with prills of AN-oil, the density of which is $P = 0.9$ and the weight strength $s = 0.9$.

8. It has been found in an investigation of pratical results in a tunnel that an increase of 20 % in the concentration of charge at the bottom of the holes makes it possible to increase the advance per round 10 %. Discuss the influence of the temperature of the dynamite cartridges on the advance, if the cartridges are stored so as to be at the outdoor air temperature and this varies from $-5°C$ to $+25°C$ during the year.

9. Discuss the difference in advance that can be obtained when loading with tamping rod as compared with the pneumatic cartridge loader.

5. BENCH BLASTING WITH AMMONIUM NITRATE EXPLOSIVES

In recent years reports from all over the world have dealt with blasting with different kinds of explosive mixtures in which ammonium nitrate (AN) forms an essential part. The conclusions drawn from these reports seem sometimes contradictory. In one case there is a claim that blasting with an AN-oil mixture is the most economical, in another that this holds true for an AN–TNT-slurry, and from other sources that a combination of dynamite and AN-oil would be the most profitable. It is quite clear that such physical conditions as water or streaming water in the holes can influence the results, but this cannot explain the divergences in opinion.

There remains to be made a survey of the whole field of application of these different kinds of explosives including all practical experience reported from different sources. This can in fact be done by proper calculation based on previous chapters. Here is shown how such calculations can be performed and applied in all the practical cases reported. It clarifies many of the amazing and contradictory points in this entire complex of problems and indicates potentialities for further reduction of the cost in rock blasting.

5.1. Conditions

Such calculations can be made in quite a general form, but to simplify the discussion I will start with such explosives and costs as are available at the moment. As a starting point for the cost of explosives, I have taken values that have been given by the Swedish Rock Blasting Committee, but I have increased the costs for freight, storage, transportation for the pre-fabricated explosives dynamite, Ammonit and Nabit, and decreased the values for the "do-it-yourself" explosives, AN-oil and AN–TNT slurries. In this way I have tried to favor the "do-it-yourself" mixtures. The values used in the calculations are given in Table 5:1.

Besides the costs, it is of importance to know the properties of the explosives, particularly their breaking force and density. Table 5:2 gives the energy content and gas volume of the explosives dealt with here, and in the last row also the values of the calculated strength. These figures can be compared with those obtained in experiments and given in the upper row of Table 5:3. A comparison between these two magnitudes shows satis-

TABLE 5:1 *Cost of Explosive in $/kg.*

Cost	Dynamite	Ammonit	AN–Oil	AN–TNT–slurry	Nabit
Price	0.43	0.17	0.11	0.18	—
Freight, storage, transp., mixing	0.07	0.03	0.02	0.02	—
Total $/kg ($/lb)	0.50 (0.23)	0.20 (0.09)	0.13 (0.06)	0.20 (0.09)	0.38 (0.17)

factory agreement. The weight strength has been decided with the greatest accuracy that can be attained in such experiments. The mean values are given with a tolerance of ±5 per cent. For the continued calculations the most favorable figures for Ammonit and for AN-oil have been chosen. The values have thus been assumed to be 0.90 and 0.89 respectively for these explosive mixtures.

The effective breaking force calculated per volume of a given drill hole is a product of the weight strength (s) and the degree of packing (P). There is a marked discrepancy here between the different explosives in question, but in addition there are great variations for one and the same explosive depending on how the loading has been done. For dynamite the degree of packing with tamping pole is between 1.0 and 1.25 kg/dm³, while with the pneumatic cartridge loader values of 1.65 to 1.70 kg/dm³ can be achieved as mentioned in the previous chapter. In the calculations 10 per cent lower values have been used, e.g. 1.50 kg/dm³. In practical application, densities of up to 1.18 for Ammonit and 1.20 for AN-oil have been achieved. Also in this case, a less favorable value for the pre-fabricated explosive has been assumed, viz. 1.10, while the calculations for AN-oil have been carried out with the value 1.20. For AN–TNT-slurries densities of up to 1.55 have been assumed, for Nabit, a powder explosive in cartridges, 1.00. The last

TABLE 5:2. *Content of Energy and Gas Volume.*

	Dynamite 35% NG	Ammonit	AN–Oil	AN–TNT–slurry	Nabit
Energy kcal/kg	1160 (1.00)	—	900 (0.78)	757 (0.65)	—
Gas volume dm³	850 (1.00)	—	973 (1.15)	950 (1.12)	—
Calculated strength	1.00	0.87	0 84	0.74	0.90

	Dynamite	Ammonit	AN–Oil	AN–TNT–slurry	Nabit
Weight strength s, exp	1.00	0.89 ± 0.04	0.86 ± 0.04	(0.74 ± 0.04)	0.90
Density, ϱ	1.45	0.9	0.8–1.0	1.4	1.0
Degree of packing, P			1.0		
a) without loader	1.0–1.25	1.0	(0.9 prills)		
b) with loader	1.5	1.1	1.2	1.55	1.0
P. s	1.0–1.5	1.02	1.08	1.21	0.90

row in Table 5:3 gives the product of the degree of packing and weight strength for the different explosives. How deviations from these assumed values influence the results of the calculations will be discussed in part 5.4 "Analysis of Results".

5.2. Various alternatives

Thus, the technical conditions are given for an analysis of blasting with different types and combinations of AN explosives. It is not sufficient merely to deal with those cases where only one and the same explosive, such as AN-oil or dynamite, is used throughout the drill hole. The properties of greatest importance in favor of the AN-explosives are the low cost per kg of the explosive itself as well as of the loading operation as compared with the more expensive dynamite which has the advantage of greater resistance to water and a higher density. To arrive at the most favorable manner of application, the possibilities of combining the different explosives at the bottom and in the pipe charge must be taken into account. Here another question comes into the picture, and that is to find the optimum conditions for the cost as a function of the height of the bottom charge.

The alternatives that have been investigated are given in Table 5:4 and have been indicated A–I (fig. 5:1). The first alternative covers blasting with only AN-oil at the bottom as well as in the pipe charge. The volume of rock excavated per drill hole is proportional to the $P \cdot s$ value 1.08 given in Table 5:3. As is indicated by the values for the density in Table 5:3, one cannot get as high a density with ammonium nitrate in prills as with crystalline AN loaded with a loader. The degree of packing obtained in loading with prills is about $P = 0.9$ and the corresponding value for the volume of excavated rock per drill hole is 0.81. These two cases are given as upper and lower limits in Table 5:4 for alternative A.

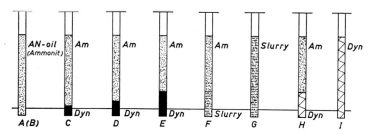

Fig. 5:1. Various alternatives for loading in bench blasting.

Alternative B refers to blasting with Ammonit only; this is a pre-fabricated, free-running AN-explosive. Alternatives C, D and E are for a bottom charge of dynamite, loaded with the pneumatic cartridge loader and with a height of $0.4\,V$, $0.6\,V$ and $1.0\,V$ meters for the bottom charge, V indicating the burden in meters (or ft). For these alternatives the pipe charges have been indicated as Am in the table, which means either AN-oil or Ammonit. Separate calculations for one or the other group of cases have been made in the following tables. With increasing height of the bottom charge, the volume of rock excavated per drill hole increases; the relative value for the last mentioned alternative E is 1.50.

How the volume excavated per drill hole is influenced by the degree of packing in the bottom charge can clearly be seen by comparing H with E, both with the same height of bottom charge but with a different way of loading. For H and G with lower degree of packing the volume excavated per drill hole is 1.00 in relative value. Note, that here a slightly lower volume excavated per drill hole is to be counted on than for A and B with crystalline AN. As can be seen from the calculations, this is fully explained by the low density in loading dynamite with tamping pole, and that this alternative is compared with blasting with AN explosives loaded with loading machines. Thus, the reason is not, as has often wrongly been assumed or stated, that the AN explosives have the same strength as the dynamite Such an explanation has no theoretical foundation and is simply not required.

Finally, such cases have been dealt with as blasting with what is known as slurries of AN, TNT and water in the proportions 65.5/20/14.5. This mixture detonates in holes of 100 mm diameter or more. Maximum energy output is obtained with a slurry of 71/18/11, but then a larger diameter is required. If slurries are to be used in smaller holes, the percentage of TNT must be increased to 30–35. As reported by Farnam, it is in such cases possible to get detonation in diameters down to $1\frac{1}{2}$ inches. This also applies if some part of the AN is substituted by sodium nitrate (SN), in proportions

120

Alternative	Explosive bottom charge			Column charge	Relative broken volume/hole
	V	$0.6\,V$	$0.4\,V$		
A	AN + oil			AN + oil	0.81–1.08
B	Ammonit			Ammonit	1.02
C			Dyn	Am	1.28
D		Dyn		Am	1.40
E	Dyn			Am	1.50
F	Slurry			Am	1.21
G	Slurry			Slurry	1.21
H	Dyn			Am	1.00
I	Dyn			Dyn	1.00

30/30/30/10 of $AN/SN/TNT/H_2O$. The cost of the explosive is then increased about 35 per cent; the density is greater and can be up to 1.70.

For G, there is slurry at the bottom as well as in the pipe charge, whereas the other alternative F has a bottom charge of slurry to a height which equals the burden and the rest of the drill hole is loaded with AN explosive. It can be seen from the last column in Table 5:4 that a definitely greater volume can be excavated per drill hole with slurry than with AN-oil or Ammonit. Nor does this mean that the weight strength of the slurry is greater (it is actually 13% less); it depends entirely on the greater density of the slurry in the drill hole.

5.3. Calculation of the costs

The costs to be counted on are the total costs for explosive and drill-hole. For the calculations the following relations shall be used in the metric system (M) where l is in kg/m, d in mm, V in m. They are derived from the formulas (3:1 M) and (3:2 M) with regard to the influence of the degree of packing P, the weight strength of the explosive s and the relation E/V if they deviate from the values $P = 1.27$, $s = 1.0$ and $E/V = 1.25$

$$l = P\ 0.0008\,d^2 \qquad (5\!:\!1\ \text{M})$$

$$V = 0.046\,d\sqrt{Ps\,(V/E)} \qquad (5\!:\!2\ \text{M})$$

V indicates the maximum burden and has to be diminished by 0.1–1.0 m for collaring faults and 0.02–0.15 m/m for alignment faults to get the practical burden V_1. Of special interest here is the case with drilling machines

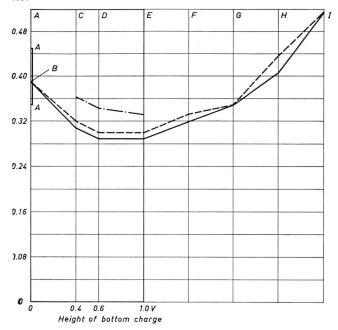

Fig. 5:2. Total relative cost for drill hole and explosive in different alternatives A–I acc. to fig. 5:1. Drilling cost 0.60 $/dm³. —— blasting with AN-oil, – – – with Ammonit, – · – · – · with a powder explosive in cartridges (Nabit).

on a jumbo, or otherwise in a fixed position. In this case the values 0.2 m and 0.02 m/m could be used if the arrangement for the drilling is carefully made. With $E = V$ we get

$$V_1 = 0.046 \, d\sqrt{Ps} - 0.2 - 0.02 \, K \qquad (5:3 \text{ M})$$

$$V_1 = 3.8 \, d\sqrt{Ps} - 0.7 - 0.02 \, K \qquad (5:3 \text{ E})$$

The equation (5:3 E) is given for English units, V_1 in ft, d in inches.

To make the best of the hole it should be drilled beneath the required bottom. As much as $0.3 \, V$ can give a satisfactory effect and will be used in the following calculations. For vertical holes the depth is

$$H = K + 0.3 \, V \qquad (5:4)$$

The height of the bottom charge is $0.4 \, V_1$, $0.6 \, V_1$ and V_1 for the alternatives investigated. The top of the hole is uncharged. In the calculations this part has been assumed to be equal to V_1 and the bench height $K = 4 \, V_1$. For the sake of comparison $K = 2 \, V_1$ has also been investigated.

122

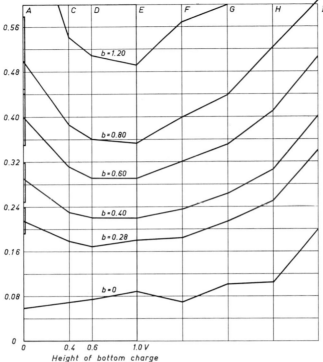

Fig. 5:3. Total cost of drill hole and explosive for different alternatives acc. to fig. 5:1 and with Am = An-oil. The curves 1.20–0 refer to the drilling cost per volume of the hole in $/dm³.

The volume of the drill hole per m is $8(d/100)^2$ dm³, the cost in $/dm³ is denoted b, the depth is $H = 4.3 V$ and the cost of the drill hole

$$C_1 = 4.3 V_1 0.0008 d^2 \cdot b \qquad (5:5)$$

The costs of the explosive are the sum of the bottom and column charges and are for alternative E, for instance:

$$C_2 = 0.0008 d^2 (V_1 P_b 0.5 + 2.3 V_1 P_c 0.13) \qquad (5:6)$$

where the degree of packing at the bottom is $P_b = 1.5$ and for the column $P_c = 1.2$.

The calculations are carried out for $b = 1.2$, 0.6, 0.4 and 0.28 $/dm³. These figures shall include *all* costs, for instance for salaries, machines, equipment depreciation, maintenance, administration, vacations, social expenses etc.

123

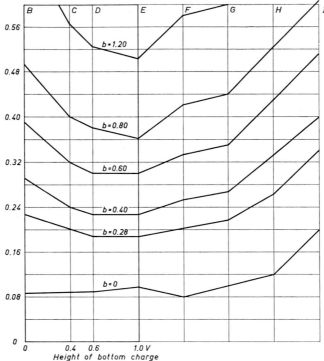

FIG. 5:4. Total relative cost of drill hole and explosive for different alternatives according to fig. 5:1 and with Am = Ammonit. The curves 1.20–0 refer to the drilling cost per volume of the hole.

For a width of the bench B the volume of rock in a row is $B V_1 K_1$ and the number of holes is $B/V_1 + 1$. The costs for the drilling in \$/m³ are then

$$c_1 = C_1(B/V_1 + 1)/B V_1 K = C_1(1 + V_1/B)/V_1^2 K \qquad (5:7)$$

and the costs for the explosive i \$/m³ are

$$c_2 = C_2(1 + V_1/B)/V_1^2 K \qquad (5:8)$$

The total cost of drill-hole and explosive is $c_1 + c_2$ \$/m³. If the width of the bench is disregarded we get relative values for the costs $c_1 = C_1/V_1^2 K$ and $c_2 = C_2/V_1^2 K$.

Example 5:1. The figures in alternative A are calculated for $K = 4\ V_1$ and $H = 4.3\ V_1$ as follows:

The cost of the *explosive* is for A, where bottom and column charges have the same degree of packing $P_b = P_c = 1.2$ and the same cost 0.13 \$/kg

124

		Relative cost per m³ rock			
			Drill hole		
Alternative	Explosive Am = AN–oil/ammonit	1.20	0.60	0.40	0.28 $/dm³
A	0.06/–	0.78–0.59	0.39–0.29	0.26–0.19	0.18–0.13
B	–/0.09	0.61	0.31	0.20	0.14
C	0.07/0.09	0.47	0.24	0.16	0.11
D	0.07/0.09	0.43	0.22	0.14	0.10
E	0.09/0.10	0.40	0.20	0.13	0.09
F	0.07/0.08	0.50	0.25	0.17	0.12
G	0.10/0.10	0.50	0.25	0.17	0.12
H	0.10/0.12	0.62	0.31	0.21	0.14
I	0.20/0.20	0.62	0.31	0.21	0.14

$$l_2 = 0.0008 \ d^2 V_1 P(0.13 + 2.3 \cdot 0.13)$$

and the relative cost per m³

$$c_2 = C_2/(V_1^2 \cdot 4 \ V) = 0.000103 \ (d/V)^2.$$

According to (5:3 M) $d/V_1 = 24.5$, 23.5 and 23.0 mm/m, when $d = 50$, 100 and 200 mm. For the middle value $d/V_1 = 23.5$ we get $c_2 = 0.057$ abbreviated to 0.06, the value given in table 5:5.

The relative cost of the *drill hole* is

$$c_1 = 4.3 \ V_1 \ 0.0008 \ d^2 b/4 \ V_1^3 = 0.00086 \ (d/V)^2 \ b.$$

Thus the relation $c_1/c_2 = 8.3 \ b$ irrespective of the d/V-value. For $b = 0.6$ $/dm³ we get $c_1/c_2 = 5.0$ i.e. the cost of the drill hole is 5 times as great as the cost of the explosive in the case investigated. The value obtained is $c_1 = 5.0 \times 0.057 = 0.29$. This is the lowest relative cost for the drill hole as the highest density for AN, $P = 1.2$, is assumed. For prills with $P = 0.9$ and for crystalline AN with this lower degree of packing the value $c_1 = 0.39$ is obtained. In table 5:5 the relative cost for the drill hole ($b = 0.6$) is given as 0.39–0.29 which indicates that the practical value is between these two upper and lower limits.

The other alternatives have been calculated in the same way for drilling costs of $b = 0.28$–1.20 $/m³ (fig. 2–4). It is worth mentioning that even the *difference* in drill hole cost per m³ for, say, A and E may amount to a greater value than the whole cost of the explosive for the alternative with AN-oil. Even if the explosive ingredients are received free of charge, it is not certain that this gives the cheapest alternative, as the relative figures for

TABLE 5:6. *Relative total cost of drill hole and explosive for alternatives A–I at different drilling costs 1.2–0.28 $/dm³ (0.40 $/dm³ = 1 $/ft in 4″ hole) V = K/4. Am = AN + oil.*

Alternative	Relative cost of drilling and explosive			
	1.20	0.60	0.40	0.28 $/dm³
A	0.84–0.64	0.45–0.35	0.32–0.25	0.24–0.19
B	0.70	0.39	0.29	0.83
C	0.54	0.31	0.23	0.18
D	0.51	0.29	*0.22*	*0.17*
E	*0.49*	*0.29*	0.22	0.18
F	0.57	0.32	0.23	0.18
G	0.60	0.35	0.27	0.22
H	0.72	0.41	0.31	0.25
I	0.81	0.51	0.40	0.34

alternative A are 0.05 units lower. A further discussion is given in the following section (5.4).

For the bottom charge of dynamite in E the degree of packing can also be less than that assumed if for one reason or another there are difficulties in loading the holes, if the temperature is below 0°C (32°F) or if the holes are greatly extended or are more than 100 mm (4 in) in diameter. However, disregarding too special cases, the deviations in the degree of packing are less than those indicated for A. Furthermore, as mentioned in the previous chapter, the P-value in E for dynamite is not even a maximum value.

TABLE 5:7 *Relative total cost of drill hole and explosive for alternatives A–I at different drilling costs 1.2–0.28 $/dm³. V = K/4. Am = Ammonit.*

Alternative	Relative cost of drilling and explosive			
	1.20	0.60	0.40	0.28 $/dm³
A	0.84–0.64	0.45–0.35	0.32–0.25	0.24–0.19
B	0.70	0.39	0.29	0.23
C	0.56	0.32	0.24	0.20
D	0.52	0.30	*0.23*	*0.19*
E	*0.50*	*0.30*	0.23	0.19
F	0.58	0.33	0.25	0.20
G	0.60	0.35	0.27	0.22
H	0.74	0.43	0.33	0.26
I	0.81	0.51	0.40	0.34

TABLE 5:8. *Relative cost of drill hole and explosive for a bench width* $B = 3\ K$ *and burden* $V_1 = K/2$ *and* $K/4$. *Drilling cost 0.28 and 0.40* $/dm^3. Am = AN + oil. Ex:B = 48\ m\ (160\ ft)\ K = 16\ m\ (53\ ft),\ V_1 = 8.0\ m\ (27\ ft)\ and\ 4.0\ m\ (13.5\ ft).$

Alternative	Relative cost of drilling and explosive	
	0.28 $(V = K/2)$	0.40 $(V = K/4)$
A	0.28–0.22	0.34–0.27
B	0.26	0.31
C	*0.23*	0.25
D	0.24	*0.24*
E	0.27	0.24
F	0.23	0.25
G	0.24	0.29
H	0.33	0.34
I	0.36	0.43

When the influence of the bench width B is taken into account the actual cost in $/m³ is one or some tenths higher than the relative costs in tables 5:5–5:7 and figures 5:2–5:5.

Example 5:2. The cost for alternative B with $V_1 = K/2$, $b = 0.28$ $/dm^3$, $B = 3 \cdot K = 6\ V_1$ is to be calculated.
The depth of the holes is $H = 2.3\ V_1$, height of the charge $h = 1.3\ V_1$. The cost for a drill hole is

TABLE 5:9. *Relative cost of drill hole and explosive for a bench width* $B = K$ *and burden* $V_1 = K/2$ *and* $V_1 = K/4$. *Drilling cost 0.40 and* $0.40\sqrt{2}$ $/dm^3. Am = AN + oil. Ex: B = 6\ m\ (20\ ft),\ K = 6\ m\ (20\ ft),\ V_1 = 3m\ (10\ ft).$

Alternative	Relative cost of drilling and explosive	
	0.40 $(V_1 = K/2)$	0.57 $(V_1 = K/4)$
A	0.50–0.38	0.50–0.41
B	0.43	0.47
C	*0.38*	0.39
D	0.38	*0.36*
E	0.41	0.36
F	0.38	0.40
G	0.40	0.42
H	0.53	0.51
I	0.57	0.61

127

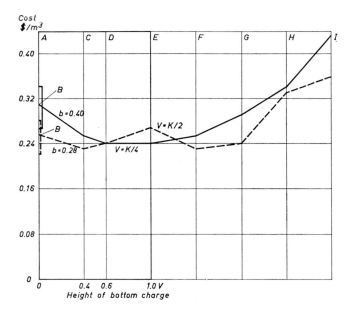

FIG. 5:5. Total cost for drill hole and explosive at a bench width equal to three times the bench height, $B = 3\,K$ and burdens of $V_1 = K/4$ ———— and $V_1 = K/2$ ——— (table 5:8). Ex: $B = 48$ m (160 ft), $K = 16$ m (53 ft), $V_1 = 8$ m (27 ft) and 4 m (13.5 ft).

$$C_1 = 2.3\ V_1\ 0.0008\ d^2\ 0.28 = 0.00052\ d^2 V_1$$

and for the explosive in a hole

$$C_2 = 0.0008\ d^2\ P\ V_1\ (0.20 + 0.3 \times 0.20) = 0.00023\ d^2 V_1.$$

Thus $C_1 = 2.26\ C_2$ and $C_1 + C_2 = 3.26\ C_2$.

According to (5:8) the cost for the explosive per m^3 is

$$c_2 = 0.00023\ d^2 V_1\ (1 + 1/6)/2 V^2 = 0.000134\ (d/V_1)^2$$

With $d = 100$ mm (4 in) according to (5:3 M) we have $d/V_1 = 24.3$ and $c_2 = 0.079$ $/m^3$ and $c_1 = 2.26\ c_2 = 0.177$ $/m^3$. The total cost is $c_1 + c_2 = 0.26$ (table 5:8).

The costs for the other alternatives have been similarly calculated (fig. 5:5). Corresponding cases with a narrower bench width have been included (fig. 5:6). They indicate conditions in underground mining where the geometrical limit largely determines the blasting procedure. The total cost will be greater here; the minimum is obtained for D and E with the smaller burden. This is, however, one of many factors, such as deviation of drill holes, ground vibrations and demands for strength in the remaining rock, which play an important part; the two last mentioned favor smaller burdens.

128

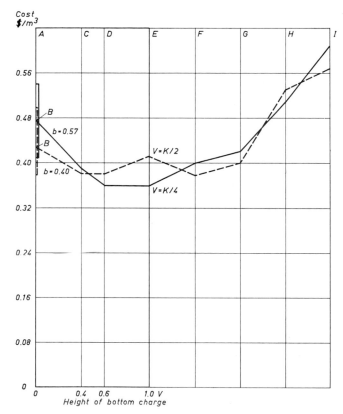

Fig. 5:6. Total cost for drill hole and explosive for a bench with $B = K$ and burden $V_1 = K/2$ and $V_1 = K/4$ (table 5:9). Ex: $B = 6$ m (20 ft), $V_1 = 3$ m (10 ft) and $V_1 = 1.5$ m (5 ft).

The fact that lower drilling costs are obtained with a larger drilling diameter implies that the choice of diameter of the holes, and with that the burden, as well as the bench height affect the total cost. This cannot be judged, however, without including the cost of digging and transportation, and is then a function of the fragmentation, which in its turn depends on the arrangement of holes, loading and ignition. There is not as yet a basis for a satisfactory calculation of those factors so that the influence of bench width, diameter of hole and drilling cost is only accounted for here.

These relations can be given in a general form, but it will be sufficient to compare in some cases the conditions which apply for the burdens $K/4$ and $K/2$. The latter represents an ordinary, but, with due regard to the fragmentation, an inappropriate procedure.

It has been assumed that the cost of the drill hole decreases with increasing diameter according to the relation

Total cost in $/dm³ for drill hole and explosive as a function of bench height with different drill hole diameters. Width of the bench B – 40 m (130 ft). Alternative D, Am = AN + oil.

Diameter d		Drilling cost		K = 3	5	10	15	20	25	30 m
m	in	$/dm³	$/ft	10	17	33	50	67	83	100 ft
0.05	2	0.56	0.38	(0.39)	0.34	0.33	0.34	0.37	0.40	0.45
0.075	3	0.46	0.69	(0.39)	(0.32)	0.29	0.28	0.20	0.30	0.32
0.100	4	0.40	1.06	(0.33)	(0.33)	0.28	0.26	0.26	0.27	0.27
0.200	8	0.28	3.00			(0.30)	(0.25)	0.24	0.23	0.23
0.300	12	0.23	4.20			(0.34)	(0.29)	(0.26)	0.24	0.23

$$b = 10b_1/\sqrt{d} \qquad\qquad (5:9\ \mathrm{M})$$

$$b = 2b_1/\sqrt{d} \qquad\qquad (5:9\ \mathrm{E})$$

where b_1 is the cost in $/dm³ for a drill hole with $d = 100$ mm (4 in). Calculations have been made for $b = 0.28$ and 0.4 (fig. 5:5) and for 0.4 and 0.56 $/dm³ (fig. 5:6). They show the influence that a halving of the diameter and burden has on the total cost of drill hole and explosive.

With larger diameters of holes the height of the bottom charge will also be greater; the depth of hole increases as the under-drilling is in proportion to the burden and a greater relative consumption of drill hole as well as of explosive results from limiting the bench laterally. All this increases the total cost of blasting with a greater burden so that a minimum of costs can be lower for $V = K/4$ than for $V = K/2$ (figs. 5:5 and 5:6). This is a surprising conclusion of great practical importance.

Quite irrespective of this it should be emphasized that the smaller burden gives better fragmentation, increased removal capacity and, in certain cases, a marked reduction in the need of repairs for the digging machines. Investigations made by the Swedish State Power Board, and reported on by Bäckman and Bjarnekull, have shown, in comparing 100 and 50 mm (4 in and 2 in) drill holes, that the removal capacity is up to 50% greater for the smaller diameter of holes.

It is known from this and other investigations that the minimum of costs largely depends on the conditions for removal being made as favorable as possible. This has, on the other hand, led to a difficulty in calculating the most suitable procedure as a definite relation between loading disposition, quantity of the charge, ignition arrangement of the holes, fragmentation and removal capacity is not known at present. It is of special interest to note that the results obtained render it possible to overcome this difficulty in cases of great practical interest.

In the regions where the methods of blasting suggested give about the same calculated costs for drilling and blasting the remaining factor which influences the result, the fragmentation, will be decisive when choosing the hole diameter.

5.4. Analysis of results

Figures for the drilling costs vary considerably from various working places. This often causes uncertainty in the discussion on the final costs for the excavation. It is not unusual that the contractors calculate with costs for the drilling two to three times as high as those estimated in mining operations. To overcome this difficulty, the calculations have been made for drilling costs between 0.23 $/dm³ and 1.20 $/dm³, which may cover the ordinary figures in question. For a 100 mm (4 in) hole this corresponds to values between 0.55 $/ft and 2.9 $/ft.

The conclusions can in every case be compared as far as different drilling costs are concerned. It will be found that certain general conclusions to be drawn apply independently of drilling costs. It is found (Table 5:5) that for alternative A with AN-oil the cost for the drill hole, even at the lowest value, is 2.3–3 times as high as that for the explosive. For the drilling cost 0.60 $/dm³, which is not an unusual value, the corresponding figures are 5–7.

Changing from blasting according to alternative E, with a bottom charge of the same height as the burden, to alternative A with AN-oil only, the cost for the explosive decreases approximately 50 per cent, while the drilling cost increases 45 per cent. The cost for the drill hole (Table 5:5) is 0.04, 0.06, 0.09 and 0.10 units higher for the blasting with AN-oil only, which can be compared with the entire cost for AN and oil given in the same units, viz. 0.05. This means that reference only to the cost of explosive may be quite misleading. The choice and the distribution of the explosive also influence the drilling cost. That it is not possible either to refer only to a certain explosive can clearly be seen by comparing alternatives E and H, both with a bottom charge of dynamite. The latter has the lower density obtained in loading with a tamping pole or, for a larger diameter than 100 mm (4 in), with dynamite that is simply dropped into the hole without tamping. Here the cost of the explosive increases without decreasing the drilling cost.

Regarding the total costs for blasting with "do-it-yourself" AN-oil, on

131

the one hand, and pre-fabricated Ammonit on the other, Tables 5:6 and 5:7 contain a survey of the costs for different alternatives. Changing from blasting with dynamite loaded in the ordinary way (alternative G), which is the most expensive of the alternatives investigated, to alternative A with AN-oil, an essentially reduced cost is obtained if crystalline AN is used and packed with a loading apparatus. The decrease in the total cost is, for a drilling cost of 0.60 \$/dm³ as much as 30 per cent, which must be regarded as rather considerable. However, if one compares this with the figures for AN in prills, loaded to a density of 0.9 in the drill hole, the total cost is only 10 per cent lower, despite the very low costs for the explosive.

It should be emphasized, that even if alternative I has been included here because this case has often been used as a reference, it is not appropriate for use as a starting point for a comparison in a discussion about costs. I have long ago shown in calculations the advantage of using a differentiated charge with different bottom and pipe charges, which represents a more suitable procedure from the technical point of view. The simplest way to make this clear is if the pipe charge in alternative I is replaced by Ammonit as in alternative H. The density at the bottom is the same. A reduction of 15 per cent in the cost is obtained and a lower value than that which applies for blasting with AN in prills and fuel oil is thus achieved.

With a bottom charge of dynamite with a higher density (table 5:6), a further reduction of the cost is attained. A minimum is reached for alternative E, with a height of the bottom charge equal to the burden. The full line (fig. 5:2) gives a graphical representation of the values for alternatives A to I according to fig. 5:1 with Am = AN-oil, and for a drilling cost of 0.60 \$/dm³. The curve of dashes gives the same relationship when Am = Ammonit. For the alternatives C, D and E a part of the curve has also been given for a powder explosive in cartridges, Nabit. The diagram shows that the essential reduction in the costs does not presuppose a change from the pre-fabricated to a "do-it-yourself" AN-explosive, but depends on a correct choice of the relation between the strength of the bottom charge and that of the pipe charge. Even with Nabit in cartridges according to alternative E, a lower total cost is obtained than for the most favorable case with "do-it-yourself" AN-oil, a fact which it may be of interest to observe.

To estimate what influence a somewhat lower degree of packing will have on the bottom charge, it can be mentioned that with a density of 1.35 as a possible practical value for diameters close to 100 mm, instead of 1.50 kg/dm³, values are obtained which are about 2 per cent higher than in the curve of dashes (fig. 5:2) for the alternatives C, D and E. Consequently, it does not essentially influence the general relations.

The values that represent the lowest costs have been underlined for each of the assumed drilling costs in the tables. With lower drilling costs this minimum shifts from alternative E to D, but the difference in costs between these alternatives is naturally insignificant, as one is in both cases evidently in the vicinity of the minimum where the curves are rather level. The diagrams in figures 5:3 and 5:4 show a graphic survey of the whole range of use of AN explosives at different drilling costs.

In curve $0.40 \times$ (fig. 5:3) one can follow step by step the whole development in the U.S.A. and Sweden during recent years. Starting from alternative I with dynamite, or some other pre-fabricated explosive of similar type, and a relative cost of 0.40 per volume of rock excavated, it was found a number of years ago that in trials with "do-it-yourself" explosives, the costs could be cut down to a relative value of 0.32 for AN in prills with oil. This led to a re-arrangement of the whole procedure in bench blasting.

Some years later, Farnam reported that it was possible to attain still lower costs when blasting with slurries, e.g. of the type developed by Cook. The corresponding relative value for the cost per volume in the diagram is 0.27. Later on, works carried out by Ivansen, Ahlman, Pousette, Hoberstorfer and others have shown that even lower values for the costs could be obtained with crystalline nitrate. The relative value for the costs can in this case be as low as 0.25. The last step in the development is evidently, according to the diagram, to use correctly a differentiated charge in order to combine the advantages of an explosive at the bottom with a high volume strength and a cheaper and more voluminous explosive in the rest of the hole. This involves in connection with slurries a change from alternative G to F, which cuts the cost down to 0.23, or using a bottom charge of dynamite according to alternatives D and E, where the minimum value 0.22 is obtained. At present, the use of a high density dynamite charge at the bottom presupposes a drill hole diameter not exceeding 100 mm (4 in). The slurries have until now been used in diameters larger than 100 mm so for the moment the three main groups, C, D and E, on the one hand, and G and F, on the other, complete one another. It can be predicted, however, that slurries suitable for diameters less than 100 mm can if necessary be used as well as methods for loading dynamite in holes of bigger diameter. Regarding the proper choice of the diameter of the hole, it is found that for ordinary bench heights and widths the lowest total costs for drill hole and explosive, are between 100 and 200 mm when undifferentiated charges are used in the holes, (as in alternatives A, B, G and I). The differentiated charge alternatives (D, E, F and H) however, give lower costs and will give minimum values for diameters of 75–100 mm (3–4 in) in benches with $K = 7$–14 m (23–46 ft) as is evident in fig. 5:7. If

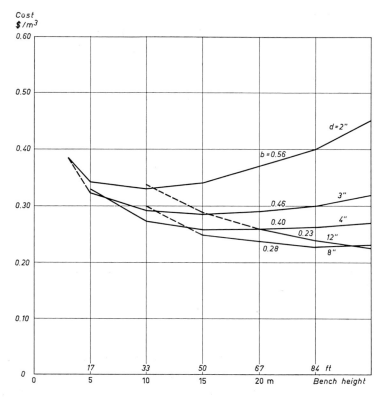

Fig. 5:7. An example of total cost according to alternative D for drill hole and explosives as a function of bench height for different diameters (2 in–12 in) and a drilling cost per volume proportional to $1/\sqrt{d}$. One row of holes. Width of the bench 40 m (130 ft).

the cost of haulage and transport also is taken into account the minimum as a function of drill hole diameter moves from 100–200 mm (4–8 in) to values definitely not exceeding 100 mm (4 in) (fig. 5:8).

As mentioned before the curves and the final value for the minimum cost are only very slightly influenced if AN-oil is exchanged for pre-fabricated Ammonit. For the alternatives D and E which represent the most favorable ones the final gain is in fact so small in mixing the explosive itself that it is almost symbolic.

Ordinarily in bench blasting there is a given bench height, and a suitable diameter and burden have to be chosen. A comparison between two different cases, one of which has been chosen equal to one half and the other to one fourth of the bench height, has been made to show how this choice influences the final cost of drill hole and explosive. As the cost of the drill

134

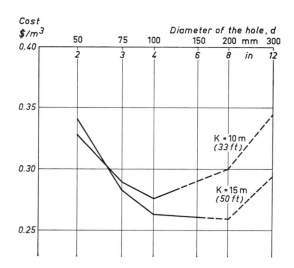

Fɪɢ. 5:8. An example of total cost for drill hole and explosive as a function of drill hole diameter (d) according to alternative B for bench heights $K = 10$ m (33 ft) and $K = 15$ m (50 ft) and a drilling cost per volume of hole proportional to $1/\sqrt{d}$. One row of holes. Width of the bench $B = 40$ m (130 ft).

hole per volume in the case of the simpler alternatives A and I decreases with a bigger diameter, lower cost is obtained for bigger burdens and diameters of the holes. The conclusion that it should be cheaper to choose a burden $V = K/2$ instead of a smaller one, e.g. $V = K/4$ is often drawn but is nevertheless not correct. For the alternatives where the *real minimum* of costs is obtained an inversed relationship partly applies (fig. 5:5).

This is even more pronounced for a smaller bench width, which has a wide range of application for various types of bench blasting in mining. Here there is even less reason to hesitate in taking a smaller diameter and burden (fig. 5:6 and table 5:9), and using the alternatives D or E.

In planning a bench blast for a quarry, or in contracting work one is often free also to choose the bench height. Then the interaction of bench height, diameter of the hole, burden and costs should be known. An example of the relationship in this connection is shown in Fig. 5:7 where the drilling costs have been assumed to be in inverse proportion to d. At a bench height of 5–8 m (17–27 ft) the cost for the drill hole and explosive for a 75 mm (3 in) hole and one of 100 mm (4 in) is about the same. For bigger bench heights up to 14 m (46 ft) blasting with a 100 mm (4 in) hole is the cheapest alternative. In this case and with this cost of the drill hole it is definitely unsuitable to use a diameter of, say, 150 mm (6 in). There is no point at all in using a diameter of 200 mm (8″) or more.

The analysis of costs made here does not include the fact that with a smaller burden a better fragmentation, and thus also a lower cost for unloading the material, for repair and maintenance of the machine equipment, is obtained. The cost for unloading the broken material is about the same as the minimum costs found here, and it is evident that if the costs for drilling and blasting together are about the same in a choice between two cases with different burdens, the final decision will depend on the cost for the digging and thus favor a smaller diameter. This means that in practice it would be advisable to use diameters not exactly up to the minimum in the figures, but to choose smaller values.

Other factors that influence the choice of burden and the diameter of the hole are the cost of loading the explosive, regard to the ground vibrations and the risk of damage in the surroundings and, finally, the time needed for the work. With the intense development of methods for loading the drill holes with explosives in cartridges, one can expect that the difference in costs for different types of loading will be cut down to very small values. If it is supposed that the cost which may be added for loading the dynamite charge in alternative E is 0.05 $/kg, a highly exaggerated value, this will raise the figures for the relative cost by less than 0.006 units in tables 6–8.

If the cost for loading the holes is scarcely a decisive factor, the problem of ground vibrations on the other hand will in many cases play an important part not only with regard to houses and structures in the vicinity, but also when it is desired to diminish the forces in the rock close to the round.

For such cases smaller diameters of the holes make it much easier to carry out the blasting operations according to requirements.

Example 5:3. In a choice between 75 mm (3 in) and 150 mm (6 in) holes for a bench height $K = 10$ m (33 ft) and a bench width $B = 40$ m (130 ft) the cost for the drill hole and explosive according to fig. 5:8 is about the same. But with the 75 mm holes the ground vibrations were reduced to 25 % of those with 150 mm holes, and the capacity of the digger was increased. The cost for the excavation as a whole was lowest with the 75 mm holes.

The calculations allow for the cost of the drill hole being higher per volume of smaller diameters. A drill hole of 200 mm drilled by one drilling-machine can be substituted by four holes of 100 mm drilled by four smaller machines at a cost assumed to be about 40% higher. However, the calculations have not taken into account the fact that this higher cost at the same time is connected with a higher drilling speed. In contracting a saving of time ordinarily implies great indirect savings, a factor that should always be borne in mind. If the capacity of the diggers cannot cope with the greater drilling capacity then two drilling-machines can be installed instead of four to replace one drilling-machine for 200 mm. Besides lower investment costs the smaller diameter drill holes also give a greater flexibility in the operations, as the blasting can then be more easily adapted to meet other requirements, and the work can be performed technically in a more satisfactory way and with better economy.

5.5. Summary

The cost of blasting largely depends on the choice of explosive, its loading and distribution in the drill hole. The minimum cost will be achieved with a concentrated bottom charge of dynamite and a column charge of ammonium nitrate explosive. With bench heights of up to 14 m (46 ft) diameters between 50 and 100 mm should be chosen for drill holes. The choice of these depends on the capacity of the removal arrangements, the importance attached to the work being rapidly carried out, etc. The costs in addition to drilling and blasting are 0.20–0.30 $/m³ rock for digging, 0.4–0.6 $/m³ for transportation (maximum 1000 m) and an additional 30% for administration, in all 1.20–1.50 $/m³ for rock blasted and removed.

Blasting with a free-running AN-explosive only gives a considerably lower cost (relative figure 0.35–0.45) than with loosely packed dynamite alone (0.52). This comparison, which is often made to illustrate the reduction in cost when blasting with AN-oil, for example, is entirely misleading, however. If AN-explosives can be used with regard to water in the holes and other conditions they ought to be compared with other simple, low-cost powder explosives, a type of which is called Nabit, in tables 5:1–5:3. This explosive in ordinary cartridges gives, with a bottom charge of dynamite, a comparatively low cost (0.33).

The gain in mixing the explosive on the working site is very small when compared with a prefabricated AN-explosive (Ammonit). It is at the minimum cost about 0.01 $/m³, that is less than 1% of the final cost. This is unimportant compared with other factors in the picture, for instance the reduction in cost (0.08–0.21 $/m³) obtainable by using the right explosive correctly tamped with the proper loader and correctly distributed in the hole.

5.6. Problems

1. The price of AN-oil is assumed to be 0.11 $/kg (0.05 $/lb) in table 5:1. Calculate the total costs of the explosive and the relative cost in table 5:5 for Alternative A if the price is a) 0.13 $/kg (0.06 $/lb), b) 0.08 $/kg (0.36 $/lb), c) 0.045 $/kg (0.02 $/lb).

2. Calculate the total relative cost of explosive and drill hole with prices of AN-oil as in Problem 1. Compare with the other values in table 5:6.

3. Calculate the relative cost of the AN–TNT-slurry in Alternative G in table 5:5 if the price in table 5:1 is 0.22 $/lb) instead of 0.18 $/kg.

4. Calculate the relative cost of the drill hole for Alternative G if another slurry with 10% greater weight strength is used.

5. Calculate the total relative cost according to problem 4 and compare the values with those of other alternatives in table 5:5.

6. In a bench blasting with $d = 63$ mm holes the degree of packing for the bottom charge is found to be $P = 1.35$. How does this influence the final costs for Alternative E in table 5:6? Compare with the other alternatives.

7. At the place mentioned in problem 6 the height of the bottom charge is $h_b = V$. Calculate the relative cost for the explosive, if a column charge of Nabit (an ammonia dynamite) in prefabricated cartridges of $d = 50$ mm in 1 m lengths is used in a length of $h_c = 2.3$ V instead of AN-oil as in Alternative A.

Calculate the cost if the height of the column charge is increased to $h_c = 2.8$ V.

8. A bench height of 11 m (37 ft) and a diameter $d = 150$ mm (6 in) are used in a quarry and charged according to Alternative A. The cost of mucking is 50% of the total cost of drill hole and explosive. It is found that blasting with half the diameter gives better fragmentation which reduces the mucking cost by 30%; this is to a large extent due to lower repair and maintenance costs. The excavating work can obviously be rationalized. Make a suggestion and discuss various alternatives, also taking into consideration such factors that influence the result, but which are not specified in the calculations (height of muck pile, etc.).

6. SHORT DELAY BLASTING
MULTIPLE-ROW ROUNDS

The introduction of short delay blasting with delays of the order of 1–100 ms has caused a revolution in the technic of rock blasting. It has made it possible to master the problems of ground vibrations and increase the size of the rounds. In multiple-row benching with short delays the operations of drilling and mucking can be performed independently of one another, which greatly facilitates the use of mechanical equipment. Together with an accurate calculation of the charge, it has made possible the planning of operations according to a carefully prescribed working schedule, thus transforming the rock blasting from a mere manual operation into an industrially standardized procedure with increased safety.

There is still a desire to extend the use of shorter delays into the range of ordinary ones and to adopt delays of the order of 0.1–0.25 secs (100–250 ms) instead of second and half-second ones, and also to increase the number of intervals available. Such detonators could be used in all cases instead of ordinary delays but in addition, they could in many important special instances increase the number of delays and the total time extension of multiple-row rounds even if they should not be called short delay detonators.

6.1. Ignition

Short delay rounds can be fired electrically with short delay caps, with detonating fuse and so-called MS connectors or with non electric firing, the so called Nonel system (described in Chapter 12).

a. Blasting machines

Short delay blasting has greatly increased the demands on the blasting machines, and it has been necessary to develop new apparatus with a capacity for rounds from some hundreds of detonators up to several thousands.

Modern, low weight and efficient condenser blasting machines (fig. 6:1) are now available as standards for 50, 600, 2400 and 6000 detonators and even for 25,000 when this is specially demanded. They consist of condenser units and a hand-operated inductor, and are independent of dry cells or net voltage. Their reliability is within wide limits not influenced by heat or cold.

Fɪɢ 6: 1. Condenser blasting machines for 2, 15, 100, 275 and 700 + 800 VA-high safety detonators and a test-instrument.

The machines can be supplied with a special instrument for testing, which is of importance in the case of extremely big rounds. In table 6:1 are given figures for voltage and number of detonators for NAB blasting machines.

The detonators are connected either in a single series (fig. 6:2a) in a number that should not exceed the one given in the table or, for bigger rounds, the detonators are connected in two or more parallel series so that there is roughly the same electric resistance in each series. The series may coincide with the rows (fig. 6:2b), but in many cases they do not (fig. 6:2c). The resistance of a single detonator including the wires is supposed not to exceed 2.5 ohm.

The condenser machines up to 2400 detonators are usually rated at less than 1100 volts. Ordinarily, one third of the potential drop is in the blasting cable when connecting in several parallel series. This means that the round itself will be provided with about 700 volts. Experience has shown that in underwater blasting in salt water, in mining ore with good conductivity, and in other cases with a low resistance to earth, there is with increasing voltage an increasing risk of current leakage to the earth. Therefore, voltages at the round not greater than 700 volts are recommended when possible.

140

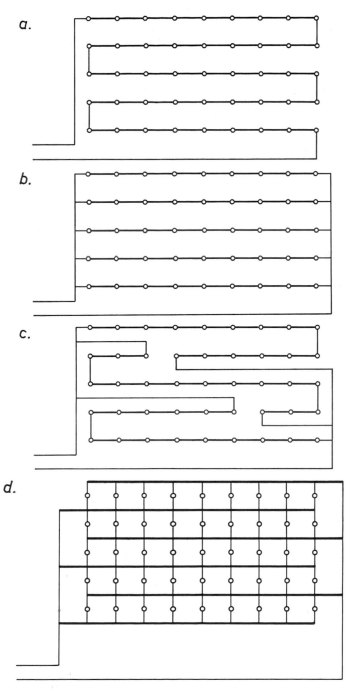

Fig. 6:2. Various ways of connecting in multiple-row blasting. In big rounds the type shown in c is the most common one. a) Series circuit. b) Parallel-series circuit. c) Parallel-series circuit. d) Parallel circuit.

Blasting machine type	Ignition cable resistance ohm	Number of series	Largest mumber of deto-nators type VA	
			Per serie	Total
CI 2 VA	2	1	2	2
	5	1	2	2
	10	1	2	2
CI 15 VA	10	1	15	15
CI 100 VA	5	1	50	50
	2	4	30	120
	5	4	25	100
	10	4	20	80
CI 275 VA	5	1	120	120
	2	6	50	300
	5	5	55	275
	10	4	60	240
CI 700 VA	5	1	130	130
	2	14	50	700
	5	11	50	550
	10	9	50	450
CI 700 VA +	2	25	60	1500
C 800 VA	5	20	50	1000
	10	10	50	750

Even so it is important, when blasting big rounds, to have insulated joints and to make a careful check of earth faults (see 6.9).

b. Electric MS–detonators

Short delay caps are provided with a built-in timing element (fig. 6:3). The most common have 15–30 delays with intervals ranging from 5–100 ms. For underground operations the number of delays can be extended by some 10–15 ordinary delays, with intervals of 0.1–1.0 secs.

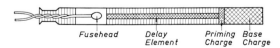

Fusehead   Delay   Priming Base
              Element   Charge   Charge

FIG. 6:3. Section through an electric delay detonator.

TABLE 6: 2. *Firing times in milliseconds (1000 ms = 1 sec) for millisecond delay detonators.*

Delay no	Nitro Nobel Sweden		Atlas USA Rockmaster	Du Pont USA MS Delay	Hercules USA Millidet	ICI Great Britain	CIL Canada MA+ ECHA	Dynamit Nobel Germany	
	VA-MS Det	Nonel							
0	8		1		12	—	8		
	25		8	25	25	30	30	30	20
2	50		25	50	50	55	50	60	40
	75	75	50	75	75	80	75	90	68
4	100	100	75	100	100	105	100	120	80
	125	125	100	125	135	130	128	150	100
6	150	150	125	150	170	155	157	180	120
	175	175	150	175	205	180	190	210	140
8	200	200	175	200	240	205	230	240	160
	225	225	200	250	280	230	280	270	180
10	250	250	250	300	320	255	340	300	200
	275	275	300	350	360	280	410	330	220
12	300	300	350	400	400	305	490	360	240
	325	325	400	450	450	335	570	390	260
14	350	350	450	500	500	365	650	420	280
	375	375	500	600	550	395	725	450	300
16	400	400	550	700	600	425	800	480	320
	425	425	650	800	700	455	875	510	340
18	450	450	750	900		485	950	540	360
	475	475	875	1000		515	1025		
20	500	500	1000			545	1125		
			1125			575	1225		
22			1250			605	1350		
			1375			635	1500		
24	600	600	1500			665	1675		
			1625			695	1875		
26			1750			725	2075		
			1875			755	2300		
28	700	700	2000			785	2550		
			2125			815	2800		
30			2250			845	3050		
			2375						
32	800	800	2500						
			2625						
34			2750						
			2875						
36	900	900	3000						
			3125						
38			3250						
40	1000	1000							

Table 6: 2 (*continued*)

De-lay no	Nitro Nobel Sweden		Atlas KSA Rock-master	Du Pont USA MS Delay	Her-cules USA Millidet	ICI Great Britain	CIL Canada MA + ECHA	Dynamit Nobel Germany
	VA-MS Det	Nonel						
44	1100	1100						
48	1200							
50		1250						
52	1300							
56	1400	1400						
60	1500							
62		1550						
64	1600							
68		1700						
72	1800							
74		1850						
80	2000	2000						

Tables 6:2 and 6:3 give the delay numbers, firing times (τ) and scattering in firing time ($\pm \Delta\tau$) for some types of delay detonators. These figures must be known for the type of detonators used when calculating the ground vibrations. Note that one or two of the lowest numbers in the ordinary intervals detonators may detonate before the highest numbers of the short delays.

The fact that there is a certain scattering effect in the firing time is often neglected, but has important consequences. The charges in a row ignited by one and the same delay number will not be delayed at precisely the same interval. Better fragmentation and less ground vibration are obtained as the scattering gives a short delay effect in the row. On the other hand the scattering is a drawback in smooth blasting if it exceeds some milliseconds; one does not get as smooth a face as in instantaneous ignition. It is generally desirable to have a scatter of $\frac{1}{3}$–$\frac{1}{2}$ of the intervals, so that they do not overlap too much but, at the same time, do not come too close to one another if several are fired in the same nominal delay. If the scattering exceeds ± 50 ms it is no longer a question of short delay effect.

c. Checking the electric system

A check of the resistance of the series and of the round as a whole should always be made when using electric firing. For this purpose only special *resistance meters* should be used, for instance the NAB type shown in

TABLE 6:3. *Firing times in seconds for ordinary electric detonators.*

De. lay no	Nitro Nobel Sweden VA-HS Det	Atlas USA Time-master	Du Pont USA Acudet Mark V	Hercules USA Super-det	ICI Great Britain Long Delay	CIL Canada Half second delay series	Dynamit Nobel Germany
Zero = 0	0.008	0.008	0.025	0.012	0.157	—	
$\frac{1}{4}$				0.24			
$\frac{1}{2}$		0.25	0.3	0.5			
$\frac{3}{4}$				0.7			
1	0.5	0.5	0.5	0.9	0.490	0.5	0.5
$1\frac{1}{2}$		0.75	0.7				
2	1.0	1.0	1.0	1.5	0.800	1.0	1.0
3	1.5	1.5	1.5	2.1	1.125	1.5	1.5
4	2.0	2.0	2.2	2.9	1.400	2.0	2.0
5	2.5	2.5	3.0	3.7	1.675	2.5	2.5
6	3.0	3.0	3.8	4.5	1.950	3.0	3.0
7	3.5	3.5	4.6	5.4	2.275	3.5	3.5
8	4.0	4.0	5.5	6.5	2.650	4.0	4.0
9	4.5	4.5	6.4	7.6	3.050	4.5	4.5
10	5.0	5.0	7.4	8.8	3.450	5.0	5.0
11	5.5	5.5	8.5	10.0	3.900	5.5	5.5
12	6.0	6.0	9.6	11.2	4.350	6.0	6.0
13		6.5	10.8	12.5	4.850	6.5	
14		7.0	12.1	14.0	5.350	7.0	
15		7.5	13.6	15.6	5.900	7.5	
16		8.0	15.2		6.550	8.0	
17					7.250	8.5	
18					8.050	9.0	
19					9.950		
20					10.000		

fig. 6:4. This instrument is easy to handle and read, and covers the whole range needed.

When blasting in conducting ore bodies, in wet shale or clay and in underwater blasting, especially in salt water, the current leakage to the ground may cause misfire in big parts of the round. To prevent or reduce such leakage, an *earthfault tester NAB JP-3* (fig. 6:5) is used. The apparatus has no battery and can be used when loading the hole to check if the conducting wires become damaged during this operation. If this happens the detonator should be replaced immediately. If the detonators are undamaged during the loading, there is a very low frequency of insulation faults immediately after the loading. Sometimes the insulation of the wires may

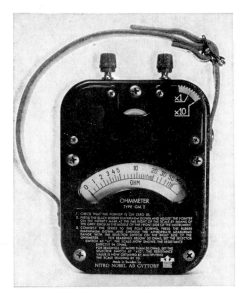

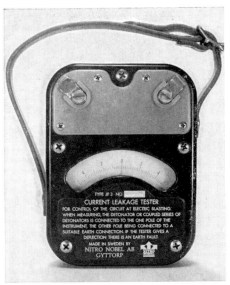

FIG 6:4. Blasting resistance meter.　　FIG. 6:5. Earth-fault tester JP-3 ("Current leakage tester").

deteriorate with time when in contact with ingredients of the explosive. Therefore for holes that have to remain loaded for several days, the use of detonators specially designed for the purpose (so-called OD-detonators) is definitely recommended.

In those cases where it is necessary to know the earth fault, the series can be checked one by one with an *insulation meter*, but it is then necessary to be at a safe distance as this instrument has a high electric potential (500 V). The current should of course be too low to cause ignition of the detonators if checked one at a time. Even so, in the event of extreme variations, a rupture or partial rupture of the circuit inside the detonator or other similar faults may cause ignition of the detonator in question.

A new instrument a so called *earth-fault meter* is constructed by B. Kihlström. With this instrument the insulation fault can be measured at low voltage (low amperage is not enough!) without polarisation. As this instrument is more informative in checking detonators in loaded holes it can replace the insulation meter.

According to investigations made by N. Lundborg, a total earth fault

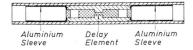

| Aluminium | Delay | Aluminium |
| Sleeve | Element | Sleeve |

FIG. 6:6. Sectioned MS connector (detonating relay).

146

in a series that equals or exceeds twice the resistance of the series can be permitted. In a round with several parallel series the same holds true for each individual series. When there is bad insulation in the detonators or the wires and joints, it may, for small rounds, be recommended to connect only one detonator in every "series", as lower values for the earth faults can then be accepted. For big rounds such parallel connection requires extremely low resistance in the conducting wires and the ignition cable and doubles the number of joints that have to be made.

d. Detonating fuse

Short delay ignition can also be obtained by detonating fuse and so-called MS connectors. They consist of a tube with a delay element (fig. 6:6). The connector transmits the detonation from one fuse to another with a delay of about 10–50 ms (0.01–0.05 s).

Firing with detonating fuse is used when there is risk of premature ignition of electric caps by lightning or other electric risks. It is comparatively easy to connect and arrange, but if the fuse is torn off by the first shots in a round a misfire will occur. This cannot occur in electric firing.

The total number of delays is practically unlimited with MS-connectors, and shorter delays (5 ms) can be used than with the electric MS-detonators. The detonating fuse itself provides a very low delay time, 0.15 ms per m.

In order to ensure that one charge will be ignited even in the case of a cut-off in one part of the system, it is advisable to connect it in such a way that every charge can be ignited from two separate directions. Some examples of such connections are given in fig. 6:7. The principles shown can be used also in other ignition schemes. The figure indicates the sequence of ignition.

In a round with fixation for the edge holes (fig. 6:7a) the holes in the central portion should be ignited first. With one or two free sides the ignition should start at the outer holes (fig. 6:7b and c). A row without MS-connectors can be regarded as rather close to an instantaneously ignited row. The short delay effect occurs only between the rows, not inside them in a round connected as in fig. 6:7b. If the same round is ignited as in fig. 6:7c a better fragmentation and less ground vibrations are obtained. There is a short delay effect in every one of the original rows; the more instantaneous effect in the diagonal rows can, according to what is said in Chapter 1, be expected to favor good fragmentation.

6.2. Delay interval and fragmentation

For a discussion of the relationship between delay time and fragmentation, benching with a single line of bore-holes is the simplest case. Malmgren

Relay No	5 ms	10 ms	15 ms	Per #1	35 ms	40 ms	Per #2
Delay CIL	5	10	15	25	35	40	50
Primacord Colour	G	O	Y	B	BW	BW	R
Per = Period   G = Green   O = Orange   Y = Yellow   B = Blue   R = Red   V = Violet
All Colours Have One Black Tracer Thread
Relay No	1	2	3	4	5
Delay ICI	10	17	25	35	45

and Berglund counted the number of boulders larger than 40 cm in a series of blasts with different delays; the results were as follows.

Although only two blasts were fired with each delay, the figures are considered to be representative, since the results in the duplicated tests were quite consistent. Measurement of the rest of the débris according to size (less than 40 cm) showed that 10 ms gave average-sized fragments, 20 ms gave large and small fragments, and 30 ms gave mainly large fragments. Thus, 10 ms gave the best result in this case. Similar investigations by Mecir and Valek have proved that within the range of 300–15 ms the best fragmentation was achieved with the shortest delay periods. For loosening rock with burdens of 1–2 m (3–6 ft), interval times of 20 ms or more do not in blasting one single row give a short delay, but an ordinary interval effect. In open air blasting a delay time of 20 ms is, however, still sufficient for a reduction of the throw. If the broken rock acquires a velocity of 15 m/s after 15 ms it has moved 0.3 m and can still shelter the surroundings when a subsequent shot is ignited.

The results from blasting with burdens between 0.5 and 8 m indicate that there is linear relationship between the burden and the delay time giving the best breakage. They can be summed up in the simple but important formula

$$\tau = kV \qquad (6:1)$$

where τ is the delay time in ms. V is the burden in meters, and the value of the constant k is 3 to 5 ms/m.

Experience gained from large blasts in limestone quarries with a burden of 5 to 8 m shows that the best fragmentation is obtained with delay times of 25 to 40 ms. The size of the fragments is comparatively constant, and if the burden and the distance between the holes are chosen so that the average

TABLE 6:5. *Tests with different delays.*

Delay interval (millisecs)	10	20	30	1000
Number of boulders	10	26	37	25

65 ms	Per #3	Per #4	Per #5	Per #6	Per #7	Per #8	Per #9	Per #10
65	75	100	128	157	190	230	280	340
BW	BW	V	BW	BW	BW	BW	BW	BW

BW = Buff White

size does not exceed the capacity of the loading and crushing machines, the necessary secondary blasting becomes small.

Some interesting figures about the consumption of explosives in instantaneous and short delay blasting, given by Janelid, are reproduced in Table 6:6. They refer to a limestone quarry with benches 15 to 20 m in height, 5 m burden 6 m distance between the holes, and a delay time of 20 to 25 ms.

The saving in total consumption of explosives in short delay blasting compared with instantaneous blasting amounts to 10 per cent. The number of boulders which had to be blasted was 10 per cent greater for short delay, but nevertheless the consumption of explosives for this purpose decreased from 29.0 to 12.4 gram, per ton of rock. This means that the short delay blasting gave better fragmentation but that the average size of the boulders was somewhat too large for the arrangements available for loading and crushing.

6.3. Drilling and ignition patterns

The relationship between diameter of the hole and the practical burden has been given as $V_1 \approx 40d$. In ordinary cases a burden that is not more than 50% of the bench height should be used, and not less than 20% of it. This gives an approximate relationship that the diameter should be between 0.5 and 1.25 % of the bench height. At a bench height of $K = 10$ m (33 ft), it

TABLE 6:6. *Consumption of explosives in grammes per ton of rock. Average values over year.*

Blasting method	Year	No of rounds fired	Consumption of explosive		
			Round	Secondary blasting	Total
Instantaneous	1947/48	23	95.5	29.0	124.5
Short delay	1948/49	48	100.1	12.4	112.5

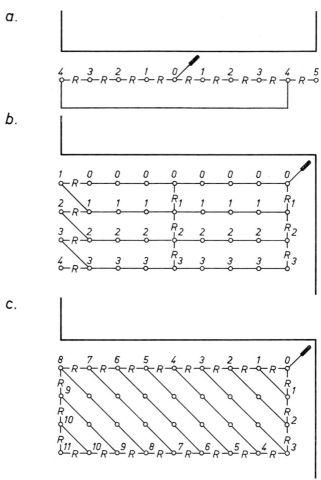

FIG. 6:7. Example of connection for firing of short delay rounds with detonating fuse and MS connectors (R).

means that a suitable diameter is between 50 and 125 mm (2–5 in). For $K = 20$ m (66 ft) the corresponding values are 100–250 mm (4–10 m). For bench heights below 6 m (20 ft) the diameter should consequently be 30–75 mm (1.2–3 in); the value 30 mm (12 in) being a lower limit for the ordinary drilling equipment. In practice when varying bench heights are involved, one equipment with one and the same drilling series is often used for the whole job.

Other factors can affect the diameter in the direction of lower values: a wish for better fragmentation and to employ lighter drilling equipment, consideration of ground vibrations or the fact that the lateral extension of the round is so small that it justifies a smaller drilling diameter.

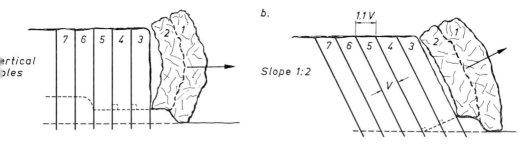

FIG. 6:8. With inclined holes the influence of stumps in the bottom is eliminated after one or two rows.

a. Slope of the holes

The holes are normally driven about $0.3\,V$ below the grade. With vertical holes the marking of their position and the drilling are easier than for inclined ones. In return, the risk of stumps and incomplete tearing at the bottom is considerably greater than with inclined holes. This is particularly the case in large multiple-row rounds, where there may be stumps from earlier rows. One must then expect the possibility of the bottom layer behind such a stump being incompletely torn off. This may continue all the way to the last row of holes due to the fact that the tearing at the bottom takes place at right angles to the direction of the holes. There is then very little or no tendency of the bottom to work its way down to a lower level than in previous rows. Fig. 6:8a shows an example of how a vertically drilled bench round which, owing to incomplete ignition in row No. 2, is incompletely torn off at the bottom not only in this row but also in all the successive ones. Crevices in the rock close to the bottom level moreover can cause the bottom level to rise, which is not uncommon. This fact should be especially borne in mind when blasting very large rounds.

With sloping drill holes the round will work itself down to the grade level whenever there happens to be a stump. This also means that faults in the bottom level cannot influence the round as a whole. If one row leaves a height of 1 meter (3 ft) at the bottom, the following rows will, according to the principle of rectangularity, work their way down at a slope of $1:2$ if the slope of the hole is $2:1$. Thus the intended bottom level (fig. 6:8b) is again reached 2 m (6 ft) farther back in the round.

As there with sloping drill holes is a less degree of fixation at the bottom, more rock can be blasted with the same number of drill holes than with vertical holes. This amounts to about 10–15%, that is to say the burden and distance of the holes can both the increased 5–7.5% at a slope of $3:1$–$2:1$.

Improved performance at the top of the round near the free surface is also an advantage with sloping drill holes. The risk of tearing in the round

151

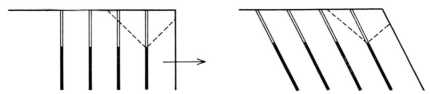

FIG. 6:9. The back break is reduced with inclined holes.

and of overbreak in the remaining rock is considerably reduced (fig. 6:9) a fact which proved of importance in the infancy of short delay blasting, when it was necessary to prevent the electrical wires in the row immediately behind from being torn off at the detonation. This aspect still applies when ignition is carried out with a detonating fuse.

When using short delay detonators, on the other hand, this condition does not exist as all the detonators in the round receive their ignition impulse simultaneously, and the delay is built into the actual detonator. It therefore does not matter if the ignition wires are torn off before the detonation of the cap. The greater risk of tearing is, however, unfavorable for the fragmentation. The tearing around a charge will often follow actual cracks and fissures, and screens off sections of the rock from the blasting effect from subsequent neighbouring holes.

b. Ignition sequence in a row

The simplest type of short delay blasting is shown in fig. 6:10, top picture, in which a series of charges in a row is initiated at successive intervals. In this case better fragmentation is obtained when the distance of the holes is 10–50% greater than the burden; $E = 1.25 V$ is indicated as an ordinary case. Another type of ignition which in certain cases implies an improved short delay effect and improved fragmentation is with rising and falling intervals according to the example in the middle picture of fig. 6:10. An important case is indicated by the lowermost picture with short delay detonators of one and the same number for all the holes in a row. This may not seem to give any short delay effect. But in view of what has already been mentioned concerning spread in the delay time, it is clear that here we get a difference in the ignition time between adjacent holes depending on the size of the spread. This of course does not give a defined interval between two holes but one which varies from close to zero up to ± 5 ms, ± 10 ms, or whatever is the actual value for the detonators in question according to table 6:2. The interval between two holes in a row can be 2 ms, for example, between two others 15 ms, and so on. With this type of ignition pattern too, the best fragmentation is obtained if the distance between the holes is greater than the burden.

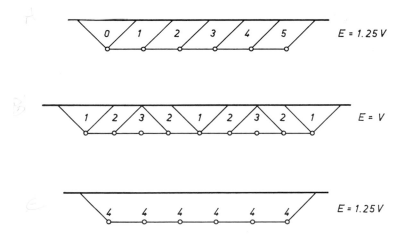

FIG. 6:10. Some ignition patterns for one row of holes.

c. Multiple-row rounds

First, attention should be drawn to the fact that the reduced throw in short delay blasting depends on the delay time between charges close to each other beeing sufficiently small. However, the definition of the ideal time depends on conditions. A delay time of 100 ms gives a short delay effect in a big bench with a burden of $V=8$ m (10 ft), but not in a small bench, of for instance, $K=0.6$ m (2 ft) and $V=0.7$ m (2.3 ft) if the rock gains a velocity of 20 m/s (70 ft/s). Here the result may be an unexpectedly heavy throw of stones.

It is desirable that the rock has not moved more than a distance 3–30% of the burden when the next charge in the row detonates. Thus for the example above with $V=0.7$ m, the delay time should be $\tau=2$–20 ms, and shorter if the round is more heavily charged. If the same delay number is used for one row, the delay between the holes is the scattering in the delay time ($\Delta\tau$).

When one separate delay is used for every hole it must be observed that the time difference between one hole and the nearest hole in the row behind may be so big that there is no longer a short delay effect between the rows. Serious damage has often been caused by such too long delay times being used close to buildings. A heavy covering material, anchored behind the round, must be used if the delay time in question cannot be reduced.

The three basic types of short delay blasting indicated for one row can be applied and combined in various ways in multiple-row blasting. Here the mutual relation between the rows and the adjustment of the ignition pattern for the edge holes in a laterally constricted round also comes

153

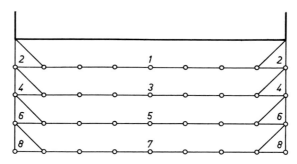

Fig. 6:11. Ignition pattern in multiple-row blasting, 2 delays per row.

into the picture. The simplest type of accurate plan for the ignition of a laterally constricted multiple-row round is given in fig. 6:11. All the holes in a row, except the edge holes, can be ignited with one and the same delay number. This means that the hole that happens to be ignited first will have a perfectly free breakage whichever it may be. This would not be the case if the edge holes were ignited with the same interval. One would then have to count on a probability (75%) of one of the edge holes being ignited before its immediate neighbor, and only in 25% of all the rows both of the edge holes having entirely suitable breakage conditions. According to fig. 6:11 the main part of the row is first blasted and successively thereafter the edge holes row upon row. There is then always the least constriction possible for every single hole. This type of ignition calls for twice as many intervals as rows.

With large multiple-row rounds the number of intervals available for applying an ignition sequence as in fig. 6:11 is not sufficient. The ignition sequence can then be modified as in fig. 6:12, where all the holes of a row, except the side ones, are ignited in the same interval as the side holes in the previous row. Note that there is a greater constriction for the two holes

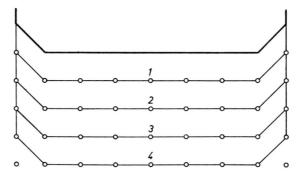

Fig. 6:12. Ignition pattern in multiple-row blasting, 1 delay per row.

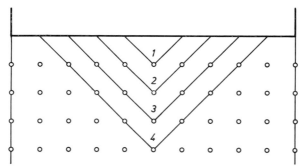

Fig. 6:13. Ignition pattern for better fragmentation, less overbreak at the walls and a concentrated muck pile. However, the pattern provides bad conditions for loosening in the central part.

which are next to the side holes, because the bench face created by the row in front is not straight but curved. This gives increased tearing in the vicinity of the two holes and may also result in a greater tendency to create stumps at the bottom.

An example of a faulty sequence of ignition is given in fig. 6:13. After blasting interval No. 1, which has free breakage, the two holes on both sides in the same row are ignited with delay No. 2 as is the hole straight behind the first one. As a result of the ignition scatter, there is a probability of 75% of the holes in the second row being ignited before at least one of the two in front. These holes will then at their moment of ignition be rather constricted. When charged in the same way as the other, it is more difficult to break the rock and complete breakage may be delayed till the two other holes are ignited. If this happens so late that too much of the energy released in the detonation has escaped there will be incomplete breakage.

The problems are not eliminated if a sequence of ignition is adopted as in fig. 6:14, but this type of ignition is usually better than the one given in fig. 6:12. The two holes which are slightly more constricted than the others have been drawn farther into the row so that the greater tearing in these surroundings does not affect the final contour of the wall.

Furthermore, the ignition in rows which run obliquely to the middle line of the round implies that the throw which occurs at right angles to the ignition rows does not proceed straight ahead, but is directed at an angle of 45° towards the elongation of the middle line. This gives less throw and consequently greater possibility for a more powerful charge and better fragmentation.

The original quadratic drilling pattern would, when ignited according to 6:12, give $E = V$, but the burden will be smaller and the distance of the holes greater if the ignition takes place along diagonal rows as in fig. 6:14.

155

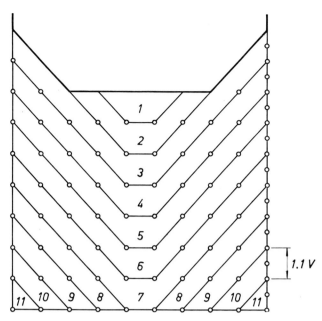

Fig. 6:14. Ignition and drilling pattern with less fixation for the central holes than according to fig. 6:13 and with partial smooth blasting of two holes at a time in the walls.

There we have $E' = E\sqrt{2}$ and $V' = V/\sqrt{2}$ and consequently $E' = 2V'$. This too can be expected to affect the fragmentation favorably. The ignition may be made with a detonating fuse so that all the holes with one and the same ignition number ignite almost simultaneously or with short delay detonators.

When it is necessary to leave the remaining rock walls undamaged and with as smooth a surface as possible, the side contours of the round could be blasted with presplitting or ordinary careful blasting as indicated on the right side of the round in fig. 6:14. The charge per m is 0.2–0.5 of that in the other holes. The ignition of the rows in a V-form, as in the preceding example, is an advantage since the broken rock will not lie so heavily on the two side rows. This makes it possible to charge them with small charges. The holes closest to the side holes should not be too heavily loaded; light charges if necessary in the upper part can extend right up to the top of the holes in order to reduce tearing. The stemming should not be more than $0.5\,V$. When smoothing is carried out as presplitting before blasting the main round, up to two times as many holes is required, as compared with the left-hand side of fig. 6:14. The ignition has to be made simultaneously, either separately before the round or in the round ignited with instantaneous detonators and/or a detonating fuse. Another

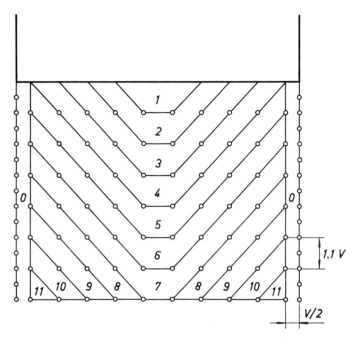

Fig. 6:15. Ignition and drilling pattern with presplitting of the walls. (Spacing 0.3–0.55 V.)

way is to use trimming, leaving the last part at the sides to be smooth-blasted after the muck pile from the main part of the round has been un-loaded. It is then possible to perform smooth blasting with light charges almost down to the bottom of the holes and—which essentially improves the result—to have all the charges ignited at the same moment. It is an important advantage with the presplitting method (see Chapter 10) that this type of ignition can be performed without an extra-digging operation as in trimming.

Another ignition pattern which has proved to give good fragmentation is given in fig. 6:16. It is based on a drilling which has been carried out in a quadratic net and has, like fig. 6:11, a free burden in front of every hole at the moment of ignition. Here too, however, twice as large a number of intervals as rows of holes is required. Rounds of this type can be suitably combined with smoothing. Ordinary smooth blasting has been indicated at the right contour of the round. As in previous cases, the desire to have smooth blasting holes of one row ignited simultaneously, or with as small a spread in the ignition time as possible means that the best result is ob-tained if all the charges in one side row are ignited in one and the same interval, preferably connected with a detonating fuse. The lower part of

157

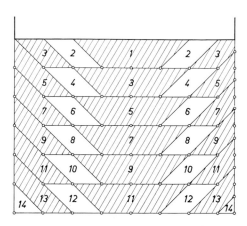

Fig. 6:16. Example of a multiple-row round with almost every hole in the row on a separate delay. Partial smooth blasting of the right wall.

the holes must be charged with the same amount of explosive per m³ (less per hole) as for the rest of the round in order to ensure a satisfactory breakage at the bottom if the above mentioned technique of trimming is not used.

In certain cases when incomplete loosening causes formation of stumps at the bottom, it has been considered advantageous to adopt an alternate hole pattern instead of one with the holes placed straight behind each other in rows using a rectangular drilling pattern. Ignition and drilling pattern for a typical round with alternate position of the holes is shown in fig. 6:17, where a certain smooth blasting of the walls has also been assumed.

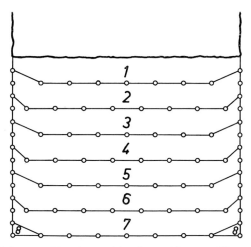

Fig. 6:17. Multiple-row blasting with alternating positions of the holes and smooth blasting.

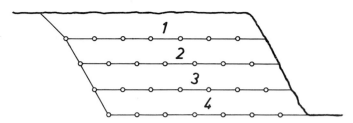

FIG. 6 : 18. Multiple-row round with two free faces.

d. Free faces

With great widths of benches or in other instances where there is no lateral geometric limitation, there are much simpler means of arriving at a simple and accurate ignition pattern. Even when one interval per row is employed it is possible to avoid the constricted effect previously mentioned in connection with laterally constricted multiple-row rounds (fig. 6:18).

This example shows a suitable and accurate type of multiple-row blast with two free faces. The simple basic type can be varied largely by employing

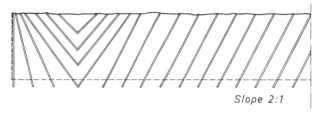

Slope 2:1

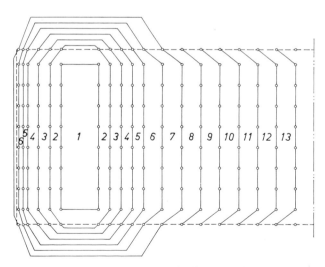

FIG. 6 : 19. Multiple-row round including a V-cut opening.

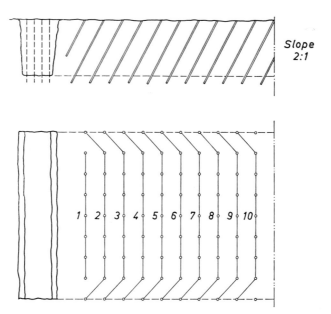

FIG. 6:20. Opening for multiple round by ditch blasting.

principles indicated above. The throw can to a certain extent be directed
with the help of the ignition. In the example given in fig. 6:18, the débris
moves straight forward. If, on the other hand, the sequence of ignition is
laid in oblique rows from left to right, the rock will have its weight trans-
ferred farther to the right after blasting.

e. Rounds including opening cut

In blasting house foundations, basins and shaft sinking with a large cross-
section, free faces as in the above mentioned multiple-row rounds, are not
available to aid blasting as in benching or stoping. In such instances the
round can be opened with a plough or fan cut, which in contrast to their
application in tunneling can be given a much greater width. Fig. 6:19
gives an example of a multiple-row round where the opening is made with
a plough cut.

The opening can also be made with a trench round as in fig. 6:20. For
blasting of the actual trench see paragraph 6:8. In subsequent blasting
of the main round it must be specially observed that the swelling can take
place. This implies a greater need to slope the drill holes. This type of
opening cannot be employed for depths greater than 4 m (12 ft) unless it
is combined with a fan type for the first rows of the main round perpendicular to the trench.

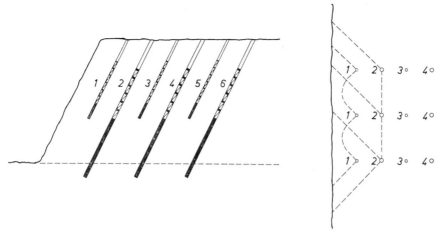

FIG. 6:21. Example of improved fragmentation in a large hole round with reduced throw.

f. Improved fragmentation

All the drilling and ignition patterns indicated have been designed to render the best and most uniform fragmentation. Some special types of blasting deserve to be mentioned, however. In many cases the problem of attaining the best fragmentation is to distribute as great a specific charge as possible per m³ in the rock without having the effect of an overcharge in which a part of the excess energy is transformed into throw. Fig. 6:21 gives an example of a blast where there was a line 15 m (50 ft) from the round beyond which the throw was not permitted, at least not of heavy stones. This problem was solved by drilling the round with sloping holes 50 mm (2 in) in diameter to full depth, the holes being located and charged only to ensure breakage and swelling and including the necessary technical margin. In addition to these main holes, smaller ones, 32 mm ($1\frac{1}{4}$ in) in diameter and about half the depth, were placed in the upper part of the rock. The charges in them were adjusted so as to be near the limit for full breakage of the rock in front of them, that is to say these charges did not give any throw. The small holes were located immediately in front of the large ones, not on the diagonal between them where cracks from the large holes are apt to meet. With this type of blasting a fragmentation corresponding to the small burden for the small holes could be obtained, without using small holes throughout the round. Small diameter holes would mean twice the number of holes and, in addition, driving all the holes to the very bottom.

The possibilities of improving fragmentation are closely associated in yet another respect with the risk of throw. In open air blasts the upper part of the drill holes must often be left uncharged in order to avoid throw and scattering of small pieces of rock. This means that the large boulders resulting from a round will be produced from the upper part of the round. A small concentration of charge which is distributed farther up in the upper part of the column can help to break up these boulders and improve the fragmentation without increasing the number of holes and without appreciably increasing the charge per m³, calculated as a mean value of the round as a whole.

In quarrying with large burdens and bench heights it has been possible to improve fragmentation by introducing short delay ignition in one and the same hole so that two or more parts of the charge are separated by sand plugs and initiated with different delay numbers. In multiple-row blasting, it should be noted that the stemmings must not be placed at the same level in two rows in sequence as this will give a section in the round with inferior fragmentation. The stemmings should consist of dry sand with a length of about 20 times the diameter of the drill hole. This means 1.0 m (3.3 ft) in a 50 mm (2 in) hole.

6.4. Bottom bench in tunnels. Canals

For tunneling with a so-called top heading followed by separate blasting of the bottom part, the introduction of short delay blasting with multiple-row rounds has made it possible to carry out this part of the work as a continuous operation in which drilling, blasting and loading can be done quite independently of one another. We have here an even, uniform bench height and consequently also favorable conditions for mechanizing the drilling. Fig. 6:22 shows examples of such highly mechanized work at Stornorrfors Power Station, where 13 chain-fed Atlas Copco BBC-21 machines have been set up on a specially constructed drilling wagon. When such drilling can be done with few or no changes, the drilling capacity per machine can be very high, while two or three machines can at the same time be managed by one man. The meters (footage) drilled per man and shift have amounted to more than 250 m (800 feet). As drilling patterns figs. 6:14, 6:15 or 6:16 apply here. The last mentioned one corresponds nearly to an ignition pattern which, according to Beckman, gave the best fragmentation and distribution of the broken rock.

As the broken part of the round cannot swell laterally, the blasting of canals and bottom benches in tunnels requires a greater heaving of rock

FIG. 6:22. Rationalized drilling for multiple-row round acc. to fig. 6:16 in a bottom bench in a water power tunnel (Stornorrfors). (Courtesy Swedish State Power Board.)

to ensure swelling. Tests with holes with a slope of 3:1 and continuous blasting have in such cases been unsuccesful with bench heights of 7 to 10 m (23–33 ft), and it has been necessary to alter the slope to 2:1 or 1.7:1. Both these values gave satisfactory results. As long stretches can be drilled before blasting takes place, problems may be encountered at the joints between the rounds on account of tearing into the part behind the round which has already been drilled. The sloping hole pattern and a minor concentration of charge extending to the top of the holes in the last row of holes will reduce such problems.

As swelling is a more pronounced problem in blasting bottom benches in tunnels it may be advisable to drill the holes farther below the bottom than is otherwise recommended. If under-drilling is increased from 0.3 V to 0.6 V, the part of the charge which contributes to swelling in the bottom section is also increased.

6.5. Low bench blasting

Low benches are nowadays blasted with holes drilled from above instead of horizontally. This makes it possible to blast rocks continuously over large surfaces without mucking, which compensates for the increased number of holes. Drilling is also simplified, and it is easier to avoid scattering of stones from the blast. The slope of the holes should be 2:1.

Low benches mean in this connection heights lower than 1.8 V. The reason for choosing this limit is that in lower benches there is not room for a bottom charge corresponding to the maximum burden for the drill hole diameter (given in table 3:1). The bottom charge that can be accommodated in a hole will therefore depend not only on the diameter of the hole but also on the height of the bench. This will be seen from table 3:3. The position of the hole is marked directly on the rock. The intended zero level can be marked, as well as the one meter level, for example. The first row of drill holes can be collared behind the contour indicating the zero level at the distance indicated as practical burden (V_1) in tables 3:4–3:6 and in fig. 6:23. The term bench height is used to designate the average height of that part of the rock to be detached by the charge. The burden shall be measured as the shortest (perpendicular) distance between two lines with the same slope as the holes, the one as a continuation of the drill hole, the other which aims at the indicated zero level. For the subsequent rows the burden is measured and similarly indicated as the minimum distance between the elongation of the holes. When the placing of the holes is checked at the working sites, it will practically always be found that the distance between two holes has been measured along the rock surface

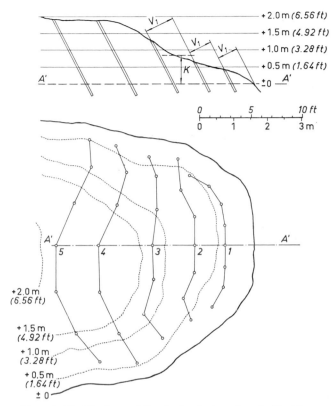

Fig. 6 : 23. Placing of the holes in low bench blasting.

instead of at right angles between the one hole and the elongation of the other. The reader can amuse himself by estimating in fig. 6:23 the difference this will make in the arrangement of the holes.

With this type of blasting, rows are generally free laterally, and every row can be ignited at one and the same interval. When large surfaces with low benches are concerned, the number of holes can be reduced by lowering the level of the bottom of breakage.

Example: With a bench height $K = 0.30$ m (1 ft) and a diameter of the drill hole $d = 50$ mm (2 in), the practical burden is $V_1 = 0.70$ m (2.3 ft). If the bottom level is lowered 1.0 m (3.3 ft) the practical burden can be increased to $V_1 = 1.40$ m (4.7 ft) when only a quarter of the number of holes is needed. A detailed discussion is given in connection with underwater blasting (Chapter 11).

6.6. Road construction

In such constructional projects as highways, airfields and other open air excavations of a similar type the conditions differ from mining and quarry-

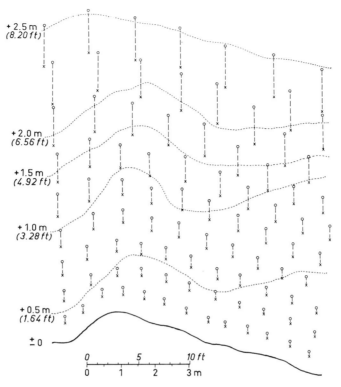

+2.5 m
(8.20 ft)

+2.0 m
(6.56 ft)

+1.5 m
(4.92 ft)

+1.0 m
(3.28 ft)

+0.5 m
(1.64 ft)

±0

0   5   10 ft

0   1   2   3 m

× the position of the hole at the intended bottom of breakage
∘ the collaring of the hole on the rock surface

FIG. 6:24. The distribution of holes in blasting a road cut.

ing in that the bench height changes continously from zero up to the maximum value. This means that the drilling patterns given in fig. 6:11 have to be modified with varying burdens and spacing, which makes the construction of drilling and ignition patterns more complicated in a similar way to that in low bench blasting.

If the height (depth) of the cut is less than 10 m (33 ft) it can be excavated in one lift. With higher values the operation is done in two or more benches. The best conditions for drilling, blasting and digging taken as a whole are obtained with bench heights of 5–8 m (16–26 ft). The diameter of the drills when blasting in hard rock is $d = 32$ mm (1.25 in) with hand-held machines, $d = 50$–75 mm (2–3 in) with heavier equipment. Larger diameters should in this case not be used. In looser rock and with big mucking equipment, diameters of up to 125 mm can be employed. However, even with big shovels it has been shown that smaller diameters of the holes may be more econo-

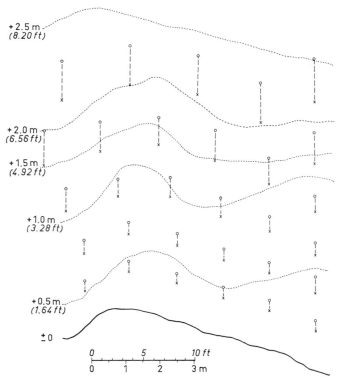

+2.5 m
(8.20 ft)

+2.0 m
(6.56 ft)

+1.5 m
(4.92 ft)

+1.0 m
(3.28 ft)

+0.5 m
(1.64 ft)

± 0

0   5   10 ft

0   1   2   3 m

x *the position of the hole at the intended bottom of breakage*

o *the collaring of the hole on the rock surface*

FIG. 6:25. The above cut with 63 mm holes, bottom loaded with the pneu-matice cartridge loader.

mical. This increases the digging capacity. Some drilling patterns are shown in fig. 6:24 for hand-held machines. In 6:25 the same cut is drilled with 75 mm (3 in) holes.

In the blasting of road cuts special attention should be paid to the remaining wall of the rock being left in a slope that will eliminate the risk of slides and with smooth undamaged faces (see Chapter 10). It is no longer necessary to accept rough rock faces after the blasting.

6.7. Foundation of buildings

In a discussion on the foundations of buildings a contractor summarized the change in attitude that has taken place during the past decade in the words: "In ancient days we did all we could to avoid constructional work in or on rock, now we go all out for it." Rock blasting can now be carried

167

FIG. 6:26. Rock blasting of the foundation for a city office.

out quickly and highly mechanized. Founding buildings on rock eliminates all problems concerning the stability of the ground.

With short delays and multiple-row blasting big rounds can be fired close to buildings. Preference is given mainly to small diameters with jack hammer drilling machines, as the problems of vibrations and throw often play an important part. Examples of blasting foundations for big buildings are given in fig. 6:26 and in Chapter 9, where the disposition of the blasting job is also discussed.

As the throw is reduced or eliminated in short delay blasting, there is no general need to use blasting mats if the spacing, loading and ignition are correctly made. The mats are placed only on special parts, for the lowest bench heights and where the explosive for one reason or another comes close to the upper face.

FIG. 6 : 27. Drill holes ready for continuous trench blasting of round after round over the whole new suburban area in the making (Bagarmossen, Stockholm).

It is important for the control of throw to avoid too long delays between the holes in a row and between the rows, especially for low benches with a great specific charge (kg/m³) and small burdens. Here the broken rock moves forward at a speed of 10–20 m/s. A delay of 50 ms means that the

169

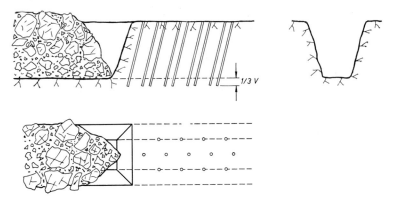

FIG. 6:28. Trench blasting with short delay ignition.

rock has moved 0.5–1.0 m when the next charges detonate. This is too much to prevent throw when the bench height and the burden are about this size. The delay time for burdens 0.5–1.0 m can be 3–10 ms. For longer delays attention must be paid to increased throw, and more mats must be used.

6.8. Trenching

Trenches for water supply, drains, electrical conduits and pipe lines are made as narrow as possible to allow enough space for the pipe. With short delay blasting the drilling can be completed over great areas before blasting and digging are carried out (fig. 6:27).

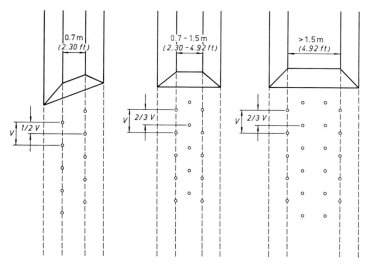

FIG. 6:29. Drilling patterns for rock trenches, $d = 30$–40 mm (1.2–1.6 in).

170

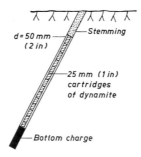

FIG. 6:30. Example of distribution of the charge in trenching.

Figures 6:28–31 show placing of the holes and ignition sequence for rock trenching. Fig. 6:32 shows how the connection should be made with detonating fuse and MS-connectors. The holes are placed in rows without sloping to the sides, at a distance that is for simplicity, but not quite correctly, called burden and denoted by V. The number of rows depends on the width of the trench at the bottom. The holes should have a slope of 2:1–3:1 in vertical planes through the rows (fig. 6:28), and be drilled 10% of the trench depth, but at least 0.2 m below the intended bottom.

In populated areas drills with diameters 30–35 mm are preferred, otherwise 50 mm and 62 mm are widely used. Drilling patterns are given in figs. 6:29 and 6:30. The corresponding values for the burden (V) are given for 32 mm drill holes in table 6:7, and for 50 and 62 mm in table 6:8 for some actual values for the width (B) and for varying depths. The charges per hole and their distribution are given in the same tables. The bottom charge is packed at the bottom, which means a concentration of about 1 kg/m for $d=32$ mm, 2.0–2.5 kg/m for $d=50$ and 3.0–3.9 kg/m for $d=62$ mm. The column charge is 0.25, 0.60 and 0.90 kg/m for the three diameters mentioned and the stemming 0.6, 0.9 and 1.1 m respectively. Examples of the distribution of the charge are given in figs 4:1 and 6:30.

FIG. 6:31. Ignition sequence in trenching.

FIG. 6:32. Connection with detonating fuse in trenching.

171

Fig. 6:33. Presplitting in trenching (compare p. 178).

The ignition sequence recommended is shown in fig. 6:31. Delays down to 3 ms can be used when $d = 32$ mm (1.25 in), and 5–6 ms for the 50–62 mm (2–2.5 in) holes. The ordinary delays of about 25 ms can be and are often used, but give a greater risk of throw and scattering of stones for lower depths and smaller burdens. They give, on the other hand, sometimes (but not generally) smaller ground vibrations.

The practical directions for connecting the holes if detonating fuse and MS-connectors are used are given in fig. 6:32. With this system it is easier to have shorter delays and practically an unlimited number of holes in a round.

Even with the holes placed in vertical planes there is an overbreak that gives a slope of at least 4:1 on the sides of the trench. In order to give the final contour a more definite finish, the excavation may be combined with presplitting of the upper part of the trench as shown in fig. 6:33 (see Chapter 10).

TABLE 6:7. *Trench blasting with 32 mm holes with a slope of 3:1 according to fig. 6:29.*

Depth of ditch K		Depth of holes H		Burden V		Bottom charge per hole, kg.		Column charge per hole (0.25 kg/m)
						$B_2 = 0.7$ m or $B_3 = 1.0$ m	$B_3 = 1.5$ m or $B_4 = 2.0$ m	
m	ft	m	ft	m	ft			kg
0.3	1	0.5	1.7	0.4	1.3	0.05	0.05	—
0.6	2	0.7	2.3	0.6	2	0.10	0.10	—
0.9	3	1.1	3.7	0.8	2.7	0.12	0.20	0.08
1.2	4	1.4	4.7	0.8	2.7	0.15	0.25	0.15
1.5	5	1.8	6	0.8	2.7	0.20	0.30	0.25
1.8	6	2.1	7	0.8	2.7	0.25	0.35	0.30
2.4	8	2.7	9	0.8	2.7	0.30	0.40	0.40
3.0	10	3.5	11.5	0.8	2.7	0.40	0.55	0.60
3.6	12	4.2	14	0.8	2.7	0.50	0.70	0.70
4.2	14	4.8	16	0.7	2.3	0.60	0.90	0.80

TABLE 6:8. *Trench blasting with 50 mm and 62 mm holes with a slope of 3:1 according to fig. 6:39.*

Depth of ditch K		Depth of holes H		Burden V				Bottom charge per hole, kg.		Column charge[a]	
				d = 50 mm		62 m		d = 50 mm B_2 = 1.1 m B_3 = 1.5 m	d = 62 mm B_2 = 1.4 m B_3 = 2.0 m	d = 50 mm	62 mm
m	ft	m	ft	m	ft	m	ft			kg	
0.6	2	0.8	2.7	0.6	2	0.7	2.3	0.25	0.35	–	–
0.9	3	1.1	3.7	0.9	3	1.0	3.3	0.30	0.50	0.10	–
1.2	4	1.4	4.7	1.1	3.7	1.2	4	0.40	0.50	0.20	0.75
1.5	5	1.8	6	1.2	4	1.5	5	0.50	0.80	0.45	0.40
1.8	6	2.1	7	1.2	4	1.5	5	0.60	0.80	0.60	0.70
2.4	8	2.7	9	1.2	4	1.5	5	0.80	1.00	0.90	1.0
3.0	10	3.5	11.5	1.2	4	1.5	5	0.90	1.25	1.35	1.8
3.6	12	4.2	14	1.2	4	1.4	4.7	1.10	1.50	1.70	2.3
4.2	14	4.8	16	1.1	3.7	1.4	4.7	1.40	1.75	2.00	2.8
4.8	16	5.4	18	1.1	3.7	1.3	4.3	1.60	2.00	2.30	3.2

[a] 1 = 0.6 kg/m
1 = 0.9 kg/m

6.9. Safety measures in electric ignition

With electric ignition one must consider the risk of haphazard ignition by current leakage, lightning, static electricity, induction (capacitive influence) from power lines or ignition by radio frequencies. The two first mentioned cause casualties every year in underground work. New high safety electric detonators, the Swedish VA, the German HU and others give a much higher degree of safety in all these respects.

A thorough investigation of the factors influencing the risk of premature ignition has been made by a team headed by Y. Hagerman in which technical specialists on electricity, blasting, control and working safety have taken part. The results are to be published but some results and recommendations will be given here.

a. Leakage of current

Contact with the rock has not been found to give any measurable current. The chief practical risk seems to be that rails or other conducting material acquire electric tension due to insulation faults in electric engines or equipments somewhere in the mine or at the working place. The electric wires of the detonators should not at any time during the operation of charging the drill holes or connecting the round come in contact with metallic conductors.

b. Static electricity

Electric detonators may be ignited by static electricity generated when blowing dry powderous substances such as sand, ammonium nitrate (AN) etc. through tubes of material that is not antistatic. The same type of mechanism has caused casualties from charges generated by dust and snow storms.

With a low relative humidity in the air, it is possible for a man to become charged with 15 kV at a capacity of 300 pF and to deliver an electric impulse of not more than 2 amp^{2}ms. Detonators with a lower impulse should be avoided, at least when working in a low air humidity.

Even for detonators requiring a higher ignition impulse than 2 amp^{2}ms through the heating filament, a lower impulse can ignite the cap if there is a spark flashover inside the cap. The best modern detonators are constructed so as to prevent spark ignition by static ignition.

c. Lightning

The area in which lightning can cause premature ignition by strong electric fields in the rock is proportional to

$$R_1^2 = c/K_t$$

where R_1 is the radius of the area, c a constant and K_t the electric impulse (amp^{2}ms) required for ignition. According to investigations made by I. Olsson, the working site should be evacuated when there is thunder at a distance of

$$R = R_1 + R_2 + R_3$$

where R_2 is the spread of the flashes of lightning and R_3 is the distance the thunder can move in the time for evacuation.

Example. If a detonator with $K_t = 3$ amp^2 ms gives $R_1 = 5.0$ km (3 miles), the spread $R_2 = 5$ km (3 miles) and the speed of the thunder $v = 1$ km/min, we get with an evacuation time of 1 minute

$$R = 5 + 5 + 1 = 11 \text{ km}$$

For detonators with $K_t = 100$ amp^2 ms the distance R_1 is 0.9 km and

$$R_{100} = 0.9 + 5 + 1 = 6.9 \text{ km}$$

and with $K_t = 1000$ amp^2 ms, $R_1 = 0.3$ km we get

$$R_{1000} = 0.3 + 5 + 1 = 6.3 \text{ km}$$

174

When using ordinary detonators precautions should be taken to evacuate the site when thunder is at a distance of 11 km (7 miles); this is about the greatest distance at which thunder can be heard. For high energy detonators with $K > 100$ amp² ms, the safety radius is reduced to 7 km (4.5 miles).

In underground work in thunder areas or seasons, there must be an effective warning system. The special safety detonators such as the Swedish type VA, the German typ HU and others of equal quality, with an ignition impulse $K_t > 100$ amp²ms, not only reduce the safety radius but also give within this radius a much lower risk than ordinary detonators. They are not, however, "lightning proof" as electric primary or secondary pulses may enter rails, pipelines or other conducting material with one end close to the working front.

An effective measure to eliminate the risk during the whole procedure of loading the round is to avoid ignition wires hanging down on to the soil. A flash of lightning striking the rock gives a high voltage inside the rock and a high potential between the cap and its wires if these terminate on the ground near conducting material (rails etc.) of approximately the same potential as a distant point. With the wires rolled together and placed in the holes there is no possibility of a great drop in potential. This method can also be practically regarded as safe for a high voltage in the rails and zero potential in the rock. There still remains the moment when the round is to be connected. Then the above mentioned condition should be fulfilled when the thunder is 11 or 7 km away, depending on what kind of detonators is in use.

TABLE 6:9. *Safety distances from power lines.*

Power line voltage kv	Distance[1] in m (*ft*) for detonators		
	type VA	type HU	Ordinary
10	—		
20	—	10 (*33*)	
50	—		
70	—	17 (*56*)	20 (*66*)
100	—		
130	10 (*33*)	22 (*73*)	30 (*100*)
220	10 (*33*)	22 (*73*)	40 (*130*)
400	16 (*53*)	22 (73)	60 (*200*)

[1] With regard to the risk of ignition from ground potentials no wires or joints of the connection system should at any moment of the work be in contact with the ground closer than 50 m (165 ft) from the power lines (60 m for ordinary detonators).

TABLE 6:10. *Safety distances in meters from radio antennae.*

Detonators		Effect, kW								
		0.100	1	4	10	40	100	200	400	2000
Ordinary		65	200	400	650	1300	2000	2800	4000	8600
Hu, VA	$\lambda <$ 190 m	0			30	70	100	130	200	600
	$\lambda <$ 560 m	0		40	60	150	220	300	500	1600
	$\lambda <$ 3000 m	0	70	150	200	500	700	900	1500	5000

d. Power lines

According to Y. Hagerman, the VA and the HU detonators can be used in the vicinity of power lines at the following distances. assuming that the VA-detonators have their original wire lengths and the HU wires are at least 3.5 m (11 ft) twinwire lengths.

e. Radio frequencies

There seems to be very little risk of ignition from radio waves in ordinary blasting practice, but in the event of special unfortunate combinations which obviously can occur, it can be proved that the radiowaves may cause ignition. This aspect therefore, cannot be ignored.

The safe distance between the antenna and the nearest wires of the blasting circuit are given in table 6:10 for ordinary electric detonators with minimum ignition current $I_{\min} \geqslant 0.20$ amp. and for more insensitive caps type VA and HU with $I_{\min} \geqslant 1.3$ amp. λ denotes wave length for the radio waves.

Example. A blasting job is to be performed in the vicinity of an antenna, effect 100 kW. The HU or VA detonators can be used at greater distances than 220 m if it is a question of medium wave lengths.

The effect values (kW) refer to a uniform distribution. It should be observed that many stations have a concentration of the energy in one main direction. The station of Stockholm (Nacka), for example, for medium wave lengths has an effect of 150 kW, and is concentrated in the direction of the mainland so as to correspond to a higher value, not exceeding 300 kW. This value can be used in the table, which gives a distance of about 400 m. Thus, in using the table, it is necessary to know not only the wave length and the effect but also if the effect is locally increased in the direction of the blasting job.

The distances given in the table assume for the VA detonators that the round in question includes at least 10 detonators.

There must be no insulation faults or earth faults in the connection system.

6.10. Planning the work

With the help of the methods described, blasting operations can now be carried out in accordance with a prearranged plan. For this purpose it is necessary, when planning, to be able to calculate both the amount of the charge required and the anticipated fragmentation (Table 2:4). In order to make use of the tables provided the blasting properties of the rock can be determined by trial blasts. This applies particularly when blasting is carried out in the neighbourhood of easily damaged buildings so that throw must be prevented. Before blasting, the burden, the distance between holes and the charges must be determined with regard to the average fragmentation required, the latter being decided in turn by the capacity of the machines available. Where special consideration has to be given to ground vibrations and the charge in each interval must be restricted on this account, it may be found necessary to reduce the burden and the distance between holes. More infrequently, the depths of the holes are reduced as this factor is usually determined in advance from given dimensions. On the other hand, it is not necessary, as a rule, to limit the total consumption of explosives in a round on account of ground vibrations. It is usually possible to arrange the ignition so that the same maximum values of the vibrations from a shortdelay round are obtained as when one or some of the holes are fired instantaneously.

The planning of the work must also include the necessary safety measures. Since blasting with large rounds is now carried out in the center of housing areas, greater precautions must be taken to protect the working site and surroundings from the throw of stones and fragments from the round. At the same time, however, it has actually become possible to provide adequate protection. With larger rounds their number is reduced, which in itself has proved greatly to increase the safety. This also means that a suitable time may be selected and, if necessary, the surrounding area evacuated. This was not formerly possible when blasting continued throughout the entire day. Provided that the charge is calculated correctly, the reduction of throw obtained with the short delay method enables the blasting results to be judged with such accuracy that protective material is frequently unnecessary, or need only be placed at such points of the round as are exposed to special risks from throwing, as in the case of low benches. In some cases indirect protection may be employed. For this purpose protective planking, wooden screens or ordinary shielding mats may be placed in front of nearby walls, windows or doors which are particularly exposed.

When using electric ignition all the recommendations of the preceding paragraph must be observed.

The planning of the work follows the schedule:

To be determined	*with regard to*
1. Rock factor by trial blast (if necessary)	
2. Average size of boulders	capacity of shovels, loading and crushing machines
3. Diameter of the drill	above, drilling equipment, ground vibrations, risk of scattering of stones, bench height
4. Burden and spacing	diameter, desired throw and fragmentation, bench height
5. Maximum charge in each hole and interval	ground vibrations
6. Depth of holes, bench height	greatest efficiency for the drilling and the shovel, ground vibrations
7. Delay time	fragmentation, risk of throw
8. Firing sequence	free angles of breakage, desired direction of throw
9. Size of round	the blasting machine, ground vibrations, number of intervals available and their delay time.
10. Safety measures: Shielding, indirect protection, evacuation,	the surroundings
safety against premature ignition	sensitivity of the detonators

A result of being able to plan and calculate the work as a whole will be that the blasting operations can be organized by technically qualified personnel so as to give maximum efficiency and minimum costs for all operations combined.

FIG. 6:34. Trench with and without presplitting. (The Royal Swedish Fortifications Administration. Sweden).

6.11. Problems

1. Discuss how to connect a round with a) 35 holes and 7 delays b) 240 holes and 12 delays c) 1000 holes and 25 delays.

2. A row of 20 holes in an open bench is to be blasted with short delay detonators of delay No. 5. Suppose that the rock is thrown away at a velocity of 10 m/s (33 ft/s). a) How far away is the rock from the hole that detonates the first, when the rock at the last one begins to move? b) The same question with ordinary delay detonators No. 5.

3. A round consists of three rows with 7 holes in the first one and the delay numbers 1–7, 6 holes in the next with delay 8–13 and 5 in the last row with delay 14–18. Suppose that the rock moves at a velocity of 10 m/s. a) How far has the rock in front of delay No. 5 moved when No. 6 detonates? b) How far has the rock in front of No. 1 moved when the charge in the hole behind it in the third row detonates?

4. In checking a round in a conducting ore body with 1400 detonators with a resistance of 5 ohm each, connected in 14 parallel series it is found that the resistance to earth is 10 kohm for ten of the series. For the remaining 4 series the earth resistance is found to be 5 kohm, 1 kohm, 0.8 kohm and 0.5 kohm. Can the round be ignited without reducing the earth faults? What measures are to be taken?

5. A bottom bench in a tunnel has a height of $K = 5$ m (16.5 ft) and a width of 10 m (33 ft) and is to be drilled with 50 mm (2 in) holes. Make a drilling pattern and discuss various ignition patterns for a round of ten rows. Slope of the holes $2:1$.

6. A round in a road cut, width $B = 16$ m (53 ft) and height $K = 0$–3 m (0–10 ft) is to be blasted with 20 delays, holes $d = 50$ mm (2 in) in diameter, loaded with dynamite at the bottom with $s = 1.0$, $P = 1.35$ kg/dm³.

Indicate as an example on a plane the contours for the heights, 0, 0.6, 1.2, 1.8, 3.4 and 3.0 m (0.2, 4, 6, 8, 10 ft) above the 0-level. Indicate also on this plane the position of the holes at the bottom and at the collaring. Make an ignition pattern for the round.

7. A trench is to be blasted to a depth of $K = 2.4$ m (8 ft) and a width at the bottom of $B = 1.5$ m with 50 mm (2 in) holes. Make a drilling and ignition pattern. Slope of the holes $3:1$. (The indexes for B in the tables 6:7 indicate the number of holes in a "row". Thus $B_3 = 1.5$ m means that rows of 3 holes in each are required).

8. The same trench as above is to be blasted with $d = 32$ mm. Make a pattern for a) the drilling, b) the loading, c) the ignition.

9. A road cut is to be blasted below a power line of 100 kV voltage. The rounds are ignited with electric detonators. When (and if) the safety distance is too small for the detonators the round has to be ignited with a denotating fuse. How close to the power line can the electric detonators VA, HU or ordinary ones be used?

10. A round of 200 holes is to be blasted at a distance of 750 m from a radio station, an effect of 250 kW and a wave length of 500 m. What is the safety distance for ordinary detonators? If this is too big, can the round be ignited with HU or VA detonators?

7. TUNNEL BLASTING

The driving of tunnels and drifts is the subject of special interest in the technique of rock blasting. This is natural in view of the central position that tunneling occupies in mining operations, and in view of the immense scope which the programme for the subterranean construction has acquired. Investigations in recent years have shown that there is a multiplicity of unsolved problems and possibilities for ascertaining more rapid, effective and cheaper driving methods.

In order to make clear the general conditions for such a development an account will be given here of principles and calculations for tunnel blasting, terms for the application of these calculations in practice, analysis of various types of cuts and how the construction of drilling patterns is to be carried out. Finally some practical drilling patterns will be given as examples for various tunnel cross-sections. In the following Chapter 8 special problems connected with tunnel blasting with parallel holes are dealt with and examined.

In principle, tunnel blasting is carried out by perforating the rock at the front of the tunnel with drill holes in which explosive is placed together with detonating fuse or electric detonators. The holes and ignition sequence are arranged according to a plan which is drawn up in advance and decides how the rock is to be broken. The first holes in the sequence aim at creating an opening towards which the rest of the rock is successively blasted. This opening, the cut, is the key that opens the rock to a depth depending on the features and success of the cut. The holes of the cut are arranged in a series of wedges or in a fan, cone or cylindrical pattern. The phases of the blasting dealing with the remaining contour of the tunnel should be performed so as to leave the rock with the intended contour undamaged and at its full strength.

7.1. Breakage in the case of free burden

The main part of the rock in a tunnel round is, or should be, broken with a more or less free burden which means with an angle of breakage greater than 90°. In such cases the conditions for every single hole are in principle the same as in benching, and the calculations in Chapter 3 will apply. From

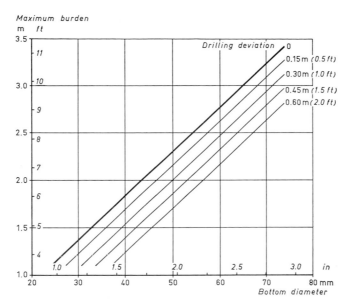

FIG. 7:1. Burden as a function of diameter of drillhole and deviation in the position of the drill hole at bottom. Rock constant $c = 0.4$. Weight strength of explosive $s = 1.0$, degree of packing $P = 1.25$ kg/dm^3.

this chapter the values in fig. 7:1 have been taken for maximum burden as a function of drill hole diameter for ordinary rock constant $c = 0.4$, explosive with weight strength $s = 1.0$ and a degree of packing of $P = 1.25$ kg/dm^3. The value $c = 0.4$ for the rock factor includes a normal variation in the rock.

In non-homogeneous rock the angle of breakage varies if the blasting is made in a direction parallel or perpendicular to the stratification of the lines of weakness. This has normally no great influence in benching and in stoping of rows with several holes, when the new face of the rock which has been created is mainly defined by the row itself. When blasting the holes close to a narrow free face only one or some few holes can be ignited at a time. With a tendency in the rock to smaller angles of breakage than 90°, the burden for the rear holes will be greater (up to some 30%) than expected. With a tendency to greater angles the fixation is, on the other hand, greater for the hole concerned, if it has a free opening of only 90°.

For these reasons the practical rock constant in homogeneous rock may well be $c = 0.4$, but in blasting hole by hole in unhomogeneous and irregular rock the influence of the greater fixation in certain directions can be compensated in practice by increasing the value to $c = 0.6$–0.7, in special cases to even higher values.

This is only one of the reasons for the higher consumption of explosive in tunnel blasting. Others are the greater influence of drilling scatter, a desire for small pieces of rock in the muck pile combined with the fact that a heavy overcharge is not so noticeable and disastrous as it can be in open-air excavation where a higher precision in the calculation of the charge is necessary.

7.2. Drilling precision

The scattering of the drill holes as a quantitative factor is often disregarded. It is included quite indefinitely in the technical margin together with the rock factor. In discussing blasting as a whole it would be a great advantage if attention could be paid to the drilling precision in calculating the charges and in constructing the drilling pattern; for the blasting of the cut it is essential.

The deviation in the position of the holes should be studied at the bottom of the round. Just as a rifle bullet does not always score a bull's eye, a drill hole may not strike the intended point but scatter around this, varying from one hole to another as is shown in fig. 7:2, where the pattern of the hits for 36 holes is given in two cases with two different types of distribution. In the space between two of the circles $\frac{1}{6}$ (17 %) of all the drill holes are expected to end at the depth in question. The left-hand picture shows a normal (N) distribution when the total deviation is regarded as the variable; in the right hand picture the distribution is denoted by F and gives a normal distribution for each of the horizontal and vertical components. Both these distributions occur in rock blasting, depending on how the drilling equipment is arranged. In principle the F-distribution is elliptical, as the spread of the vertical component does not necessarily equal that of the horizontal one. For convenience, only the circular F-distribution is given. If it is not specially mentioned it will be supposed that it is a symmetrical N-distribution. The radius R of the outer circles in fig. 7:2 include $\frac{5}{6}$ of all the holes. With increased depth the radius increases. If the deviation per m including $\frac{5}{6}$ (83%) of all the holes is denoted σ we have $\sigma = R/H$.

Faults occur in the collaring (R_c), in the alignment of the holes (R_d) and additional deviations inside the rock (R_r) owing to unevenness of the rock and to the effect of gravity. All these faults sum up as vectors in a three-dimensional space. If the radii including $\frac{5}{6}$ of the distributions are denoted R_c, R_d and R_r the resulting distribution is characterized by

$$R = \sqrt{R_c^2 + R_d^2 + R_r^2} \tag{7:1}$$

A consequence is that the greatest component dominates more than in a linear relation. If, for example, R_d is more than twice as big as the two

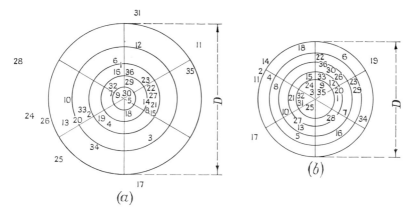

FIG. 7:2. Distribution of 36 holes in relation to the intended position (center).
a) N-distribution, b) F-distribution. For $H = 4.0$ m (13.4 ft) and $\sigma_{5/6} = 2$ cm/m
$D - 16$ cm (6.4 in). With $\sigma_{5/6} = 5$ cm/m, $D = 40$ cm (16 in).

other deviations then an approximation $R \simeq R_d$ gives values that are only
20 % lower than the proper ones. Ordinarily the resulting deviation is only
known within a tolerance of $\pm 20 \%$ and the approximation may be quite
satisfactory.

The deviation R_c is a constant, the two others increase with the depth
proportional to H and $H^{3/2}$ respectively. With the lowest practical values in
ordinary tunneling and with H given in meters we have

$$R_c = 2 \text{ cm}$$
$$R_d = H \text{ cm} \qquad\qquad (7:2)$$
$$R_r = 0.6\, H^{3/2}$$

The values at various depths and the resulting total deviation are given
in table 7:1.

The table is given as an illustration of how the components combine in
one special case and shows how surprisingly small the practical influence
of the collaring faults is, even with the extremely low deviation faults
(R_d and R_r) assumed.

The analysis of driving results with parallel hole cuts in the following
chapter indicates that the influence of the deviation faults has been
about 15% greater than in (7:2). This means that if the collaring faults
don't exceed the value $R_c = 2$ cm, they contribute less than 10 % to the
resulting deviation at depths greater than $H = 3$ m and (7:1) can be
simplified to

$$R = H(\sigma_d^2 + \sigma_r^2)^{\frac{1}{2}} \qquad\qquad (7:3)$$

183

H	m	1	2	3	4	6	8
R_c	cm	2	2	2	2	2	2
R_d	cm	1	2	3	4	5	6
R_r	cm	0.6	1.7	3.1	4.8	8.8	13.6
R	cm	2.3	3.3	4.7	6.6	11	16
σ	cm/m	2.3	1.6	1.6	1.6	1.8	2.0

where σ_d and σ_r represent the deviation per m in the two components in question, the first one being a constant and the second a linear function of $H^{1/2}$. This approximation with $\sigma_d = 1.3$ cm/m and $\sigma_r = 0.78$ cm/m$^{3/2}$ gives at depths greater than $H = 2.4$ m the same values as the relationship in table 8:2 for depths less than 8 m. If the region of application is still more restricted an even simpler approximation applies. At depths between 2.5 m and 5 m the deviation per m, within $\pm 10\%$, can be regarded as a constant, $\sigma = 2.0$ cm/m.

Practical results

In drilling parallel hole cuts with hand-held machines a skilled worker can attain as good a precision as in (7:2). There has so far been no evidence that this is obtained with machines mounted on a fixed support, even if

FIG. 7:3. With smooth blasting the deviations in the drilling are exposed. They depend on the individual worker.

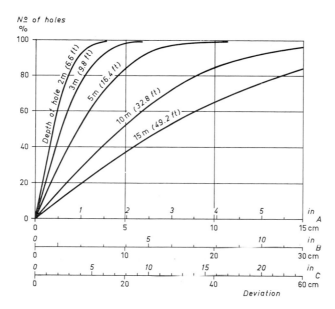

No of holes
%

Depth of hole 2m (6.6 ft)
3m (9.8 ft)
5m (16.4 ft)
10 m (32.8 ft)
15 m (49.2 ft)

Deviation

FIG. 7:4. Number of holes in percent with less deviation than that read on the horisontal scales $A–C$ (N-distribution). A = very careful drilling $\sigma_{5/6}$ = 1 cm/m, B with $\sigma_{5/6}$ = 2 cm/m, C = good drilling $\sigma_{5/6}$ = 4 cm/m.

they are specially designed for great accuracy. The material from parallel hole driving (Chapter 8) seems to indicate that the deviations are of the order of $R_d = 1.14 H$ cm and $R_r = 0.8 H^{3/2}$ cm$^{3/2}$, which gives $\sigma = 2.0$ cm/m for $H = 4$ m (13 ft). Sometimes in accurate drilling the alignment faults amount to $R_d = 2.0 H$ cm, which gives $\sigma = 2.6$ cm/m.

These values cannot be counted on in stoping, not even in the ordinary cuts such as V- and fan cuts, where $\sigma_d = 5$ cm/m can be regarded as a comparatively good result. In this case σ_d quite dominates for ordinary depths of the holes and the resulting drilling precision can be described as $\sigma \approx \sigma_d$. In ladder drilling the value of $\sigma_d \approx 5$ cm/m can be assumed. With drilling jumbos as well as with hand-held equipment $\sigma_d = 3–5$ cm/m can be regarded as rather ordinary values if the personnel is *specially* trained. The deviation in alignment (fig. 7:3) can easily be demonstrated after smooth blasting.

It is evident from the figures mentioned that the way to keep the deviations in the drilling at low values is to be careful when collaring and to use stiff bars for the drills. If this is done practically all deviation will be due to alignment faults and can be reduced almost in direct proportion to the reduction of these faults as long as they exceed $\sigma_d = 2$ cm/m.

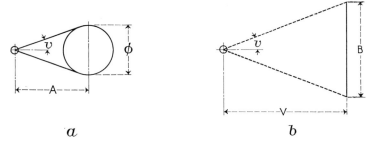

FIG. 7:5. Blasting towards a narrow opening. a) circular, b) rectangular.

7.3. Calculation of the charge for small angles of breakage

Two typical examples have been examined. In both, the pipe-charge depends upon the fixation angle and the distance to the free face, which in one case is a circular hole, and in the other a rectangular opening, as shown in fig. 7:5. If the charge per meter is indicated by l, half the aperture angle by v, the distance between the centers by A (m), the diameter of the empty holes by $\varnothing$ (m) and the distance to a free surface by V (m), charges that give full breakage are given in kg/m by the following equations:

$$l_a = 0.55\,(A - \varnothing\,/2)/(\sin v)^{3/2} \qquad\qquad (7:4)$$

$$l_b = 0.35\,V/(\sin v)^{3/2} \qquad\qquad (7:5)$$

in which indexes a and b refer to figs. 7:5a and b respectively. The relation (7:4) is shown in the next chapter (fig. 8:4); the relation (7:5) is given in table 7:2 below.

The equations differ only in the value of the constant, which in blasting towards a circular opening is 60 % greater than for the rectangular one, owing to the greater fixation in the first mentioned case.

The relationships given cover the present experimental data, but blasting towards a rectangular aperture must be the subject of further theoretical consideration. The experimental part of the investigation has, however, reached a stage that gives a basis for discussion of some of the main points involved. These are: the determination of the optimum diameter of the uncharged hole in parallel cut blasting, construction of the drilling pattern and calculation of the charge with regard to the placing of the holes and the actual drilling scatter.

The charges given in the table are sufficient for breakage but not necessarily close to limit values as the basic experimental investigation has mainly covered charges that give full breakage. Many of the values, especially those for capital B:s, may be some 50% higher than for limit charges.

TABLE 7:2. *Concentration of the charge in kg/m for various burdens (V) and extension (B) of the free face (Preliminary table).*

Max. burden V		Concentration of the charge, kg/m l											l_b kg/m limit charge with free burden
m	ft	B = 0.10 / 0.3	0.15 / 0.5	0.20 / 0.7	0.25 / 0.8	0.30 / 1	0.35 / 1.2	0.40 / 1.3	0.50 / 1.7	0.60 / 2	0.80 / 2.7	1.4 m / 4.7 ft	
						kg/m							
0.10	0.3	0.12	0.08	0.06									
0.15	0.5	0.30	0.18	0.13	0.11	0.09							
0.20	0.7	0.60	0.35	0.24	0.20	0.16	0.14	0.12					
0.25	0.8	1.0	0.60	0.35	0.30	0.26	0.22	0.18					
0.30	1	1.3	0.9	0.60	0.50	0.35	0.31	0.26	0.22	0.18			
0.35	1.2		1.2	0.9	0.65	0.45	0.40	0.35	0.30	0.25			
0.40	1.3		1.6	1.2	0.9	0.7	0.6	0.50	0.40	0.30	0.24		
0.50	1.7			2.0	1.6	1.3	1.0	0.7	0.60	0.50	0.36		0.13
0.60	2				2.2	1.9	1.6	1.3	1.0	0.7	0.52		0.17
0.70	2.3					2.5	2.2	1.8	1.3	0.9	0.7		0.25
0.80	2.7						3.2	2.4	1.8	1.4	1.0	0.6	0.32
1.00	3.3							4.0	3.0	2.4	1.4	0.9	0.5
1.20	4								4.4	3.8	2.5	1.2	0.7
1.40	4.7								5.0	3.6	1.6		0.9
1.60	5.3									4.8	2.4		1.1
2.00	6.7										4.0		1.9

The figures in lb/ft are 2/3 of those for kg/m

The last column gives the limit values for the bottom charge with a free burden.

In addition to the obvious requirement that the charge used should be sufficient for loosening the rock, allowance must also be made for the swell of the broken material if the opening is small compared to the volume of rock loosened. An increase of up to 60 per cent in the volume may be expected, and if the material is to be blown out through the opening it is suggested to allow for 100 per cent if jamming of the broken rock is to be avoided. If the opening after one shot is more or less cleaned out the next shot will have a greater volume for the swell. If, on the other hand, the material remains in the opening this will after one or more shots soon become blocked, the conditions for the swell can no longer be fulfilled and there can be no satisfactory loosening of the rock. This happens when using *too heavy charges* in blasting towards a cylindrical opening of great depth. The pieces of rock are then thrown against the opposite wall of the hole at such a great velocity that they become recompacted here and the explosion gases escape behind them. A series of such shots does not give a sufficient opening inside the rock and a full advance is not obtained.

The possibility of throwing out the material is also reduced if the burden is big compared to the width of the opening. When the ratio B/V or $\varnothing/V$ is less than 0.50 and the aperture angle is consequently small, such heavy charges are required that there will be a plastic deformation of the rock instead of ordinary loosening. Nor does such "burning" widen the opening sufficiently at great depth. The subsequent holes will not have room enough for the swell and the round breaks off some meters from the free face.

For a given opening the burden has to be chosen. It is worth nothing how much the charge must be increased with an increased burden, especially when the ratio $B/V<1$. With $B=0.4$ m and $V=0.5$ m the concentration of the charge should be 0.7 kg/m, but it increases to 1.3 kg/m if the burden is $V=0.6$ m.

7.4. Consumption of explosive

Compared to bench blasting with holes of $H=4$–5 m (13–17 ft) where the specific charge is $q=0.25$–0.40 kg/m³ (0.40–0.65 lb/cu.yd) it is remarkable that the consumption of explosive in tunneling is often 4–10 times this value. It has already been mentioned that one reason for this is the greater influence of drilling scatter that is often of the order of 0.1 m/m (1.2 in/ft). This reduces the practical burden (V_1) to

$$V_1 = V - 0.1H \qquad (7:6\,\text{M})$$

where V is the maximum burden at the bottom and H the depth of the holes. With $H=4$ m (13 ft) and $V=1.5$ m (5 ft) this means $V_1=1.1$ m (3.6 ft) instead of $V=1.3$ m, which could be the burden in benching. The corresponding increase in the specific charge is some 50%. Even if irregularities in the rock and in the lines of breakage of each individual hole have to be compensated for by a 50–75% higher charge, we have only got 2.5 times the value in benching and the corresponding specific charge $q_1=0.4$–0.8 kg/m³ (0.7–1.4 lb/cu.yd).

So far we have only discussed stoping without fixation. If we take a parallel hole round, the holes closest to the empty hole do not have a free angle of breakage. A central part with an area of about 1×1 m² (1.1×1.1 sq.yd) requires a much higher specific charge, depending in fact on the advance and the diameter of the empty hole, but an approximate figure of $q_0=7$ kg/m³ can be taken.

The total charge for a round of advance A and section S is in metric units

$$Q_\tau = Aq_0 + (S-1)q_1 A \qquad (7:7 \text{ M})$$

and the mean value for the specific charge is

$$q = (q_0 - q_1)/S + q_1 \qquad (7:8 \text{ M})$$

The cut is ordinarily placed more or less in the center of the section. The holes below the level of the cut have to heave the rock vertically upwards to ensure the swell and they demand a greater specific charge. The holes above the level of the cut need a lower specific charge. The value q_1 in (7:8 M) indicates the practical mean value of the specific charge outside the central section.

The empirical relation for tunnel areas from 4 m² to 100 m² is approximately

$$q = 14/S + 0.8 \qquad (7:9 \text{ M})$$

The noticeable difference between these two equations lies in the first term, and indicates that in practice a heavy overcharge is used in the cut section, where q_0 seems to be about 13 kg/m³. The reason for such an overcharge is of course that an extra margin for full breakage in and around the cut is desired. It may well be possible to reduce this margin as it is nowadays easy to ensure 100 % ignition of the detonators with a good blasting machine and by checking the electric system, to calculate the exact position of the holes and to see that the deviations in the drilling are kept below $\sigma = 5$ cm/m. These measures will also make it possible to reduce the second term to 0.6 or maybe to even lower values.

7.5. Different types of cut

Fan cut

The fan cut can be taken as an example of a complete realization of the relieving principle (fig. 1:6). From what has been previously mentioned, the uncertainty regarding the angle of breakage at the bottom is eliminated: a right angle can be expected. With a knowledge of the burden and the width of the tunnel the construction of a breakage pattern for the cut is reduced to a purely geometrical problem (top of fig. 7:6). On the other hand, we must not ignore the fact that the problem is three dimensional, so that we must pay attention to the breakage conditions also in a surface perpendicular to the drill holes at the bottom of the break and there apply the right angle principle laterally (bottom of fig. 7:6). It is only then that the consequences of placing the holes and sequence of ignition become evident. The distance E between the holes and the burden V are chosen

189

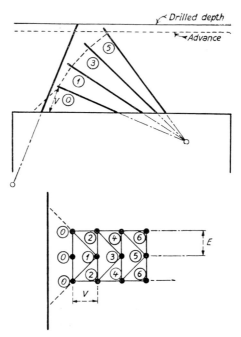

FIG. 7:6. Construction of the drilling pattern for a fan cut.

with regard to the fact that they are mutually connected (in the case shown in the figure $E = V$). This should be taken into account when the number of holes has to be increased which is sometimes necessary. The fan cut can be regarded as a form of trench to be blasted across the tunnel section.

The ignition pattern calls for special attention. A rearrangement of the sequence of ignition in a drilling pattern already given can lead to an improvement in the final result. If so, this method should of course be chosen in preference to increasing the number of holes and the size of the charge. Fig. 7:6 illustrates the situation if we compare the indicated sequence of ignition with one in which three holes are ignited in every delay number. In such a case two out of three of the side holes will detonate before the middle one. A considerably larger charge is in such a case required for breakage, as there is not a free burden for the side holes. In addition, the surrounding rock is exposed to a severe strain as a result of fixation, with a risk of tearing neighboring holes and causing cracks, which may have a bad effect on the breakage.

Plough cut (or V-cut)

There is a strange contrast between the general idea behind the plough cut and the way it actually functions in the rock. The ordinary advance

190

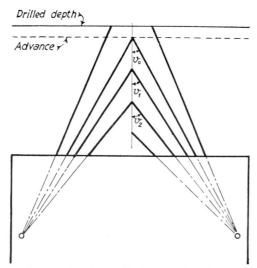

Drilled depth

Advance

v_0

v_1

v_2

o

o

FIG. 7:7. Theoretical plan of a V-cut.

here is about 45–50 % per cent of the width of the tunnel and is therefore geometrically determined as in the fan cut (fig. 7:6). This implies that the acute angle $(2v)$ at the bottom is larger than, or equal to, 58° ($\cot v = 1.8$). In principle, however, there is no such limit to the advance but there is a connection (fig. 7:8) between weight of charge and $\cot v$. For instance, at the value $\cot v = 2.3$ we have even with a moderate number of holes $(6 + 6)$ and charge (0.35 kg per m) an advance that is greater than is reached at present. The curve in fig. 7:8 applies if the drilling has been carried out so that the holes meet in pairs at the apex, and if all holes belonging to the angle of breakage in question are ignited with a sufficiently small spread in the delay time.

How does this work in practice? A comparison with the distribution curves for drilling scatter (fig. 7:4) shows that with the deviation one has reason to expect, the conditions cannot be fulfilled. The alignment faults are of the order of ± 5 % or more, which means that the holes at a depth of 5 m (16.5 ft), for example, meet the center line with deviations of more than ± 0.5 m (1.7 ft). One cannot reckon on a simultaneous detonation owing to flashover from one hole to the other. As ignition is usually arranged igni-tion scatter is so great (one or some tenths of a second) that charges ignited by one and the same delay number do not cooperate and cannot then pos-sibly tear the rock to the bottom of the V (plough) when the angle is about $2v = 60^\circ$ or less. Thus what really happens in practice is that most of the holes deeper in the V-cut will break out more or less as if ignited one at a time.

191

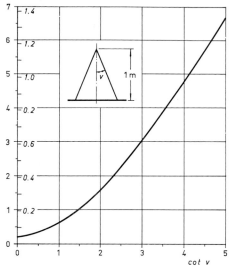

FIG. 7:8. Relation between charge per area of the cut and the angle for a single instantaneous V-cut.

In contrast to the general idea for the V-cut the individual hole will then loosen the rock in front of it in keeping with the laws of free breakage. It is then consistent to revise the drilling pattern so as to start from this fact and apply an ignition pattern which really gives free breakage (fig. 7:9). A secondary condition which must be considered is that flash-over in this case must be prevented both between opposite holes and between contiguous ones. This implies that the drill holes, where fully charged, must not be allowed to be closer than 20 cm (0.7 ft).

The construction of a cut with every hole breaking in right angle as in fig. 7:9 gives the same limitation in the advance as is attained with a V-cut. This is an indication that the tearing in a V-cut also takes place in the same way.

The way to improve the V-cut is to use detonators of good accuracy in the delay time and to reduce the influence of the deviations in the drilling. If the separate ploughs are loosened as a whole one after another it seems reasonable to expect that the ignition can be made with short delays of the order of 50 ms between every plough. If the loosening, on the other hand, takes place as in fig. 7:9 the movement of the rock is not primarily straight out of the cut but to the opposite side in a direction perpendicular to the holes where the explosion gases are acting. The subsequent holes

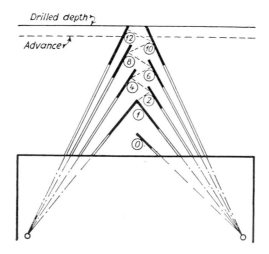

Fig. 7:9. Design of drilling pattern for a *V*-cut.

have not space enough to swell if the delay time between the intervals is too short. Experiments indicate that in ordinary *V*-cuts the delays between the ploughs can be of the order of 100 ms in the inner rows.

The influence of the deviations in the drilling is reduced when using holes with a larger diameter. Attempts to drive a tunnel with a width of 9 m (30 ft) have proved it possible to raise the advance per round from below 50% of the width to the excellent value of 60% by using ladder-drilling and better drilling precision as well as holes of 44 mm diameter instead of 34 mm (according to Lautmann).

Instantaneous cuts

The analysis of the breakage of the *V*-cut has led to the construction of new cuts characterised by the drilling and ignition being done so that all the holes really can cooperate and then break smaller top angles than would otherwise be possible. They are called instantaneous cuts as they are preferably ignited by instantaneous detonators to ensure a simultaneous detonation of all the charges in the cut. An old example of such an instantaneous cut is the pyramid cut. It is rather restricted in its use as it is not fit for deeper rounds. At the end of the holes a pyramid cut doesn't give a real opening as all the holes meet here at one point. Instantaneous cuts of the type discussed here should be so designed that at the bottom of breakage a real opening is made for the subsequent holes to break against. The following types of cuts may be mentioned.

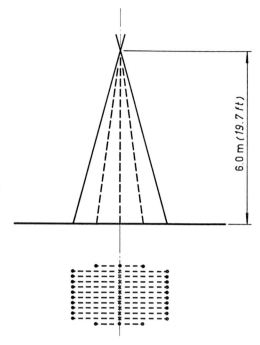

FIG. 7:10. Blåsjö-cut (instantaneous firing).

Blåsjö-cut. This is a cut with a single *V* where all the holes on one side are parallel and meet the holes from the other side at an angle that may be as low as 30°. The example shown in fig. 7:10 (10 parallel holes from each side) has been used with an average advance of 5.5 m (18 ft) with a width of only 3.3 m (11 ft) at the base of the cut. Even with rather rigid drill steels the cut can be used in tunnels with a width of down to 6.6 m (22 ft). It is a greater advance (83 %) than can be obtained with *V*- or fan cuts. At the bottom of the breakage the cut gives a rectangular opening of about 2.0×0.4 m² (6.6×1.3 ft²). A big throw is obtained even if the holes are cautiously charged in the pipes. The big throw is a disadvantage of this cut.

WP-cut. This cut is developed from the Blåsjö-cut by the same contractor (Widmark and Platzer). The same principle of instantaneous ignition with only 4–5 holes of a diameter $d = 44$ mm (1.75 in) from each side (fig. 7:11) is applied. The holes are arranged in a geometrical figure as an incomplete pyramid and are not parallel in the planes of the sides as are those of the Blåsjö-cut, and the throw has been reduced. It seems to have a somewhat lesser advance than the Blåsjö-cut, but for tunnel widths below 7.5 m (25 ft) a greater one than *V*-cuts and fan cuts.

Presplit cut. The experience with the above mentioned cuts with instan-

194

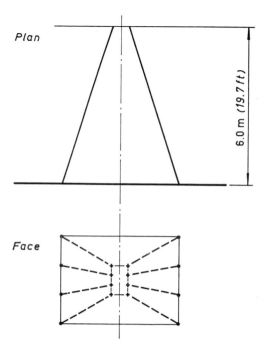

Plan

6.0 m (19.7 ft)

Face

Fig. 7:11. *WP*-cut (instantaneous firing).

taneous ignition indicates the possibility of reducing the concentration of charge and the spacing of holes to the values given in table 10:1 (p. 310). Practical tests performed with such types of cuts have, however, not yet proved successful.

Parallel hole cuts

The mechanism of breakage is entirely different in blasting cuts with parallel holes. These cuts were originally b u r n c u t s with charges and distances between the holes that give plastic deformation in the rock, but further investigations have led to the construction of a general type of cuts avoiding the burning effect, so-called c y l i n d e r c u t s, giving a cylindrical opening into the bottom of the cut. The next chapter gives an analysis of all types of parallel hole cuts including results from practical driving.

7.6. Advance per round

The driving can be done with as great advances as possible in every round, but in many cases shorter rounds give a better working cycle and more rapid advance per day. Driving with maximum advances often invites

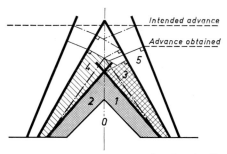

FIG. 7:12. The drillhole deviations cause reduction in the obtained advance per round. The holes 1 and 2 deviate 10 cm/m, the positions of the other holes are assumed to be correct.

trouble from over-estimation of the results. It is essential to know not only the possibilities but also the practical limitations and their causes. A blasting with a drilling pattern intended to give $A = 5.4$ m (18 ft) that actually only gives, say, 4.2 m (14 ft) is quite unsuccessful and probably not as good as one designed for $A = 3.3$ m that really gives this value.

When special interest is paid here to the maximum advance it does not mean that I regard it to be advisable to try to attain it, but mention it as a valuable and essential piece of information in tunnel driving. And of course if the limit for the advance is increased it may give better economy and greater speed in the driving.

Even where a maximum advance can be obtained one should not expect to be able to carry it out immediately. Fig. 7:13 shows a typical example

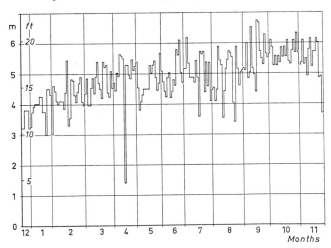

FIG. 7:13. Variation in the advances has to be counted on. This is caused by deviation in the drilling, ignition sequence and variations in the rock. (Given by K. F. Lautmann.)

196

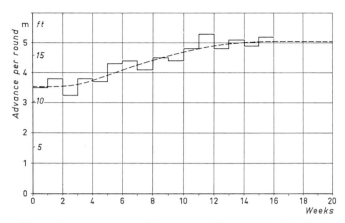

Fig. 7:14. The advance per round increases often during the first months of the job. ––– mean value.

for the advance per round in a tunnel driving. The maximum advance $A = 5.4$ m (18 ft) is only reached after 3–6 months. In the first month there was an average value of 3.9 m (13 ft). It takes time to make all the small practical corrections that the rock and other conditions may require and for the workmen to learn how to drill and load accurately.

In addition to this it is observed that the full advance is usually not obtained until after the agreement on wages is made if this agreement is to be related to the advance.

Whatever the reasons, it should be emphasized that there are practical difficulties when driving with angled cuts every time the advance is to be changed. One and the same drilling pattern is *either* being used from the beginning, or the pattern is changed in two or three stages at intervals of one or two months. This last alternative is hardly ever planned beforehand. Nevertheless when it becomes necessary it causes much trouble and work to rearrange the alignment and marking for the collaring on the rock. For the first alternative, when the drilling pattern is expected to remain unchanged, it is designed for the greatest advance aimed at, say 5.1 m (17 ft), but only the inner 4.2, 4.5 or 4.8 m (14, 15 or 16 ft) is drilled and blasted until at least 90 % of the drilled depth is obtained. Then another foot of the drilling pattern can be tried. A method that is simple in application is not to use the inner part of the drilling pattern. In this case all the collaring and the alignment are done as for a full round but the holes are not drilled to the full depth until the rounds work satisfactorily in the earlier stages.

The slope of the mean curve (–––) may be higher or lower. There are many examples of jobs where it has been so low that the curve never reached the

197

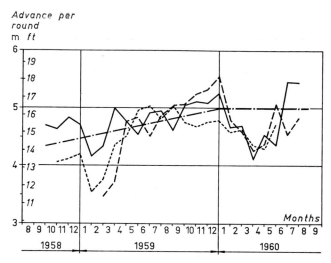

FIG. 7:15. The advance per round may depend on the temperature when this falls below zero °C (+32°F) as in the months of Dec.–March in the three examples above (Widmark o. Platzer AB).

intended level. The opposite, a curve that reaches the final value after only some few rounds is rather rare when trying to attain a maximum advance (A_m), but it can be had if the intended and final advance (A_1) per round is about 70 % of this value. Measures that increase the maximum advance p % may in this way increase the practical advance the same percentage. This is one reason for the interest in knowing the conditions for maximum advance.

With parallel hole rounds any advance below the maximum can be tried without changing the drilling pattern. This makes it easy to adjust the driving according to the advance that is possible and desirable for the moment (see Chapter 8.1i).

One factor influencing the advance obtained in most inclined cuts is the degree of packing at the bottom of the holes. A surprising result of this is that the advance is then also a function of the temperature of the dynamite if this is below $+10$°C ($+50$°F). In fig. 7:15 the advances from three separate faces at a working place in northern Sweden are shown from the end of 1958 to July 1960. In the winter season all the curves have a minimum that is quite characteristic in 1959 as well as in 1960. The advances seem to have fallen on an average more than 0.6 m (2 ft) from the summer value to the lowest value for one winter month ($A \simeq 4.3$ m). If it is correct to assume that this last value was caused by a lower degree of packing the conclusion is that the way to increase the advance in winter would be to

198

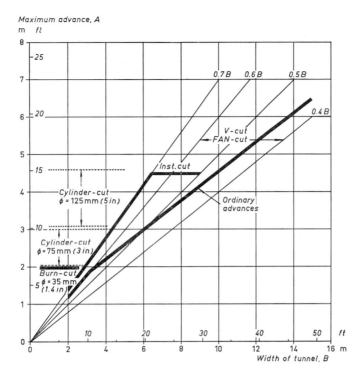

Fig. 7:16. Maximum advance per round as a function of the width of the tunnel for various types of cut. For *V*-cuts and fan cuts the values are between 40 and 60% of the width (shaded area). The percentage for the "Ordinary advances" falls with increased depth due to increased influence of the deviations in the drilling. "Inst. cut" indicates *WP*-cuts and pyramid cuts.

store the explosive at a higher temperature, inside the rock for example, or at least to arrange to do this for some hours before loading. The curves in fig. 7:15 would then also give an illustration of the importance of good packing even in ordinary tunneling. A special low-temperature dynamite has been developed. An examination of other driving results shows that an increase in the degree of packing of 20% has given a 10% greater average advance.

When talking about the advance per round it must be borne in mind that there are individual variations in every round, as shown in fig. 7:13. In one group (month 2–5) of rounds the average advance is 4.6 m (15.3 ft), but individual rounds have varied from 5.7 m down to 3.3 m (the round of only 1.4 m must have been due to some special cause apart from ordinary variation). Some months later an improved drilling pattern increased the average to 5.6 m (18.7 ft) with individual values from 6.25 m (21 ft) down to 4.4 m

(14.6 ft) and 3.7 m (12.5 ft). The variations seem to a great extent to be due to the drilling deviation. This can be assumed to follow the same distribution in some successive rounds, but in this variation it occasionally happens that the holes are unfavorably placed or unfavorable in relation to one another; it is like gambling with a number of dices, sometimes several of them give only ones, sometimes you get a lot of sixes.

This background must be clear before speaking about the maximum advance for different types of cut. It is the practical value that is of interest. It means the final value of such a mean curve, as in fig. 7:14, or the average value as received in fig. 7:13. With this definition the maximum advance is given in fig. 7:16 as a function of the width (B) of the tunnel. For angled cuts it is about 60 % of the width for small tunnels and falls to some 45 % at greater widths. The lower percentage values depend on the greater influence of the drilling scatter at greater depths of the holes. If this influence is reduced and a correct drilling and ignition pattern is employed advances of up to 60 % can be had even for a tunnel width of up to 9 m (30 ft). With instantaneous cuts 0.7 B can be obtained up to at least $B = 6.6$ m (22 ft). The advances for the parallel hole cuts are not in themselves dependent on the width of the tunnel, but the heavier equipment for the drilling of large empty holes may not be suitable in the smallest sections.

7.7. Choice of the section

Tunnels lower than 8 m (25 ft) are usually blasted in full section. For greater heights and long tunnels the lower part of the section can be left for benching with short delay multiple-row blasting. The unloading of the rock in the bench can then be done continuously and the drilling machines mounted on a jumbo so that several machines can be operated by one workman. This gives the lowest cost per m³. The height of the bench should be at least 3 m (10 ft) and as great as possible in view of the space required in the top heading for the haulage equipment. For a bottom bench higher than 10–12 m (33–40 ft) the lowest costs are obtained if it is blasted in two or more separate operations with a bench height of 5–10 m (27–54 ft) each.

In rock of poor strength the following measures can be taken

1) reduce the advance
2) divide the tunnel section into a central part and side-sections for stoping
3) smooth blasting of the roof and guniting immediately (within 30 min) after the round.
4) so-called armed pre-splitting.

The two methods 3) and 4) have made it possible to carry out tunnel blasting even in extremely bad rock without imposing such restrictions in the driving as 1) and 2) impose. A special guniting device has been developed for such purposes.[1]

When these methods are not used or are not sufficient, the advance per round must be reduced or—in very difficult cases—the tunnel section must be divided and the roof strengthened by roof bolting before the sidestopes are blasted. It is then usual to blast a pilot drift through the part of the rock where 2) is required. The drilling and ignition pattern for the side stopes is made as for benching.

7.8. Construction of stoping and ignition patterns

The object of the cut is to create a free surface towards which the rest of the blasting can be carried out. The problem in stoping is to attain just as large an advance with the remainder of the round as that effected in the cut, to get a satisfactory fragmentation and to get a suitable disposal of the broken rock. At the same time the remaining rock face should also be left undamaged and as close as possible to the planned contour.

A complete loosening of the rock depends on the breakage conditions at the bottom. The drilling pattern should therefore in the first place be based on the positions of the holes at the bottom. Starting from the surfaces exposed by the cut the stope holes should not have a greater burden at the bottom than V in table 7:2 (p. 187) when the extension B of the free face is less than twice the burden, that is to say that the angle of breakage is less than 90°. When the central opening is wide enough to provide conditions for a 90° breakage or more, the principle of rectangularity should be applied in indicating the expected lines of breakage. They decide the position of the subsequent holes as shown in fig. 7:17, right-hand stope pattern. In this way the sequence of ignition to be used is clearly defined and tearing in the surrounding rock is reduced to a minimum.

A full and consequent adoption of this system of ignition calls for a great number of delays and modern ordinary electric detonators are now manufactured with up to 24 delays. This greatly improves the conditions for a correct ignition pattern and a good fragmentation in tunneling. The short delay detonators can also be used to increase the total number of delays.

The placing of holes and sequence of ignition according to fig. 7:17, left, should be avoided. Just as much rock is blasted out with only two holes number 2, as shown by fig. 7:17, right; as with four holes number 2, as shown by fig. 7:17, left. Furthermore free breakage is not guaranteed for the subsequent holes number 3 in the pattern on the left. Because of ignition scatter every individual charge detonates separately and several will

---

[1] By AB Skånska Cementgjuteriet.

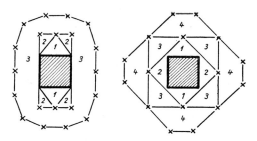

FIG. 7:17. Construction of stope pattern should be made as in the right-hand figure and not as in the left-hand one.

probably have a high degree of fixation so that there will only be breakage if they are heavily overloaded. There is also a severe strain on the surrounding rock which is torn up along cracks and crevices. This contributes to inferior fragmentation.

The maximum burden and the placing and number of holes are determined by the bottom diameter, by the explosive employed and by the degree of packing, as well as by the quality of the rock and by the drilling scatter. The effect of these different factors can be calculated according to Chapters 2 and 3, table 7:2 and section 7.3. This makes it possible to calculate the number of holes as a function of the above factors and of the area and width of the tunnel.

Some such connections for parallelhole rounds based on the values of the burden indicated in fig. 7:1 are given in fig. 7:18. It is evident from the curves that a very pronounced reduction in the number of holes is obtained for major tunnel areas when employing a drill of larger diameter. The same effect is obtained when increasing the degree of packing with the aid of the pneumatic cartridge loader or by using an explosive with increased breaking power per volume of the hole. With a section of 120 m² (144 sq.yd) and calculating according to $c = 0.6$ the number of holes can be reduced from about 160 to 110 if we go from 32 mm (1.3 in) to 43 mm (1.7 in) diameter. With the same section and 39–40 mm holes the number of holes can be reduced by 15 from 125 to 110 if the degree of packing is increased 15%.

For minor tunnel areas, on the other hand, the difference will be smaller: the corresponding numbers of holes with an area of 30 m² (36 sq.yd) are 58 and 50, which maybe does not justify the employment of a larger diameter of the bit. In addition, it is easier in small tunnels to obtain conformity with the given contour when employing small burdens. The curves in fig. 7:18 have been calculated on the assumption that this conformity has been possible so that full advantage has been taken of every hole,

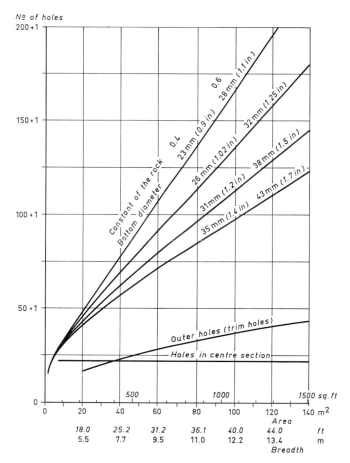

FIG. 7:18. The least number of holes as a function of the tunnel area with parallel hole rounds for different bottom diameters and for rock constants $c = 0.4$ and $c = 0.6$. Drilling deviation is assumed to be 0.25 m (0.85 ft) at the bottom. The curves include holes for smooth blasting.

with the exception, however, of the contour holes which have only been intended for an area of 0.5 m² each.

Some examples showing the construction of drilling patterns are given in fig. 7:19, where the intended position of the holes is seen at the bottom of the round for a 10 m² (12 sq.yd) and a 55 m² (66 sq.yd) tunnel. The former requires 2 holes more than the calculated minimum, because the burden for the roof holes is greater than desirable and two roof holes have been inserted. It may be mentioned that a minimum number of holes can actually be attained by a still better solution—a problem that can be left to the reader to solve.

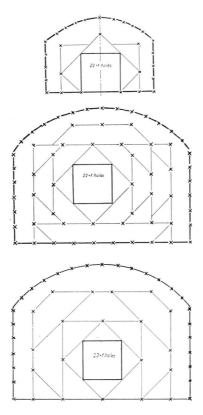

F<small>IG</small>. 7:19. Examples showing the principles in drawing a drilling pattern around a central square opening.

The two drilling patterns for a 55 m² (66 sq.yd) area illustrate the difference between one stope with small holes and one with large holes; (32 and 43 mm drill holes respectively in rock with $c=0.6$) and show the greatly simplified pattern attained in the latter case. In both cases there are fewer holes (one or two) than in the estimated minimum. This may seem surprising, but it is owing to the burden at the sides being 1.0 m (3.3 ft) instead of 0.75 m (2.5 ft), which is what was intended when calculating fig. 7:18. One cannot in general count on obtaining complete accordance. In the case of the big-hole stope it is more difficult to adjust the pattern, as has already been mentioned.

The position of the cut influences the throw of the round, the fragmentation and maybe also the number of holes. As the density of the holes is greatest close to the cut this can be taken advantage of if some of the holes are contour ones (fig. 7:20, left). This should only be done in small tunnel sec-

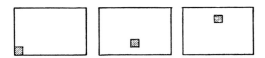

Fig. 7:20. Different placing of the cut section.

tions however. In order to secure the best throw and centering of the broken rock the cut is placed approximately in the middle of the section. Low location gives less throw, larger fragments and more compact piling of the rock. Less explosive can be used by stoping downwards.

In general the problem is the reverse; one wants to avoid employing the free fall. Note that with the given principles for drawing drilling patterns one consistently avoids downward stoping of entire rows of holes in the same interval. If the cut is placed high up in the section a less compact heap of fragments, lower and more extended with better fragmentation, is obtained.

7.9. The rhythm of the driving

The rhythm of the driving is dependent on the capacity of the mucking equipment and on the advance that can be obtained per round. For the planning of the work the data for the mucking equipment actually available should be inserted in table 7:3.

In the so-called rhythmic driving the drilling, loading and blasting are carried out in one shift and the mucking and transportation in the following one. This enables every man to specialize in his tasks and machines, which in a highly mechanized job is a necessary condition for making the best use of an expensive equipment. It also reduces or eliminates the loss of time for ventilation that in rhythmic driving is carried out between two shifts.

When the operation can be carried out in several neighboring drifts in mining it is possible even with small advances to use different and specialized crews for the blasting and mucking.

TABLE 7:3. Data for mucking equipment.

Machine	Min. section of tunnel, m^2	Width m	Height m	Volume of shovel, m^3	Capacity m^3/h

When the operations are in progress in a single tunnel the advance per round must be big enough to supply rock for the mucking equipment during one (or two) whole shift. This ordinarily means $A \approx 4$ m (13 ft) comparatively independent of the tunnel section. The rounds should also offer a high degree of certainty for full breakage in every round.

For angled cuts the sections must be greater than 50–60 m² if the advance is to be greater than 4 m. For cylinder cuts such advances can be reached in any section if the diameter of the empty hole is at least $\varnothing = 110$ mm for double spiral cuts (fig. 8:10a) and at least $\varnothing = 135$ mm for four section cuts (fig. 8:10d) if the drilling accuracy is satisfactory.

The advance per day is limited in rhythmic driving by the advance per round. Its range of application therefore increases also from this point of view with the possibilities for increased advances per round.

In *continuous driving* the same personnel do the drilling, blasting and mucking while working continuously round after round. They can in this way—except for the time for ventilation—be at work during the whole shift. Continuous driving is used when the advance per round is low and the mucking or the drilling and blasting do not need more than a part of the shift.

But continuous driving is also employed, when the highest driving speed is aimed at. Compared to rhythmic driving it is here possible to use more people at the front and to blast more frequently with rounds of the same length or shorter ones. Ordinarily the rounds are shorter than 4 m even in tunnels of great sections.

7.10. Drilling and ignition patterns

Drilling patterns with any type of cut can be designed on the principles given in previous sections and in the following chapter. Even so it has proved to be a great practical help to have details from patterns in the whole range of tunnel sections used in contracting and mining work. Such patterns are given in figs. 7:21–7:39. Figs. 7:40–7:45 show some drilling and ignition patterns specially designed and used for cautious blasting.

Every pattern is selected from a great material from various parts of the world so as to represent the best of todays technique. Some of the patterns are recommendations or given with improvement in hole placing and ignition sequence.

The drawings are given on a scale of 1:50, for a width of up to $B = 4.5$ m (15 ft), 1:100 for greater values up to $B = 9.0$ (30 ft) and 1:200 up to $B = 18$ m (60 ft). The width (B), section area (S), advance (A), diameter of the charged holes (d), diameter of empty hole ($\varnothing$), concentration of bottom

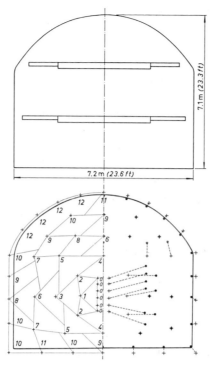

FIG. 7:21. The placing of the holes has to be adapted with regard to the drilling platforms.

$S = 38$ m² (410 sq.ft) $d = 38$–34 mm $l_b = 1.2$ kg/m (0.8 lb/ft)
$B = 7.2$ m (24 ft) (1.5–1.34 in) $q = 1.3$ kg/m³ (0.08 lb/cu.ft)
$H = 5.2$ m (17.1 ft) $\varnothing = -$ ordinary delay 0–12
$A = 4.8$ m (15.8 ft) $N = 64$

charge (l_b), number of holes (N) and specific charge (q) are indicated in a table for every figure. The patterns for cautious blasting also give the minimum distance (R) to adjacent buildings or constructions. The figures indicated at the holes give the best ignition sequence, not the delay numbers.

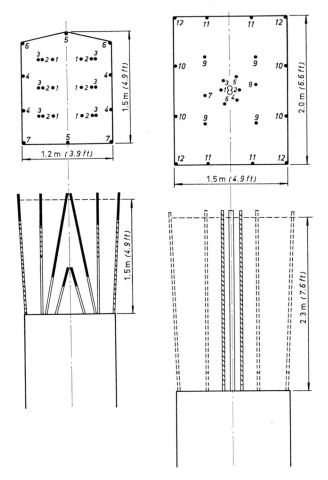

FIG. 7:22.　　　　　　　FIG. 7:23.

FIG. 7:22.

$S = 1.7$ m² (18.4 sq.ft)　$d = 32$ mm (1.25 in)　$l_b = 0.15$–0.5 kg/m (0.1–0.3 lb/ft)
$B = 1.2$ m (4 ft)　　　　$\varnothing = -$　　　　　　　$q = 5.0$ kg/m³ (0.31 lb/cu.ft)
$H = 1.6$ m (5.3 ft)　　　$N = 28$　　　　　　　ordinary delay 1–7
$A = 1.5$ m (4.9 ft)

FIG. 7:23.

$S = 3$ m² (32.5 sq.ft)　　$d = 31$ mm (1.2 in)　　$l_b = 0.3$–0.9 kg/m (0.2–0.6 lb/ft)
$B = 1.5$ m (4.9 ft)　　　　$\varnothing = 2 \times 57$ mm　　　$q = 3$ kg/m³ (0.18 lb/cu.ft)
　　　　　　　　　　　　　　$(2 \times 2.25$ in)
$H = 2.4$ m (7.9 ft)　　　　$N = 24 + 2$　　　　　hole 1–6: short delay 1–6
$A = 2.3$ m (7.6 ft)　　　　　　　　　　　　　hole 7–12: ordinary delay 2–7

208

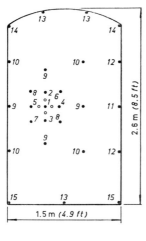

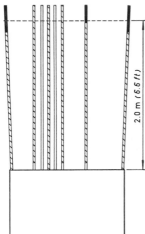

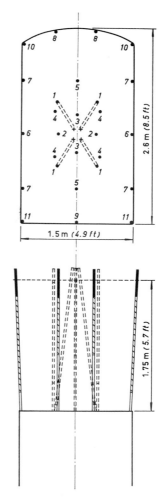

Fig. 7 : 24.   Fig. 7 : 25.

Fig. 7:24.

S = 3.8 m² (41 sq.ft)	d = 32 mm (1.25 in)	l_b = 0.4–0.8 kg/m (0.27–0.53 lb/ft)
B = 1.5 m (4.9 ft)	$\varnothing$ = 4 × 32 mm	q = 3.3 kg/m³ (0.2 lb/cu.ft)
	(4 × 1.25 in)	
H = 2.2 m (7.2 ft)	N = 31	hole 1–5: short delay 1–5
A = 2 m (6.6 ft)		hole 6–15: ordinary delay 2–11

Fig. 7:25.

S = 3.8 m² (41 sq.ft)	d = 32 mm (1.25 in)	l_b = 0.3–0.8 kg/m (0.2–0.54 lb/ft)
B = 1.5 (4.9 ft)	$\varnothing$ = –	q = 3 kg/m³ (0.18 lb/cu.ft)
H = 1.9 m (6.2 ft)	N = 30	hole 1–11: ordinary delay 1–11
A = 1.75 m (5.7 ft)		

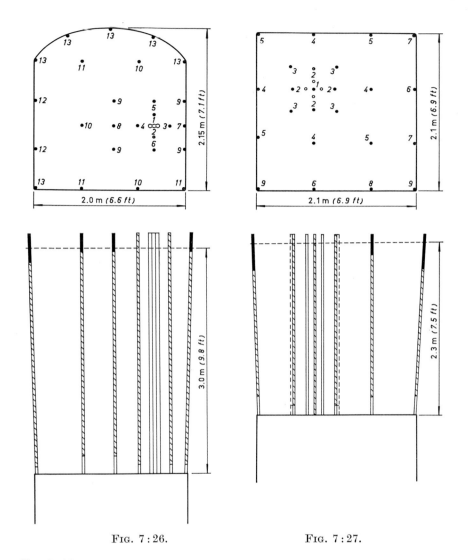

FIG. 7:26.   FIG. 7:27.

FIG. 7:26.

$S = 4.0$ m² (43 sq.ft) $d = 32$ mm (1.25 in) $l_b = 0.35$–0.9 kg/m (0.23–0.6 lb/ft)
$B = 2$ m (6.6 ft) $\varnothing = 3 \times 57$ mm $q = 3.6$ kg/m³ (0.22 lb/cu.ft)
$H = 3.2$ m (10.5 ft) $(3 \times 2.25$ in)   hole 1–4: short delay 1–4
$A = 3.0$ m (9.8 ft) $N = 26 + 3$ hole 5–13: ordinary delay 2–9

FIG. 7:27.

$S = 4.4$ m² (47 sq.ft) $d = 32$ mm (1.25 in) $l_b = 0.4$–0.8 kg/m (0.27–0.53 lb/ft)
$B = 2.1$ m (6.9 ft) $\varnothing = 4 \times 32$ mm $q = 2.9$ kg/m³ (0.18 lb/cu.ft)
$H = 2.4$ m (7.9 ft) $(4 \times 1.25$ in)   hole 1–9: ordinary delay 1–9
$A = 2.3$ m (7.6 ft) $N = 24 + 4$

210

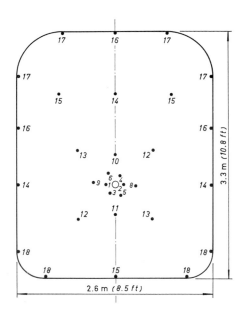

Fig. 7:28.

$S = 8.6$ m² (93 sq.ft) $d = 31$ mm (1.2 in) $l_b = 0.35$–1.0 kg/m (0.23–0.67 lb/ft)
$B = 2.6$ m (8.5 ft) $\varnothing = 83$ mm (3.25 in) $q = 2.8$ kg/m³ (0.18 lb/cu.ft)
$H = 3.2$ m (10.5 ft) $N = 31 + 1$ hole 1–11: short delay 1–11
$A = 3.1$ m (10.2 ft)   hole 12–17: ordinary delay 3–8

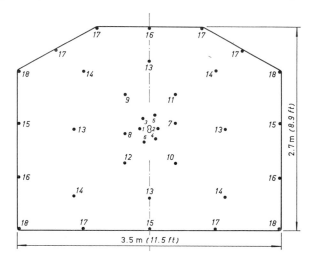

Fig. 7:29.

$S = 8.8$ m² (95 sq.ft) $d = 35$ mm (1.4 in) $l_b = 0.35$–1.0 kg/m (0.23–0.67 lb/ft)
$B = 3.5$ m (11.5 ft) $\varnothing = 2 \times 57$ mm $q = 2.3$ kg/m³ (0.14 lb/cu.ft)
$H = 4.0$ m (13.1 ft)   (2 × 2.25 in)   hole 1–8: short delay 1–8
$A = 3.8$ m (12.5 ft) $N = 36 + 2$ hole 9–18: ordinary delay 2–11

211

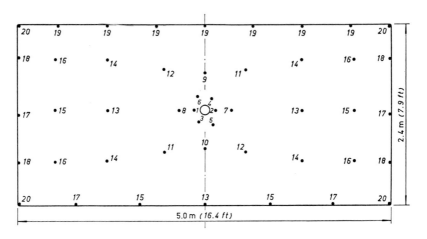

FIG. 7:30

$S = 12$ m² (130 sq.ft) $d = 30$ mm (1.2 in) $l_b = 0.35$–1.0 kg/m (0.23–0.67 lb/ft)
$B = 5.0$ m (16.5 ft) $\varnothing = 125$ mm (5 in) $q = 2.3$ kg/m³ (0.14 lb/cu.ft)
$H = 4.0$ m (13.2 ft) $N = 48 + 1$ hole 1–10: short delay 1–10
$A = 3.9$ m (12.8 ft)   hole 11–20: ordinary delay 2–11

FIG. 7:31.

$S = 14$ m² (151 sq.ft) $d = 33$–32 mm $l_b = 0.8$–1.0 kg/m (0.54–0.67 lb/ft)
$B = 3.7$ m (12.1 ft) (1.3–1.25 in) $q = 1.8$ kg/m³ (0.11 lb/cu.ft)
$H = 2.0$ m (6.6 ft) $\varnothing = -$ hole 0–5: ordinary delay 0–5
$A = 1.8$ m (5.9 ft) $N = 34$

212

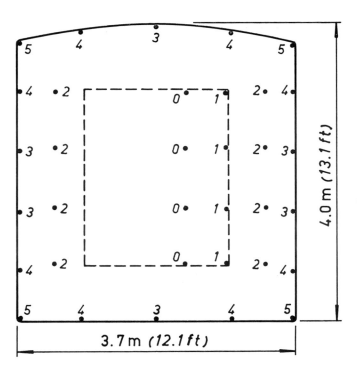

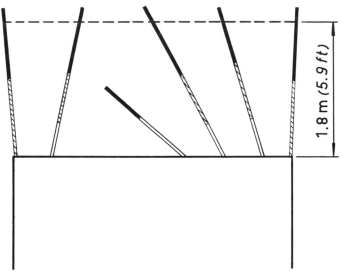

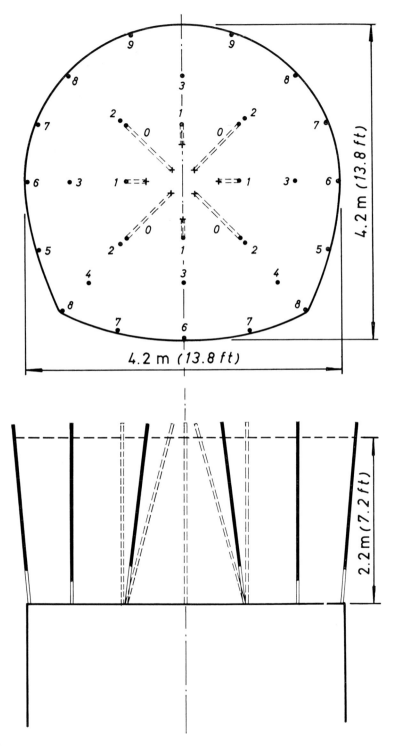

4.2 m (13.8 ft)

4.2 m (13.8 ft)

2.2 m (7.2 ft)

214

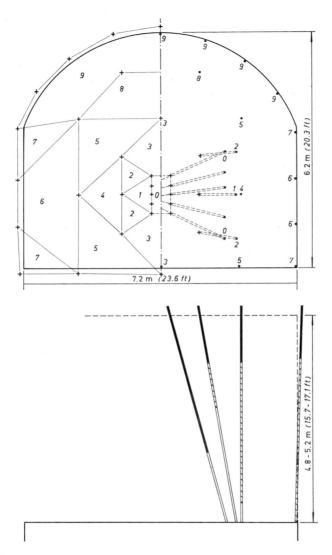

Fig. 7:33.

$S = 39$ m² (420 sq.ft)	$d = 45$ mm (1.75 in)	$l_b = 2$ kg/m (1.3 lb/ft)
$B = 7.2$ m (23.7 ft)	$\varnothing = -$	$q = 1.3$ kg/m³ (0.08 lb/cu.ft)
$H = 5.6$ m (18.4 ft)	$N = 39$	hole 0–9: ordinary delay 0–9
$A = 4.8$–5.2 m		
(15.8–17.1 ft)		

Fig. 7:32.

$S = 15$ m² (162 sq.ft)	$d = 38$ mm (1.5 in)	$l_b = 1$–1.4 kg/m (0.67–0.94 lb/ft)
$B = 4.2$ m (13.8 ft)	$\varnothing = -$	$q = 1.5$ kg/m³ (0.093 lb/cu.ft)
$H = 2.4$ m (7.9 ft)	$N = 33$	hole 0–9: ordinary delay 0–9
$A = 2.2$ m (7.2 ft)		

215

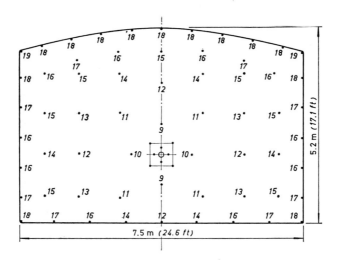

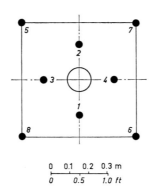

FIG. 7:34.

$S = 37$ m² (400 sq.ft) $d = 31$ mm (1.2 in) $l_b = 0.6$–1.5 kg/m (0.4–1.0 lb/ft)
$B = 7.5$ m (24.7 ft) $\varnothing = 125$ mm (5 in) $q = 1.3$ kg/m³ (0.08 lb/cu.ft)
$H = 3.8$ m (12.5 ft) $N = 70 + 1$ hole 1–10: short delay 1–10
$A = 3.6$ m (11.8 ft)   hole 11–19: ordinary delay 2–10

216

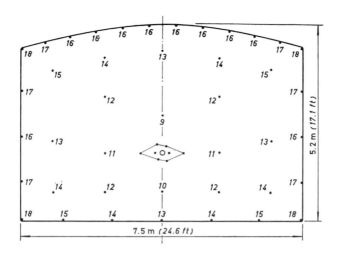

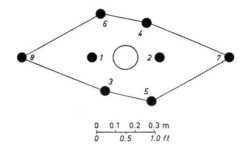

Fig. 7:35.

$S = 37$ m² (400 sq.ft) $d = 50$ mm (2 in) $l_b = 0.4$–2.5 kg/m (0.27–1.7 lb/ft)
$B = 7.5$ m (24.7 ft) $\varnothing = 125$ mm (5 in) $q = 1.3$ kg/m³ (0.08 lb/cu.ft)
$H = 4.5$ m (14.8 ft) $N = 50 + 1$ hole 1–6: short delay 1–6
$A = 4.3$ m (14.1 ft) hole 7–18: ordinary delay 2–12

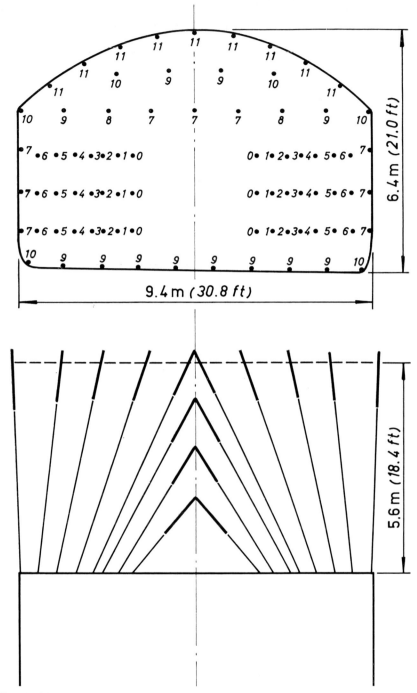

Fig. 7:36.

$S = 54$ m² (583 sq.ft)	$d = 34$ mm (1.34 in)	$l_b = 1$ kg/m (0.67 lb/ft)
$B = 9.4$ m (30.8 ft)	$\varnothing = -$	$q = 1.1$ kg/m³ (0.068 lb/cu.ft)
$H = 5.9$ m (19.4 ft)	$N = 80$	hole 0–11: ordinary delay 0–11
$A = 5.6$ m (18.4 ft)		

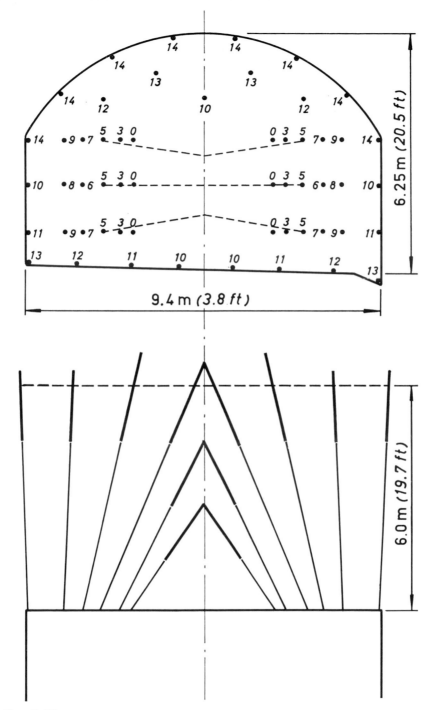

Fig. 7:37.

$S = 52$ m² (562 sq.ft) $d = 44$–37 mm $l_b = 1.5$–2.1 kg/m (1.0–1.4 lb/ft)
$B = 9.4$ m (30.8 ft)   (1.73–1.45 in) $q = 1.1$ kg/m³ (0.068 lb/cu.ft)
$H = 6.4$ m (21.1 ft) $\varnothing = -$ hole 0–5: short delay 0–5
$A = 6.0$ m (19.7 ft) $N = 55$ hole 6–14: ordinary delay 2–10

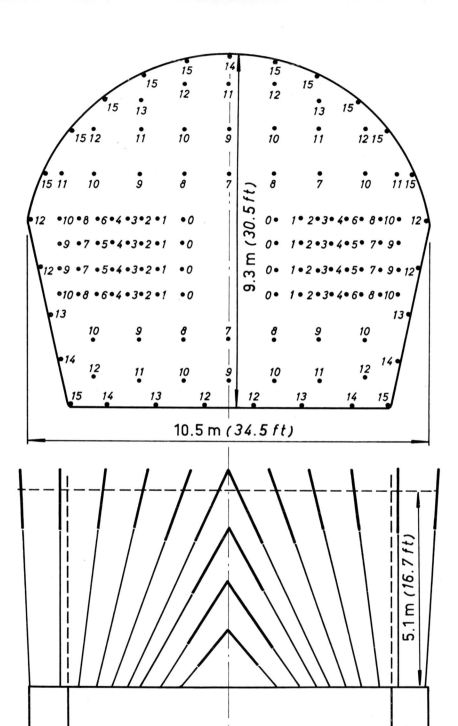

9.3 m (30.5 ft)

10.5 m (34.5 ft)

5.1 m (16.7 ft)

220

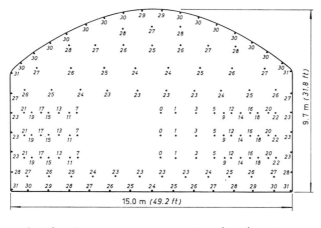

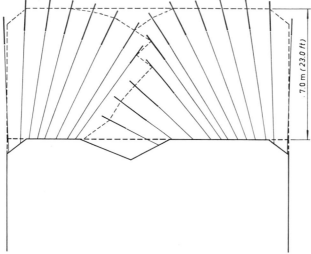

Fig. 7:39.

$S = 128\,\mathrm{m^2}\,(1380\,\mathrm{sq.ft})$ $d = 37\text{--}31$ mm $l_b = 1.2$ kg/m (0.8 lb/ft)
$B = 15$ m (49.3 ft) $(1.45\text{--}1.2$ in) $q = 0.9$ kg/m³ (0.056 lb/cu.ft)
$H = 7.4$ m (24.3 ft) $\varnothing = -$ hole 0–22: short delay 0–22
$A = 7$ m (23 ft) $N = 137$ hole 23–31: ordinary delay 4–12

Fig. 7:38.

$S = 82\,\mathrm{m^2}\,(864\,\mathrm{sq.ft})$ $d = 38\text{--}33$ mm $l_b = 1.1\text{--}1.4$ kg/m (0.74–0.94 lb/ft)
$B = 10.5$ m (34.7 ft) $(1.5\text{--}1.3$ in) $q = 1.1$ kg/m³ (0.068 lb/cu.ft)
$H = 5.7$ m (18.7 ft) $\varnothing = -$ hole 0–5: ordinary delay 0–15
$A = 5.1$ m (16.7 ft) $N = 136$

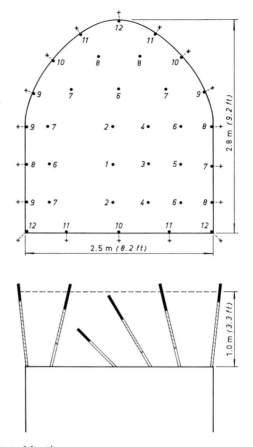

Fig. 7:40. Cautious blasting.

$S = 6$ m² (65 sq.ft) $d = 33$ mm (1.3 in) $l_b = 0.17$ kg/m (0.37 lb/ft)
$B = 2.5$ m (8.2 ft) $\varnothing = -$ $q = 0.5$ kg/m³ (0.031 lb/cu.ft)
$H = 1.1$ m (3.6 ft) $N = 35$ hole 1–12: short delay 1–12
$A = 1.0$ m (3.3 ft)

The thickness of the rock above the tunnel is only about 1 m (3–4 ft) reinforced by 1 m of concrete.

Fig. 7:41. Cautious blasting.

$R = 4$ m (13 ft) $A = 1.6$ m (5.3 ft) $l_b = 0.8$ kg/m (0.54 lb/ft)
$S = 20$ m² (217 sq.ft) $d = 34$–31 mm $q = 0.7$ kg/m³ (0.043 lb/cu.ft)
$B = 4$ m (13.1 ft)   (1.34–1.2 in)   hole 1–28: short delay 1–28
$H = 1.8$ m (5.9 ft) $\varnothing = -$ hole 29–38: ordinary delay 9–18
   $N = 71$

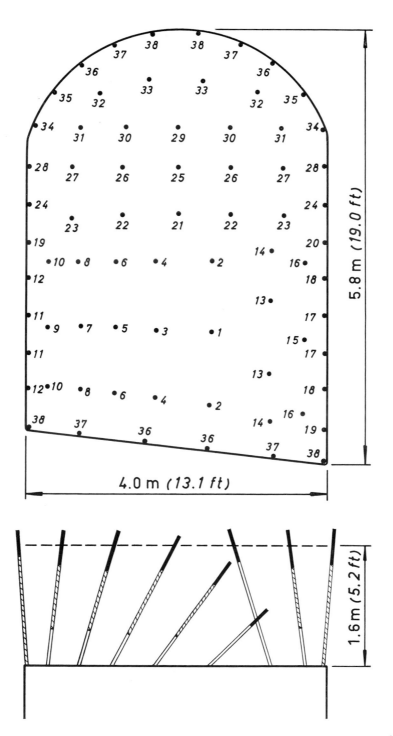

5.8 m (19.0 ft)

4.0 m (13.1 ft)

1.6 m (5.2 ft)

223

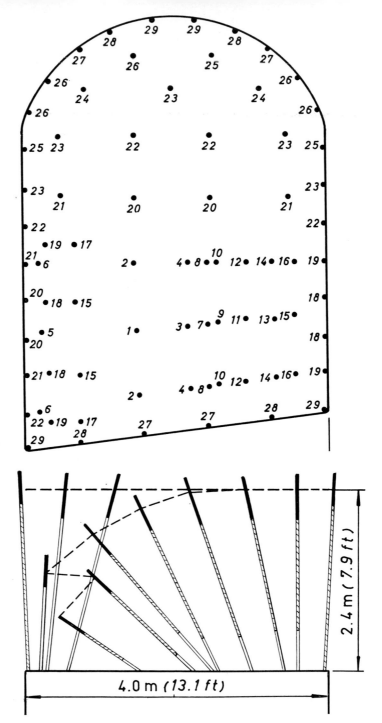

FIG. 7:42. Cautious blasting.

$R = 7$ m (23 ft) $A = 2.4$ m (7.9 ft) $l_b = 0.5$ kg/m (0.34 lb/ft)
$S = 20$ m² (217 sq.ft) $d = 33–30$ mm $q = 0.9$ kg/m³ (0.056 lb/cu.ft)
$B = 4$ m (13.1 ft) (1.3–1.18 in) hole 1–20: short delay 1–20
$H = 2.6$ m (8.4 ft) $\varnothing = -$ hole 21–29: ordinary delay 4–12
 $N = 75$ (7–15)

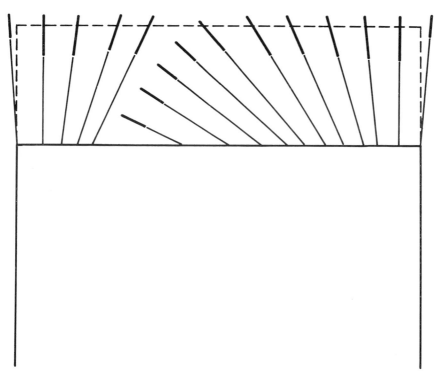

Fig. 7:43. Cautious blasting.

$R = 12$ m (39 ft)	$A = 3.2$ m (10.5 ft)	$l_b = 0.15$–1.0 kg/m (0.1–0.67 lb/ft)
$S = 57$ m² (615 sq.ft)	$d = 32$–28 mm	$q = 0.9$ kg/m³ (0.056 lb/cu.ft)
$B = 10.7$ m (35.2 ft)	(1.25–1.1 in)	hole 1–20: short delay 1–20
$H = 3.5$ m (11.5 ft)	$\varnothing = -$	hole 21–29: ordinary delay 4–12
	$N = 107$	(7–15)

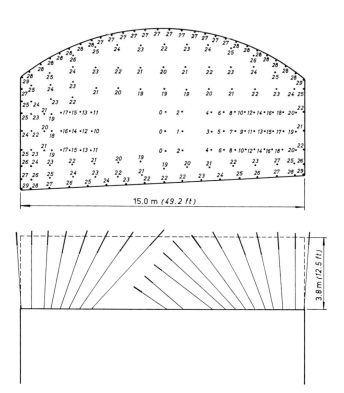

FIG. 7:44. Cautious blasting.

$R = 15$ m (49 ft) $A = 3.8$ m (12.5 ft) $l_b = 0.6$ kg/m (0.4 lb/ft)
$S = 111$ m² (1200 sq.ft) $d = 40$ mm (1.6 in) $q = 0.67$ kg/m³ (0.042 lb/cu.ft)
$B = 15$ m (49.3 ft) $\varnothing = -$ hole 0–20: short delay 0–20
$H = 4.2$ m (13.8 ft) $N = 166$ hole 21–24: ordinary delay 4–12
                                               7–15) in half delay

226

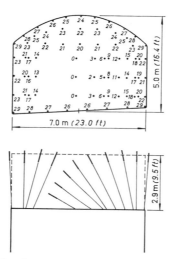

Fig. 7:45a. Cautious blasting.

$R = 15$ m (49 ft) $A = 4.3$ m (14.1 ft) $l_b = 0.55$ kg/m (0.37 lb/ft)
$S = 79$ m² (853 sq.ft) $d = 40$ mm (1.6 in) $q = 0.52$ kg/m³ (0.032 lb/cu.ft)
$B = 15$ m (49.3 ft) $\varnothing = -$ hole 0–12: ordinary delay 0–12
$H = 4.6$ m (15.1 ft) $N = 98$ in half delay

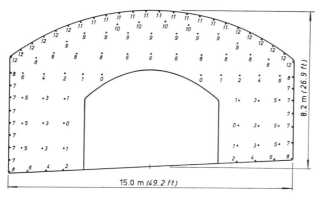

Fig. 7:45b.

$R = 8$ m (26 ft) $A = 2.9$ m (9.5 ft) $l_b = 0.25$–0.7 kg/m (0.17–0.47 lb/ft)
$S = 32$ m² (346 sq.ft) $d = 33$–31 mm $q = 0.97$ kg/m³ (0.06 lb/cu.ft)
$B = 7$ m (23 ft)   (1.3–1.2 in)   hole 0–20: short delay 0–20
$H = 3.2$ m (10.5 ft) $\varnothing = -$ hole 21–29: ordinary delay 4–12
        $N = 72$ (7–15)

227

FIG. 7:46. Underground oil storage in rock. Photo: Manne Lind. Contractor: Armerad Betong Vägförbättringar.

7.11. Problems

1. Determine the burden to be used (V_1) for a reliever with free breakage, depth of the hole of $H = 3.0$ m, diameter at the bottom $d = 30$ mm if the drill hole deviation is 5 cm/m.

2. Determine the burden to be used with $H = 5.4$ m, $d = 50$ mm and deviation 10 cm/m.

3. At what distance from a rectangular opening with width $B = 0.4$ m could a hole be placed in the drilling pattern when the concentration of the charge is a) $1 = 0.3$ kg/m, $1 = 0.7$ kg/m, c) $1 = 1.5$ kg/m.

4. Recalculate the table 7:1 assuming that the linear deviation is $R_d = 0.02 H$ cm (H in m), which corresponds to alignment faults of 2%. At what depths is this factor in itself 80% or more of the total deviation R.

5. A square opening 0.5×0.5 m² is created by a parallel hole cut. Suggest position and ignition sequence for 10 relievers with $d = 30$ mm around the opening if the deviation (R) is 0.20 m for 5/6 of all the holes.

6. Indicate in the example above the final position of the holes if they are numbered 5–15 in the N-distribution in fig. 7:2. Discuss hole by hole the conditions for breakage at the bottom.

7. Calculate according to formula (7:8 M) the consumption of explosive for a tunnel with a section a) $S = 4$ m², b) $S = 60$ m².

8. Study the effect on the breakage at the bottom of a V-cut, for example fig. 7:2, if the deviation of the holes is $\sigma_{5/6} = 10$ cm/m. Number the holes and apply the N-distribution in fig. 7:2.

9. Discuss the choice of cut for a tunnel with $B = 6$ m (20 ft). What is the greatest advance with cylinder- and with V-cut; what is the ordinary (mean) advance with V-cut?

10. How many holes have to be used in a tunnel with $c = 0.6$ diameter $d = 43$ mm, section a) $S = 10$ m², b) $S = 40$ m², c) $S = 80$ m²?

What change is there in the figures if the diameter is $d = 30$ mm?

8. TUNNEL BLASTING WITH
PARALLEL HOLE CUTS

For mechanization of the drilling the use of parallel hole cuts offers the most favorable conditions. When drilling with hand-held machines it is easier to instruct new personnel and, if desired, to supply them with light templates for drilling parallel hole cuts. Ordinary V and fan cuts may be comparatively simple on paper, but are in fact rather complicated to perform and to alter if the advance per round has to be changed.

The extended use of parallel hole cuts in USA and Europe in recent years has increased not only the knowledge but in some cases also the uncertainty of how, when and under what conditions to use these cuts. The reason is that the cut holes are often too heavily loaded ("burning") instead of the principles for cylinder cut blasting with slight charges being applied, and often with unfavorable placing of the holes. Conclusions valid for such burn cuts are drawn and supposed to represent parallel hole cuts in general.

General conditions for driving with parallel hole cuts are given in paragraph 8.1. These offer a basis for designing new cuts and calculating their advance as a function of deviation in the drilling and of the drilled depth. An examination of full-scale driving including more than 13,000 rounds will be found at the end of the chapter.

There are three different main types of cuts with parallel holes: burn cuts, cylinder cuts and crater cuts.

Burn cuts are the oldest and hitherto the most widely used type with parallel holes. The blasting takes place towards one or several uncharged empty holes, but with so large a concentration of charge that the rock sticks ("freezes") at great depths of the cut and does not give satisfactory conditions for the breakage of the round.

In *cylinder cuts* the blasting is performed towards an empty hole in such a way that as the charges in the first, second and subsequent holes detonate, the broken rock is thrown out of the cut. The opening is successively and uniformly (cylindrically) enlarged in its entire length as is shown in figs. 8:1–8:3. This implies that the advance is restricted only by the deviation of the holes and that larger advances can be obtained than with burn cuts.

Crater cuts consist of one or several fully charged holes in which blasting is carried out towards the face of the tunnel, that is to say towards a free surface at right angles to the holes.

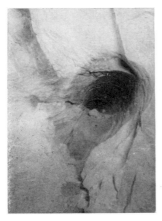

FIG. 8:1.   FIG. 8:2.

FIG. 8:1. The first hole close to an uncharged large hole gives in many cases only an insignificant increase in volume.

FIG. 8:2. Hole No. 2 increases the given empty hole volume.

Some of the advantages aimed at in rational tunnel driving are: a high degree of mechanization of the drilling, a high value for the relative advance (advance in per cent of the drilled depth), a long average advance per round, good fragmentation and a drilling pattern that can be applied unchanged

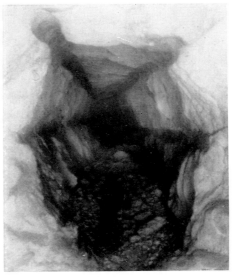

FIG. 8:3. Result of blasting a cylinder cut (acc. to fig. 8:10a). After blasting 6 cut holes the opening attained is practically cleaned to the full depth, 5 m (16.3 ft). (Photo Atlas Copco.)

231

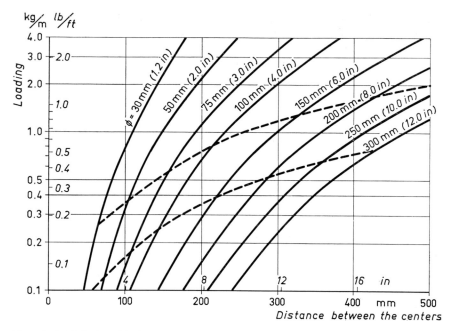

FiG. 8:4. Relationship between quantity of charge and distance between holes
when blasting towards an empty drill hole with a diameter of 30–150 mm
(1.2–6 in). ––– corresponds to the broken lines in Fig. 8:5. Diameter of the
loaded hole $d = 32$ mm.

at varying depths of drilling and various widths of the tunnels. The cylinder
cuts have been developed to meet these requirements.

For practically all tunnels and drifts with a section area of less than 50 m²
(550 sq.ft) parallel hole cuts can give a larger advance per round than
ordinary angled cuts (fig. 7:16). Also for greater sections the parallel hole
cuts may be used, when it is preferred to blast a larger number of rounds
per day instead of using the maximum advance per round.

8.1. Basic relations

a. Concentration of charge

The concentration of charge (l) required in blasting towards an empty hole
and which is given in kg/m (lb/ft) depends on the diameter of the empty
hole (ϕ), on the diameter (d) of the charged hole and on the distance (A)
between the centers of the two holes. The following relation applies if a
and ϕ are given in mm and the diameter of the charged hole is $d = 33$ mm
(see fig. 7:5 a)

$$l = 1.5 \cdot 10^{-3} \, (A/\phi)^{3/2} \, (A - \phi/2)$$

232

TABLE 8:1. *Concentration of charge* (l) *in* kg/m *for cylinder cuts and greatest distance* $\bar{a}$ *when blasting towards empty holes whith diameters between* $\phi = 2 \times 57$ *and 200 mm* (d *indicates the diameter of the loaded hole*). *The weight strength of the explosive is* $s = 1.0$.

ϕ mm	50	2×57	75	83	100	2×75	110	125	150	200
32	0.2	0.3	0.3	0.35	0.4	0.45	0.45	0.5	0.6	0.8
37	0.25	0.35	0.35	0.4	0.45	0.53	0.53	0.6	0.7	0.95
45	0.30	0.42	0.42	0.50	0.55	0.65	0.65	0.7	0.85	1.10
$\bar{a}$ mm	90	150	130	145	175	200	190	220	250	330

The relation is given in the graphs of fig. 8:4 and for some current diameters of empty holes and a greatest distance ($\bar{a}$) recommended between the holes in table 8:1.

It is remarkable, especially for small empty hole diameters, how much the charge must be increased with a growing distance between the centers. For $\phi = 30$ mm a charge of 1.0 kg/m is required for a center distance of 11 cm, and less than half the concentration of charge for a distance of 8 cm. This important fact should be observed, not so much because of the saving which can be made in the consumption of explosive, but rather for the great trouble which may be caused by heavy charges in constricted holes. The rock in the immediate vicinity of the hole is exposed to intense strain with consequent risk of tearing in the surrounding holes. The breaking capacity of these holes is impaired and there is a greater risk of faulty sequence in the ignition from flash-over.

b. Clean blasting and plastic deformation

Even if the condition for charging according to fig. 8:4 has been fulfilled the result of a blast can vary very much according to the ratio between the center distance (a) and the diameter of the empty holes (ϕ). When the distance is greater than double the diameter of the empty holes, a $> 2\phi$, there can be no real breakage as the required concentration of charge is so great that there is a plastic deformation of the rock between the two holes.

If the two holes are placed closer and the charge adjusted according to fig. 8:4 the prospects for true breakage of the rock between the holes are brighter. Breakage is not, however, the only desirable feature; as much as possible of the broken material should also be blown out through the opening by the explosion gases.

It will be seen from fig. 8:5 how the breakage conditions vary with different distances between the charged hole and the empty one. The diagram

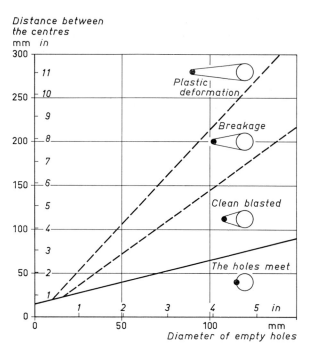

FIG. 8:5. Result when blasting towards an empty hole at different distances and dimensions of the empty hole.

presupposes that the concentration of charge is used as given in fig. 8:4 for the diameter and spacing in question. It shows the changing result with increasing distance between the holes. For $a < 1.5 \phi$ the opening is "clean-blasted". Between 1.5ϕ and 2.1ϕ there is merely "breakage" and for greater distances, as mentioned above, there will be "plastic deformation". Even for smaller distances than in the region of "plastic deformation" a burning of the rock occurs inside the opening if the concentration of the charge is too great. This is, in reality, the essential difference between so-called burn cuts and cylinder cuts correctly performed. It always seems to be difficult at the blasting sites to get the blasters to understand that too low an advance has been due to over-charging the holes nearest the empty one. This factor can obviously not be too strongly emphasized.

c. Flash-over

As has already been mentioned it is very important to avoid sympathetic detonation of charges intended to be ignited separately. For holes with a concentration of charge of 1.0 kg/m (0.7 lb/ft) there will be no flash-over at a distance of 25 cm (10 in) with ordinary dynamite (NG $\leqslant 35\%$). The corresponding values for 0.7 and 0.3 kg/m (0.5 and 0.2 lb/ft) are 20 and 10 cm respectively (8 and 4 in). These values hold good for granite and similar species of rocks. For softer species such as dolomitic limestone (U.S.A.), flash-overs can occur also for $l = 1$ kg/m (0.7 lbs/ft) at a distance of 25 cm (10 in). Flash-over can occur at greater distances than those mentioned if there are cracks and crevices between the holes.

This implies that it is difficult to avoid flash-over for the common burn cuts with normal deviations in the drilling. For cylinder cuts there is less risk of flash-over as the concentration of the charge is lower. It is more difficult to avoid flash-over in cuts with small empty holes than with large ones.

In order to reduce the risk of flash-over a relatively insensitive explosive should be chosen. Tests show that dynamites with lower percentage than 35 % NG and powder explosive of ammonia dynamite type have given favorable results in cylinder cuts. "Nabit" and ammonia dynamite entail a low risk for flash-over and have proved to be suitable from different points of view for the 4–8 holes nearest the empty one. These holes should be weakly charged.

d. The effect of the explosive

In blasting stope holes fully charged in the usual manner the detonation velocity of the dynamite can amount to 5500 m/s (18,000 ft) or more; in some types of rock this may be an advantage with regard to the fragmentation. For the 4–8 holes in a cylinder cut which are nearest the empty hole and which are only loaded with 0.3–0.5 kg/m (0.2–0.3 lb/ft) conditions differ. The detonation velocity of the dynamite is low in this case, 2400 m/s (8000 ft/sec), which is appropriate especially for one or two of the nearest cut holes where the detonator can be placed near the face of the tunnel. With low detonation velocity the charges in these two holes give better cleaning in soft as well as in hard rock.

When blasting in soft rock it has been found that a powder explosive can be compressed in adjacent holes to such an extent that it can no longer detonate (St. Joseph Lead, Missouri, USA). An important aspect in practical driving over and above what has been mentioned is how the breaking power of the explosive cartridges available correspond to the requirements for the actual cut. A powder explosive packed in 22 mm cartridges loosely piled in the hole gives a breaking power corresponding to 0.34 kg dynamite per m (0.22 lb/ft). As will be seen from table 8:1 this is the concentration of charge for cylinder cuts with empty holes of 75 mm (3 in), and for Coromant cuts with 2×57 mm empty hole. With 110 mm (4.3 in) empty holes 22 mm ($\frac{7}{8}$ in) dynamite cartridges, 25 (1 in) or 29 (1.25 in) mm Nabit cartridges can be used loosely piled and give a breaking power corresponding to dynamite of 0.48, 0.45 and 0.60 kg/m (0.32, 0.30, 0.40 lb/ft). With 125 (5 in) and 150 mm (6 in) empty holes 29 (1.25 in) mm Nabit cartridges loosely piled or 25 mm (1 in) dynamite can be employed. With a 200 mm (8 in) empty hole 32 mm (1.3 in) Nabit (a powder explosive) can be appropriately used packed loosely in the hole to the desired density in keeping with table 8:1.

e. Influence of the diameter of the charged holes

The data given in figs. 8:4, 8:5 and 8:6 refer to blasting with charges in 30–35 mm diameter holes. If larger diameters are used the effect of the explosive is diminished and the charge must be increased approximately in proportion to the diameter so as to get the same breaking power (table 8:1). At the same time the conditions for throw and ejection of the broken pieces of rock are changed. Preliminary trials indicate that the tendency for the rock to sinter decreases with larger diameter of the loaded holes.

f. Influence of the rock

The breakage conditions vary with the structure of the rock. Fig. 8:1 shows an example of how breakage for the first opening hole at times can take the form of a narrow cleft. In other cases the opening can be widened to form an angle, with the edge in the center of the charged hole and touching the contour of the empty hole. The cuts in fig. 8:10 are designed to allow for such differences as can occur in the rock.

Bullock has found that in especially soft or plastic rock a plastic deformation may occur in the upper part of the region denoted "Breakage" in fig. 8:5.

The quality of the rock can moreover affect the result by crevices and clay dikes causing flash-overs. This may interrupt the round so that it does not penetrate to the full depth. The possibilities of achieving favorable drilling precision can furthermore differ in different sections also in one and the same rock, when the direction of the crevice happens to be unfavorable.

It may generally be said, however, that a cylinder cut blast carried out correctly appears to operate in most rock. The driving in a new kind of rock could begin with a drilled depth that is 50–70% of the maximum given in table 8:4, and when the average *relative* advance for a series of rounds exceeds 95%, the drilled depth can be increased successively to the value the conditions will permit.

Burn cuts are in this respect more dependent on the rock and are some-

236

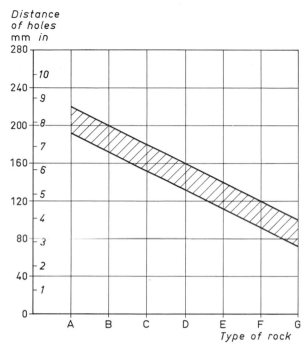

Fɪɢ. 8:6. Distance between loaded and unloaded holes with a diameter of 35 ±5 mm (1.4 ±0.2 in) when blasting burn cuts in various kinds of rock A–G acc. to Steidle, A gypsum, very ragged rock, B soft limestone, C soft sandstone, D hard limestone, E soft granite, F quartz, G hard granite.

times unsuitable. Better results can then often be attained by reducing the charges in the cut holes. Steidle has indicated for burn cuts how the most suitable distance of holes depends on the kind of rock (fig. 8:6).

g. Influence of the ignition sequence

Cylinder cuts and burn cuts appear to function well both with MS delays and with ordinary intervals. In many cases it has been ascertained that the charge in a hole can be ejected at the detonation of the charge in a neighboring hole. In igniting with short delays the charges have a better chance of igniting before ejection has begun. If the cartridge with the detonator moves with a lower velocity than neighboring cartridges an explosive with a greater flash-over distance will be more apt to ignite. This does not imply that such an explosive is recommended, but the fact should be observed.

If the detonators are placed near the free face of the rock, better cleaning of the opening can be achieved than if the detonator is placed at the bottom.

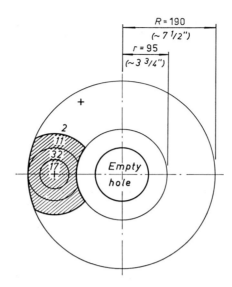

Fɪɢ. 8:7. Probable position at 5.5 m (18 ft) depth for a cut hole close to a 110 mm (4.4 in) empty hole at a deviation according to table 8:2. The figures indicate the probability of a hit within a corresponding area between two circles. 35% of the opening holes will be wrongly placed at the bottom. The figure increases to 53% if the distance between the centers of the cut hole and the empty hole increases from 140 to 185 mm.

This also applies to other cut holes, but in view of the risk of cut-offs it is advisable only to place instantaneous detonators close to the surface and the others at the bottom of the holes.

h. Deviation in drilling (Table 8:2)

When the drilling pattern is designed for a parallel hole cut allowance must be made for the fact that faulty collaring (cm), faulty alignment (cm/m) and faults from additional deviations inside the rock will occur. This will give a pattern at the bottom of the holes which may differ considerably from the original one. In fig. 8:7 $+$ has been used to indicate the intended position of the first opening hole. Figures in the rings concentric with this hole indicate the probability of the hole ending in this ring and in the area shaded between the two circles with radius r and R around the empty hole. How the total deviation R depends on the faults in collaring (R_c), alignment (R_d) and deviations inside the rock (R_r) is described by the formula

$$R = (R_c^2 + R_d^2 + R_r^2)^{\frac{1}{2}} \qquad (7:1)$$

238

TABLE 8:2. *Drill hole deviations at various depths according to the relation*
$R = (1^2 + 1.14^2 H^2 + 0.8^2 H^3)^{1/2}$.

H	m	2	3	4	5	6	7	8
R	cm	3.4	5.7	8.0	11	14	17	20
σ	cm/m	1.7	1.9	2.0	2.2	2.3	2.4	2.5

The R-values indicate the radii that include 5 out of 6 (83%) of the deviations of all the holes; the σ-values are the corresponding mean values for the deviation in cm/m (%). The table 8:2 gives the deviations for depths $H = 2$–8 m according to the formula above with $R_c = 1$ cm, $R_d = 1.14 H$ cm and $R_r = 0.8 H^{3/2}$ cm. The calculations of the advance for cylinder cuts have been performed with these deviation values. The advances obtained in driving with correct charge and placing of holes have shown a variation in proportion 0.8:1 which may well correspond to variations in R_d and R_r values at different sites. If the collaring faults are below $R_c = 2$ cm, they have a rather small influence on the resultant deviation R for depths $H \geqslant 2.4$ cm.

Still another fault may make itself strongly felt and that is when the steel, especially in drilling large holes, may as a result of the weight of the bit cause greater wear on the underside of the hole and thereby a downward deviation for the first 2–5 m (6–16 ft) of the drilling. A reverse tendency is experienced at greater depths of holes. The steel will then be so long that it is bent downwards at the middle of its length which gives the steel an upward tendency. Those types of faults are reduced by choosing a heavy diameter of the drilling steel. It is desirable that the empty hole should have the same downward deviation as other cut holes.

In the case of mechanically guided drills it must be arranged *either* that the collaring is exactly at the spot intended, *or* that the direction of the drill is adjusted to be parallel to the empty hole after the collaring. If this is not done there will be essential alignment faults, but also much greater strain in the equipment and it will last a shorter time. Conditions for accurate collaring are improved if the drill steel is supported as close to the rock as possible and by using stouter drilling steels.

i. Advance per round (Tables 8:3, 8:4 and 8:5)

One usually counts with an advance being 90% of the drilled depth of the round. For cylinder cuts, if the deviation in the drilling can be ignored, an advance (A) can be obtained as large as the depth of the holes (H), that is to say the relative advance is 100%. At greater depths, those holes which

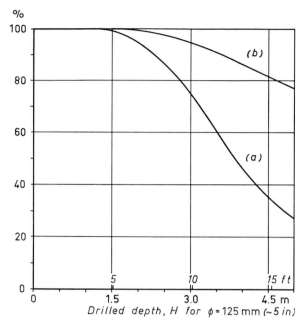

FIG. 8:8. a) Probability of breakage to various depths H when blasting a Täby cut according to fig. 8:10b with a drilling deviation of $\sigma = 2.0$ cm/m (2%). b) Average advance in % of drilled depth.

have been drilled round the empty hole will according to a probability of distribution end at such positions that the breakage cannot take place at the bottom of the round. As the placing of the holes is correct at the surface there is a depth between 0 and the full depth to which the cut will operate.

In order to make these conditions clear fig. 8:8 shows the limit for the advance of a cylinder cut resulting from the deviation of the holes. Down to a depth of 1.6 m 100% of all cuts will be correctly drilled. Of these 98% will be right also at a depth of 2.0 m, 86% at 3.0 m, etc. according to curve (a). This implies that the *mean advance* at a depth of 1.6 m is 100%, at 2.0 m 99.8% and at 3 m 98%, and so on. The mean advance is shown by curve (b) in fig. 8:8.

For burn cuts the conditions are only partly similar since the advance of these cuts is also limited by loosened rock that sticks in the opening at 2–3 m depth. For some different cylinder cuts the mean advance has been calculated for a deviation of $\sigma = 2.0$ cm/m. The values calculated are given in table 8:3 in percent of the depth (H), though in the case of the three-section and four-section cuts with a correction to 15% higher H-values than the calculation has given, since this appears to correspond better to the results obtained in practice.

240

Mean advance in percentage of drilled depth (H) *for some different cylinder cuts, with* $\phi = 125$ *mm* (*5 in*) *and a drilling deviation of* $\sigma = 2.0$ *cm/m* (%). *The advance is proportional to the diameter of the empty hole* (ϕ) *and inversely proportional to the drilling deviation per m* (σ).

H	Double spiral	Four-section	Three-section	Täby
2.5	100	100	99	97
3.0	100	98	97	95
3.4	100	96	95	92
3.8	99	94	92	88
4.2	97	92	89	84
4.7	95	89	85	80
5.1	92	86	82	76
5.5	89	83	79	72
6.0	86	80	76	68

If in practical tests with cylinder cuts a certain value of the mean advance is obtained, or if such data are received from various working sites, they can only be compared if the percentage advance is simultaneously known.

In conformity with what has been said in the preceding section the deviation of the holes per meter increase with the depth of hole. The values in table 8:3 must be corrected with reference to this. This has been done with the aid of the values in table 8:2 and means that the H-value corresponding to a certain percentage figure is increased or decreased in reverse proportion to the σ-value if this deviates from $\sigma = 2.0$ cm/m. In this way it is possible to calculate which depth of the cut will give a mean advance of 95 % of drilled depth. The values thus obtained are given in table 8:4.

In trying to find a suitable advance for a cylinder cut to be introduced at a working site, the driving should not begin with the drilled depth according to table 8:4, but with 30 % lower values. With these shorter rounds all practical details such as concentration of the charge, type of explosive, ignition sequence and drilling precision can be tried out and improved. The

TABLE 8:4. *Drilling depth for cylinder cuts at a relative advance of 95 % and a drill hole deviation according to table 8:2.*

ϕ mm	2×57	75	83	2×69	100	2×75	110	125	150	200
Double spiral	3.1	3.0	3.3	3.8	3.8	4.1	4.1	4.5	5.1	6.3
Four-section	2.6	2.5	2.7	3.2	3.2	3.4	3.4	3.8	4.4	5.4
Three-section	2.5	2.4	2.6	3.0	3.0	3.2	3.2	3.6	4.2	5.1
Täby	2.2	2.1	2.3	2.6	2.6	2.8	2.8	3.1	3.6	4.5

Type Fig 8:13	30–35 mm Grönlund "Cat hole" a	85–100 mm Michigan b	75 mm "cat hole" c	38 mm triangular d	75 mm triangular (Bullock-cut) e
H	2.2	3.0	^{a}	2.0	3.3
A_m	2.0	2.7	^{a}	1.8	3.0

^{a} The values not reported

four section cut with $\phi = 125$ mm gives a 95 % advance at $H = 3.6$ m. If tests or trial driving with $H = 3.0$ m are made with this cut the mean advance, say during a week, shall be 98 %, that is to say 2.94 m (for the cut itself). Only when this value has been approximately achieved should the depth of the holes be increased to 3.6 m, and if desired in the next step beyond that value if the drilling precision is so good that the mean advance per round is greater than 95 % of drilled depth. An important advantage of parallel hole cuts is that tests with new advances merely call for increasing the depth of the hole. This can be done from one round to the next. This is not the case with angled cuts, where every attempt towards a new advance calls for a change in the drilling pattern and a careful control of the new drilling programme so that it is carried out at the working site exactly as intended. This has proved to be very difficult.

The values given in tables 8:3 and 8:4 are calculated exclusively with regard to the drilling deviation. This means that the practial mean advance can be lower if there is flash-over or difficulty in effecting complete ejection of the debris through the drilled opening. At hole depths up to 4–5 m (13–17 ft) and possibly also for greater values the drilling scatter appears, however, to be the decisive factor. Results from full-scale driving show on an average even 5 % greater advances than according to table 8:4. In table 8:5 the mean advance for different burn cuts is given for a lower relative advance (90 %) than for the cylinder cuts in table 8:4.

j. Throw, fragmentation

Since the blasting is done in principle towards a central opening there is a comparatively small throw and the muck pile is rather concentrated. The throw is affected by way of ignition and increases considerably if the round is ignited with short delays. If the advance is increased to more than 5 m for tunnels with small sections, a great part of the round must be ignited

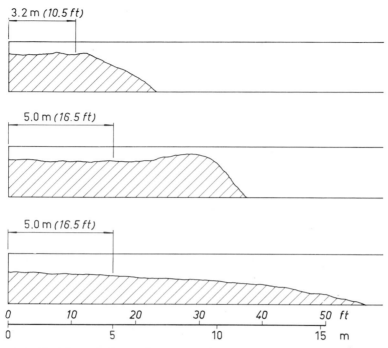

Fig. 8:9. Position of the rock masses after blasting rounds in a 5 m² drift.
a) second-delay ignition, advance 3.2 m. b) second-delay ignition, advance 5.0 m.
c) short-delay ignition, advance 5.0 m.

with short delays so that the round will have sufficient chance to swell.
Fig. 8:9 shows some examples of the distribution of the muck pile in 3.2–
5.0 m rounds in a 5 m² tunnel.

The fragmentation will be better and more uniform with cylinder and
burn cuts than with angled ones. The rock masses are all the time thrown
with great force against the opposite wall in the central opening. This means
that the energy of the throw is transformed into internal energy, which
contributes to the crushing of the rock. It is also easier to avoid tearing
inside the round as almost every hole has a free angle of breakage, and the
explosive is uniformly distributed throughout the entire body of rock
that is to be blasted.

8.2. Different types of parallel hole cuts

There is a great variety of parallel hole cuts. Even when there seems to be
only a slight difference in the appearance of the drilling pattern the results
can be very different. Only those cylinder cuts which have proved to give

Data for double spiral cuts with varying diameters of empty holes (ϕ). Notations acc. figs 8:10. The concentrations of charge, l_1 and l_2, refer to holes marked + and ● respectively.

ϕ mm	75	85	100	110	125	150	200
a mm	110	120	130	140	160	190	250
b mm	130	140	160	170	190	230	310
c mm	160	175	195	210	240	290	380
d mm	270	290	325	350	400		
l_1 kg/m	0.30	0.35	0.40	0.45	0.5	0.6	0.8
l_2 kg/m	0.65	0.75	0.85	0.9	1.1	1.3	1.7

good results in practical driving will be mentioned here. It should be emphasized that the stipulated drilling and charging patterns must be strictly observed. Every adjustment, for example towards greater distances of holes and larger quantities of charge in kg/m, lowers the mean advance appreciably. The ignition of the various holes in the cut should be carried out in the sequence given in figs. 8:10 and 8:11. When it is a question of the double spiral, Täby and Coromant cuts ignition of holes Nos. 1 and 2 with instantaneous detonators can give a slightly better result. Instantaneous detonators should be placed near the mouth of the hole. Regarding the risk of tearing off the charges the other detonators should be placed at the bottom of the holes. There should be no bottom charge in the 6–8 holes nearest to the empty hole.

a. Cylinder cuts

Double spiral cut. (fig. 8:10 a)

A spiral hole pattern gives the widest opening. When high advances are called for, however, a double spiral should be used in conformity with fig. 8:10a, with data for the spacing of holes and concentration of the charges as in table 8:6. In the double spiral pattern we gain the advantage that opposite holes can be ignited successively. This gives the best cleaning of the opening. Further, safety in the advance is increased, since one section of the double spiral can give breakage irrespective of the other. As will be seen from data of mean advances in table 8:3 the double spiral cut is definitely superior to other types of cuts, with at least 20% greater advance than the other parallel hole cuts. A drawback in practice is the fact that the two holes which are to be placed nearest to a 100–110 mm empty hole are only at a 130–140 mm distant from its center. In drilling with stationary frames and present day equipment a distance of 160 mm

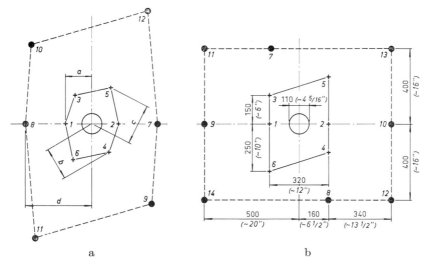

FIG. 8:10a. Double spiral cut. (Notations according to table 8:6.)

FIG. 8:10b. Täby cut.

is required. This entails a need to increase the empty hole diameter to 125 mm, when the center distance can be increased to 160 mm and, if desired, makes it possible to increase the advance 10 %. The alternative is to increase the distance for the two nearest holes to 160 mm which probably reduces the mean advance at least 10–5 %.

Täby cut. (fig. 8:10b)

This, as will be seen from the ignition sequence, is a modified double spiral cut. As regards the advance it is inferior to the proper double spiral. This

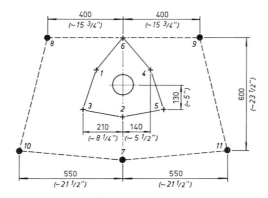

FIG. 8:10c. Three-section cut. $\phi = 110$ mm.

245

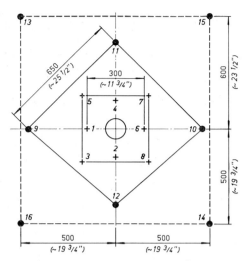

FIG. 8:10d. Four-section cut. $\phi = 110$ mm.

will be seen from table 8:3. This circumstance can thereby also serve as an example of the effect which even seemingly insignificant alterations in the drilling pattern can acquire.

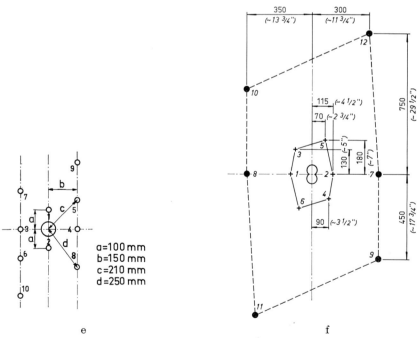

a=100 mm
b=150 mm
c=210 mm
d=250 mm

e

f

FIG. 8:10e. Fagersta cut. $\phi = 75$ mm.
FIG. 8:10f. Coromant cut. $\phi = 2 \times 57$ mm.

FIG. 8:11. Drill jig (template) with light metal sheets for Coromant-cut, weight about 15 kg (33 lbs) (acc. to Lautmann).

The benefits of the Täby cut are, however, that the holes are here located vertically below one another on one and the same line. This facilitates drilling and mechanization of the drilling.

Three-section cut (fig. 8:10 c) and four-section cut (fig. 8:10 d)

These are two other types of cylinder cuts which have been widely used. They give an advance with is 15 and 20 % respectively greater than the Täby cut. The three-section cut has above all been employed in driving with light, hand-held equipment with 75 mm empty holes. Compared with the Täby and four-section cuts it requires fewer holes. The four-section cut, which gives a satisfactory advance value, is at present the type of cylinder cut that is in most current use.

Fagersta cut (fig. 8:10 e)

This cut is also drilled with handheld equipment. The empty hole, 64 ($2\,^1/_2$″) or 76 (3″) mm is drilled in two steps, the first as an ordinary hole and the second as an enlargement of this pilot hole. The cut is something between a four-section cut and a double spiral cut.

The Coromant and the Fagersta cuts are both drilled with light equipment, which makes them suitable for use in mines and in small drifts,

247

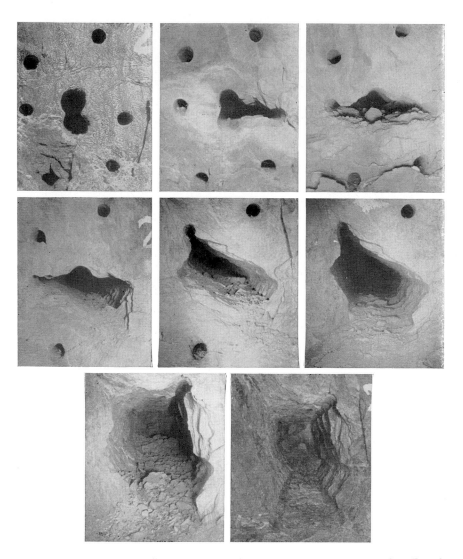

Fig. 8:12. Blasting the Coromant-cut with two 57 mm (2.25 in) holes. Depth of the holes: 3.2 m (10.5 ft), advance 3.1 m (10.2 ft), charge 0.3 kg/m (0.20 lb/ft).

Top row left to right: before blasting: after hole number 1, 2. Middle row left to right: after hole number 3, 4, 5. Bottom row, left: after hole number 6. Right: picture after cleaning the cut (acc. to Lautmann).

where drilling with heavy machines is not profitable. They are superior to all *burn* cuts with one empty hole of 75 mm, or several empty holes of smaller diameter. They give a greater and a more reliable advance and a lower consumption of explosive.

248

Coromant cut (fig. 8:10 f, 8:11 and 8:12)

The advantage of this type of cut is that it can easily be drilled with hand-held equipment. It is, in principle, a double spiral cut. Two 57 mm holes drilled together to one common 8-shaped hole are used as an empty hole. Like the charged holes in the cut, the 57 mm holes are drilled with the aid of a ready-made template (fig.8:11). From table 8:4 it will be seen that approximately the same advance is obtained with this type of cut as with one with a 75 mm empty hole.

b. Burn cuts (fig. 8:13 a–e)

The most common burn cuts operate with empty holes of the same diameter as the charged ones. The Grönlund cut (fig. 8:13a) and triangle cut (fig. 8:13d) have the great advantage that no additional equipment whatsoever is needed for the drilling. It follows that these types of cut will also in future have a range of use of their own to which they are justly entitled. Even if it is necessary here to use charges which give plastic deformation of the rock it is advisable not to load fully holes Nos. 1–5 in the Grönlund cut and holes Nos. 1, 2 and 3 in the triangle burn cut. These holes are charged with 25 mm cartridges loosely piled which give 0.65–0.70 kg dynamite per meter. The other holes are fully loaded. Ignition of hole No. 1 should be done with an instantaneous detonator close to the face of the tunnel.

There is little or no reason to include or enumerate the number of variations that can be made in these types of cuts. Only one further common burn cut, the so-called "cat-hole", will be mentioned; it can be obtained by turning the four uncharged holes of the Grönlund cut 45° round the center hole.

Michigan cut with a larger central empty hole is one of the first burn cuts that has been adopted. It is now merely of historical interest, since better and more reliable results are to be had with cylinder cuts if a large empty hole is available for the blasting.

The "cat-hole" with a 75 mm hole (fig. 8:13c) is an enlarged cat-hole cut, with the center hole charged with 1.0–1.2 kg dynamite per meter.

In conclusion an example of a 75 mm triangle cut according to fig. 8:13e can be shown. This type of cut, which has been investigated and reported by Bullock, gives a better and more reliable advance than some thirty other types of burn cuts which were tested at the same time. Strictly speaking it is not a question here of a parallel hole cut since the 6 outer holes have to be slightly angled inwards.

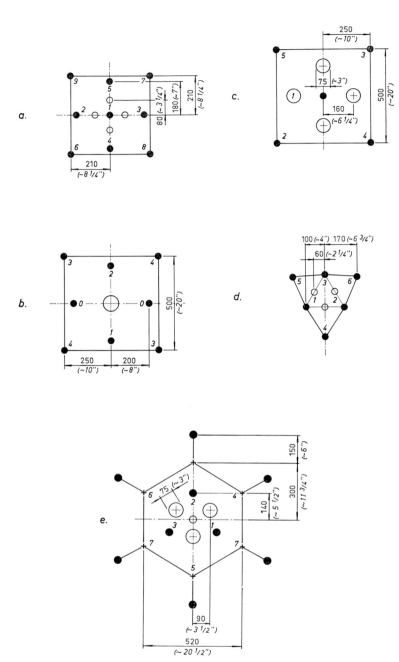

FIG. 8:13. Examples of burn cuts. a) Grönlund cut. b) Michigan cut. c) cat hole with 75 mm (3 in) holes. d) triangular burn cut with 35 mm (1.4 in) holes. e) Burn cut with three 75 mm (3 in) empty holes and six outer cut holes acc. to Bullock.

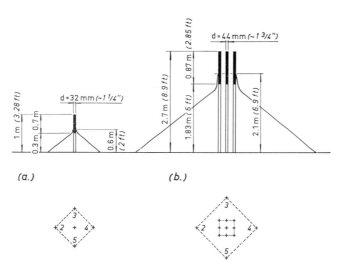

FIG. 8:14. a) Crater cut with breakage parallel to direction of holes. b) acc. to K Hino.

c. Crater cuts (fig. 8:14)

These represent in principle a completely new type of cuts and make use of the crater effect which is obtained in blasting a single hole at a free rock surface (fig. 8:14a). This type of cut has been suggested and developed by K. Hino. The possibility of a uniform enlargement can be counted on. This means that if the scale is enlarged so that the diameter and depth of hole and length of the charge are all doubled, for example, a crater of double the depth will be obtained. The number of holes can be increased instead of increasing the diameter of the holes. In fig. 8:14b, 9 small holes have replaced one with the three-fold diameter. The depth of the crater is also three times as large. Compared with other types of cuts with parallel hole pattern the crater cuts have hardly any advantage to offer in ordinary tunneling. Crater blasting can, on the other hand, be utilized as an important aid in raise blasting with long holes. In such cases crater blasting can be used to blast some meters in short rounds where the position of the holes due to the deviation renders blasting towards the empty hole impossible.

8.3. Results from driving with parallel hole cuts

The practical results obtained with these different kinds of cuts have been carefully examined in material collected during recent years in America and Europe. For all types of burn cuts with a central hole of up to 100 mm (4 in) the experience is satisfactorily described in table 8:5 for conditions

251

where these cuts can be used. For cylinder cuts the corresponding information is given in table 8:4 (p. 241).

Recent development with cylinder cuts, avoiding the burning effect, have revealed potentialties for such big advances that the drilling equipment, and, in many cases, also the training of the workmen have been able to meet the demands only in special cases. For these cuts there has been a special need to check the calculations in full scale operations. This has been done for some 8000 rounds with cylinder cuts and are accounted for together with some 5000 burn cuts in tables 8:7–8:9. A great part of the basic material was reported to the Swedish Rock Blasting Committee by different companies and was presented by T. Bjarnekull. Further material has been compiled during 1960 and 1961. The blasting has been performed in a wide variety of rock with empty holes ranging from 30 to 125 mm. The sections for the tunnels have been mainly 4–20 m² (5–25 sq.yds.).

The first column of table 8:7 gives the number of the working site concerned, the second the type of rock. It is important to know the number of rounds given in the third column, when drawing conclusions from the final results. In the next columns the size of the empty hole and the type of cut are indicated. The first four cuts mentioned are burn cuts, then come cylinder cuts Nos. 5–27. The notation 2×57 indicates that two holes of 57 mm ($2\frac{1}{4}$ in) have been used, in this specific case drilled to a joint slit. Note that No. 5 and No. 27 are different cuts. In the first mentioned denoted "2×57 Täby" the six holes closest to the empty hole are arranged on two vertical lines and not as in the Coromant-cut (No 27).

The following column gives the depth of the drill holes (H) which varies between 1.6 and 4.0 (5.5 and 16 ft). The mean value for the advance per round (A_m) is given in relation to the drilled depth as $\alpha = A_m/H$ and varies from 0.69–0.97. The working site No. 24 has had a drilled depth of 4.8 m (16 ft) and has given 0.69 (69 %) of this value as advance, thus $A_m = 3.3$ m (11 ft). No. 20 has had an advance of $A_m = \alpha H = 0.97 \times 3.1 = 3.0$ (10 ft). The advances have been practically the same in these two cases, but not the cost and the results by any means.

We cannot use figures for the average advance in comparing the results. It would be quite misleading. What we want to know is the drilled depth corresponding to 95 % advance. It can be calculated with the aid of such curves as fig. 8:8(b) and the relations given in table 8:3. In most cases it is a question of a correction of only some few per cent. In Problem 8.5 the course of the calculation is shown. The corrected values H_{95} are given in the last column of table 8:7. It shows for No. 24 that $H_{95} = 2.8$ m (9.3 ft) instead of $H = 4.8$ m (16 ft) thus indicating that with the conditions and drilling accuracy on the site the drilling should only have been 2.8 m if

252

TABLE 8:7. *Results from tunnel driving with parallel hole cuts. Tunnel section 4–20 m² (5–25 sq. yds), except No. 29 with 80 m² (100 sq. yds.).*

Site No.	Type of rock	Number of rounds	ϕ mm	Cut	H m	$\alpha = A_m/H$	H_{95} m
1	Magnetite	33	32	Grönlund	2.07	0.92	1.8
2	Leptite	118	30	Grönlund	2.00	0.97	2.3
3	Ore and granite	2500	30	Grönlund	2.00	0.92	1.75
4	Ore	2500	30	Grönlund	2.00	0.92	1.75
5	Leptite	1040	2 × 57	Täby	2.00	0.95	2.0
6	Leptite	3	2 × 57	Four section	2.40	0.95	2.4
7	Pyrite	57	2 × 57	Täby	1.60	0.97	1.8
8	"Tuffit"	123	2 × 57	Double spiral	4.0	0.88	3.3
9	Ore	2400	75	Three section	2.4	0.96	2.6
10	Leptite, ore	8	76	Four section	3.2	0.77	1.9
11	Leptite, pegmatite	190	76	Four section	2.0	0.95	2.0
12	Gneiss	900	75	Täby	2.4	0.93	2.2
13	Clay-slate	419	83	Four section	3.1	0.90	2.55
14	Granite	6	102	Four section	3.15	0.95	3.15
15	Granite	455	110	Four section	4.0	0.95	4.0
16	Leptite	550	110	Täby	3.0	0.90	2.5
17	Green-slate	242	110	Double spiral	2.0	0.80	1.7
18	Leptite	242	110	Four section	2.4	0.93	2.2
19	Lime-leptite	55	110	Täby	2.4	0.97	2.7
20	Gneiss-diabase	200	110	Four section	3.1	0.97	3.5
21		250	110	Four section	4.0	0.90	3.3
22	Granite	240	75	Four section	3.15	0.86	2.4
23	Granite	115	125	Four section	3.9	0.97	4.4
24	Granite	40	110	Täby	4.8	0.69	2.8
25		500	75	Three section	3.20	0.87	2.6
26	Leptite	30	110	Four section	3.17	0.95	3.2
27	Granite	57	2 × 57	Coromant	3.17	0.92	3.0
28		25	3 × 75	Bullock	3.6	0.93	3.3
29			225	Burn cut	3.6	0.92	3.3
30	Dol. limestone	50	3 × 75	Bullock	2.85	0.90	2.5

a 95 % advance is desired. Inversely in No. 20 the actual value for the drilled depth could have been greater as here $H_{95} = 3.5$ m (11.6 ft). The H_{95}-values make it possible to compare the experimental results from various sites correctly.

In table 8:8 the distances for the four holes closest to the empty one are indicated as *a* when equal, and as *a/b* when the two closest differ from the next two. (*a*) gives the *a*-value suggested according to the figures 8:10 and

TABLE 8:8 *Spacing and concentration of charge for the cut holes compared with the corresponding values given in the tables 8:6 or 8:1 (in brackets).*

Site No	Diameter of empty hole ϕ mm	Distance of holes a/b (a) mm	c (c)	Concentration of the charge l_1 (l_1) kg/m	Ignition	$\varkappa = H_{95}/H_{cal}$
1	32	90 (80)	210 (180)	0.53 (0.65–0.70)	fuse	1.00
2		85 (80)	180 (180)	0.78 (0.65–0.70)	HS	1.28
3		95 (80)	300 (180)	0.78 (0.65–0.70)	fuse	0.97
4		100 (80)	300 (180)	0.79 (0.65–0.70)	fuse	0.97
5	2×57	120 (120)	170 (160)	0.43 (0.30)	fuse	0.91
6		100/150 (120)	250 (200)	0.18–0.30 (0.30)	MS	0.93
7		120 (120)	170 (160)	0.45 (0.30)	fuse	0.82
8		150 (120)	150 (200)	0.25 (0.30)	MS	1.07
9	75	120 (120)	220 (210)	0.24–0.28 (0.30)	fuse	1.08
10		100/160 (120)	220 (200)		fuse	0.76
11		100/150 (120)	210/290 (200)		fuse	0.80
12		150 (125)	210 (160)	0.29 (0.30)	MS	1.05
13	83	150 (130)	280 (220)	0.45 (0.40)	MS	0.95
14	102	110/140 (140)	350 (230)	0.20–0.34 (0.40)	fuse	0.97
15	110	150 (150)	280 (250)	0.5 (0.5)	MS	1.18
16		270 (160)	440 (210)		fuse	0.89
17		190 (140)	210 (170)	0.67 (0.5)	MS	0.41
18		130/150 (150)	290 (210)	?	?	0.65
19		170 (160)	225 (210)	0.41 (0.5)	HS	0.96
20		130/200 (150)	270 (250)	0.35–0.57 (0.5)	fuse	1.03
21	110	150 (150)	300 (250)	0.40 (0.5)	MS	0.97
22	75	130 (120)	240 (200)	0.36 (0.35)	MS	0.96
23	125	140 (170)	260 (280)	0.40 (0.55)	MS	1.16
24	110	150 (160)	210 (210)	0.5 (0.5)	MS	1.00
25	75	110 (120)		0.35 (0.35)	MS	1.08
26	110	165/200 (150)	(250)	0.30 (0.40)	MS	0.94
27	2×57	90 (90)	120 (120)	0.30 (0.30)	MS	0.97
28	3×75	90 (90)	145 (145)		MS	1.23
29	200	—	—	—	—	—
30	3×75	—	—	—	—	0.86

8:13 or, for the double spiral cuts, according to table 8:6. The distances indicated as c (c) give the value used and calculated (in brackets) respectively for the next two or four holes closest to the empty one.

Actual (calculated) values l (l) are also given in kg/m in the column for the concentration of the charge. The holes in the cut have been ignited with an ordinary fuse, with HS (half-second) or MS (short delay) electric detonators.

The last column gives the relation $\varkappa$ between the experimental depth of hole corresponding to 95 % mean advance and the calculated value. The values calculated (H_{cal}) are taken from table 8:4, or for the burn cuts from table 8:5. When the results obtained are in accordance with tables 8:4–8:5 the last column gives. $\varkappa = 1.00$. With this notation the $\varkappa$-values clearly indicate if the blasting operation has been successful or not. It also makes it possible to explain some important results of the investigation as simply as follows.

The $\varkappa$-values for the cylinder cuts are arranged in four separate groups in table 8:9 according to their concentration charge and spacing of the holes. The first group is the most interesting one as it gives the results obtained with spacing and concentration of the charge according to the calculations. Even cases with a lower concentration of the charge belong to this group. We have here obtained values between 0.94 and 1.18 with an approximate average value of $\varkappa = 1.05$. One conclusion is that the values of table 8:4 for the advance are if anything slightly too low. In full-scale driving 5 % higher values have on an average evidently been obtained.

In the next group, with correct placing but an excessive concentration of the charge the average value is only $\varkappa = 0.85$. Even if expected—as previous experiments have shown the importance of avoiding excessive charges—it is here surprising that such small overcharges as 20–25 % characterize this group.

With correct charges, but excessive distances from the loaded holes to the empty hole an average of about $\varkappa = 0.97$ is obtained. "Excessive" means that the distances have been about 20 % more for the four closest holes than those recommended in tables 8:1 and 8:6 (instead of $a = 140$ mm the distances have been 160–170 mm), or 25 % for the following holes (instead of $c = 210$ mm the distances have been up to 260 mm). It is worthy of special attention that even if the distance is increased by 20–30 mm (0.8–1.2 in) for the closest holes the mean advance obtained is only 8 % less than the average advance with spacing according to the tables. Values obtained from different working places deviate ± 10 % from the mean values of every group.

Finally, with excessive distances and charges the lowest $\varkappa$-values are obtained, $\varkappa = 0.41$–0.89. For the burn cuts, not included in table 8:9, the experimental values from the cases investigated in tables 8:7–8:8 are in accordance with the general values of table 8:5. The mean advances have been about 5 % higher than those of table 8:5, which is entirely within the tolerance limits.

TABLE 8:9. *Ratio between practical mean values for the advance per round and the values for the advance calculated according to table 8:4.*

Distance to center empty hole	Conc. of charge	Ratio $\varkappa$
correct	correct	0.94 1.08 1.18 0.96 1.03 0.97 0.96
		1.16 1.00 1.0 1.07 0.97
correct	too great or unknown	0.91 0.82 0.76 0.80
too great	correct	0.93 1.05 0.95 0.97
too great	too great	0.89 0.41 0.65

Correct distance to center of empty hole means that the values in the drilling pattern have not exceeded those given in table 8:6 within 10 % for the four inner holes and 25 % for the outer holes.

Correct concentration of charge means concentrations not exceeding the l_1-values given in table 8:6 by more than 10 %.

8.4. Conclusions

The experimental results mainly confirm the general calculations and the tables stress the importance of not using excessive charges, but also that a slightly (less than 20 %) increased spacing of the holes around the empty hole may be used, if this facilitates the arrangement for the drilling. The calculations as well as the driving results show that the double spiral cuts give a superior performance with advances that are 20 % larger than for the four-section cut, 25–30 % better than the three-section cut and 50 % more than the Täby-cut. It seems reasonable to expect that the debris can be satisfactorily thrown out of the central opening up to depths of 5 m (17 ft) or even more. Table 8:4 gives the drilled depth for cylinder cuts, when the relative advance is 95 %.

The burn cuts have given about 5 % greater advance than table 8:5 which table gives the depth drilled when the relative advance is 90 %.

The fundamental investigations as to the conditions for blasting with the examined types of parallel hole cuts are thus carried out. Concerning practical application there still remain improvement of the facilities for drilling empty holes of about 125 mm (5 in) or more and for lowering the cost of this hole, to coordinate the use of the cuts with increased mechanization of the drilling, to adopt prefabricated cartridges specially designed for the purpose and to organize the driving by training the workmen to make the best use of parallel hole cuts.

8.5. Problems

1. Construct a drilling pattern for a 83 mm (3.25 in) double spiral cut. To what depth could it be drilled with a satisfactory (95%) advance?

2. Reconstruct the pattern above with the 6 cut holes at a distance that is increased 10%. Discuss the influence on the proposed drilling depth (H_{95}) and the advance according to the table 8:9.

3. Decide how to load the 6 cut holes in problem 1: a) with a semi-gelatine explosive with a weight strength of $s = 1.0$, b) with a powder explosive with $s = 0.9$.

4. Discuss the diameter of the 6 cut holes in problem 1 with regard to the explosive cartridges available.

5. When driving with a four-section cut No. 21 in table 8:7 the drilled depth was $H = 4.0$ m (13.3 ft) and the mean value for the advance $A_m = 3.6$ m (12 ft). Determine H_{95} a) supposing the deviation in the drilling to be $\sigma = 2.0$ cm/m independent of H as in table 8:3, b) with the deviation depending on the depth according to the values of table 8:2.

6. The deviation of the drill holes assumed when calculating the values given in table 8:2 is for depths greater than $H = 2.4$ m satisfactorily described as $\sigma = (1.3 + 0.64\ H)^{1/2}$ cm/m. The first term in the brackets depends on alignment faults, the second on deviations inside the rock. Calculate for double spiral cuts the drilled depth H_{95} that will give 95% advance if the fault in the alignment is 2 cm/m instead of 1 cm/m for cylinder cuts when the diameter of the empty hole is a) $\varnothing = 83$ mm (3.25 in), b) $\varnothing = 2 \times 57$ mm (2×2.25 in), c) $\varnothing = 125$ mm (5 in), d) $\varnothing = 150$ mm (6 in) and e) $\varnothing = 200$ mm (i in).

7. A tunnel driving is performed with a double spiral cut and alignment faults of 2 cm/m as in problem 6 but with the depth drilled according to table 8:4. The output would then be less than 95%. Calculate the percentage for cuts with an empty hole a) $\varnothing = 75$ mm (3 in), b) $\varnothing = 125$ mm (5 in) and c) $\varnothing = 150$ mm (6 in).

8. Construct a drilling pattern for a four-section cut with a) $\varnothing = 83$ mm (3.25 in) and b) $\varnothing = 150$ mm (6 in). c) Discuss for $\varnothing = 150$ mm the effect of increasing the distances of the holes 10%.

9. Decide how to load the 8 cut holes in problem 8 a) and b).

10. When driving a month with a double spiral cut $\varnothing = 100$ mm (4 in), and a depth of the holes $H = 3.1$ m (10.3 ft), the total advance for 24 rounds is a) 71 m (235 ft) and b) 73 m (244 ft). Calculate $a = A_m/H$ and the value H_{95} (corresponding to a relative advance of $a = 0.95$).

11. Discuss the effect of the ignition on the results shown in table 8:8.

9. GROUND VIBRATIONS

With the development of rock blasting technique the question of ground vibrations has become increasingly important. It has become more or less routine to excavate rocks close to or below buildings even in central city areas. During the 1950's big rounds were widely used; this has greatly increased the efficiency and the safety of the work. Especially in the city of Stockholm this development has been spectacular and is probably unparallelled in the world. Rounds of hundreds and thousands of cubic meters in the vicinity of dwelling areas are now regarded as ordinary ones —in special cases rounds of up to 50,000 kg (100,000 lbs) of explosive have been blasted less than 200 m (700 ft) the from the nearest houses.

This has become possible by the adoption of multiple-row blasting with short delay ignition, by new methods for calculating the charge required to ensure loosening of the rock without unnecessary throw, and by increased experience concerning ground vibrations and how these can be reduced in large rounds. When blasting work has to be carried out in the neighborhood of buildings or existing structures the ground vibration is often the factor which finally decides how the blasting is to be carried out. Experience from records and investigations which have been made by Nitroglycerin A.B., the Street Department of the City of Stockholm and Skånska Cement A.B. during a long sequence of years has been collected and analysed together with investigations in Central Europe and South America and material from reports made by Rockwell, Thoenen and Windes, Crandell, Morris, Fish, Köhler, Baule, Otto, Teichman, Westwater, Edwards and Northwood and others. This material covers practically all kinds of rock from hard bed to sedimentary rock, clay and glacial till, and the conclusions drawn in this chapter can be generally applied in any kind of rock in the planning of blasting operations and in controlling the results obtained.

A great difficulty when it comes to determining the limit values for varying degrees of damage is due to the fact that there have been relatively few cases where damage could be proved. As a rule every care has been taken to avoid damage. It is questionable, however, if this attitude should not be changed. It sometimes entails great expense to impose heavy restrictions on the carrying out of blasting operations, whereas the damage which

Fig. 9:1. The explosive correctly distributed and detonated must break up the olid rock into small pieces but leave nearby houses intact.

ss avoided by doing so, e.g. the falling of plaster or occasional cracks in repeated blasts, could have been repaired or compensated at a moderate cost. When there is no risk of personal injuries it may therefore be right to risk minor damage and to carry out the work in a more satisfactory manner from a technical and economic point of view.

9.1. Apparatus for recording ground vibrations

In order to ascertain if the ground vibration could be reduced by inter-ference in successive initiation of the charges, an investigation was begun by Nitro Nobel AB in 1946 at different blasting places. As the vibrographs then on the market could not reproduce the actual range of frequencies in a satisfactory manner, an apparatus was constructed comprising a capa-citive detector with a natural frequency of 2 c/s and a cathode-ray oscillo-graph. In order to obtain sufficient dissolution at the highest frequencies and avoid an unnecessarily large and expensive consumption of film, a special arrangement was introduced for the photographic recording. In

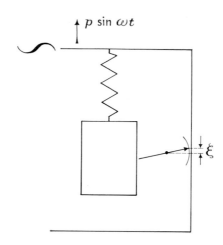

$p \sin \omega t$

ξ

Fig. 9:2. Vibrograph with frictional damping.

this the oscillograph is utilised in the ordinary manner with repeated deflections from left to right, and the film of the camera is fed downwards at a constant speed. The recording takes place with parallel lines almost perpendicular to the longitudinal direction of the film as a base line. The sensitivity and the recording speed can easily be varied within wide limits, and this has proved to be a great advantage.

The recordings obtained with this apparatus indicated that an interval of the order of centiseconds might reduce the ground vibration by interference.

The so-called Cambridge vibrograph which is independent of electric current and easily transportable, has been used to a great extent for field tests. The instrument has a swinging mass with relative damping and the vibrogram is drawn in a moderate enlargement on a celluloid strip. The proper frequency is 4 c/s and the instrument gives vibrograms with a legible dissolution up to 500 c/s.

In many cases the ground vibrations which we have determined have frequencies which are many times larger than those which have been measured elsewhere. The question of the origin of ground vibrations is so complex that we will not deal with it here. We would only point out that we have examined whether these high frequencies could be due to natural vibrations in the instrument. If so it should be possible to find one or more permanently recurring frequencies. This has not been the case. Furthermore, it should be possible to produce the eventual natural frequencies by jolts, but when a change is made in the testing conditions recurring frequencies are never obtained. In addition, when recording simultaneously

260

with two entirely different vibrating systems, as with the old and present measuring equipments, we have obtained the same graph in the actual frequency area.

In an ordinary vibrograph is fitted with a vibrating system with relative friction damping (fig. 9:2). If the base executes a constant sine oscillation with the amplitude A and the frequency f the spring-suspended mass will move slightly in relation to the stand, and there will be obtained for the elongation of the spring

$$-\xi = -p\sin\omega t + x \qquad (9:1)$$

in which $\omega = 2\pi f$ and x is the distance of the mass from the position of equilibrium. The magnitude ξ is recorded graphically on a celluloid strip which is continuously fed forward by a needle with a rounded point, the movement of which in relation to the strip is proportional to ξ. The damping force is assumed to be

$$F = f_1 \frac{d\xi}{dt} + f_2 \qquad (9:2)$$

in which f_1 represents "viscous" damping, f_2 "frictional" damping. For the damping coefficients, the relationships $f_1 = $ constant and $f_2 = $ constant apply; f_2 changes its sign with the direction of movement so that the movement is counteracted. The relative movement is determined by equating the sum of the inertia force of the mass, the damping force, and the tension in the spring, to zero.

$$-m\frac{d^2x}{dt^2} + f_1\frac{d\xi}{dt} + c\xi = -f_2 \qquad (9:3)$$

or

$$\frac{d^2\xi}{dt^2} + 2b\frac{d\xi}{dt} + \omega_o^2\,\xi = -\omega^2 p \sin \omega t + h\omega_o^2 \qquad (9:4)$$

where m is the spring-suspended mass, c the spring constant, $f_1/m = 2b$, $c/m = \omega_0^2$ and $-f_2/m = h\omega_0^2$.

To determine the magnitudes b, ω_0 and h, the natural oscillations of the system can be studied according to (9:4) with $p=0$. As long as the needle moves in one direction, for instance upwards, h is constant and the movement follows the equation

$$\frac{d^2(\xi+h)}{dt^2} + 2b\frac{d(\xi+h)}{dt} + \omega_o^2(\xi+h) = 0 \qquad (9:5)$$

which implies that $(\xi + h)$ performs an ordinary damped sine oscillation

$$\xi + h = B e^{-bt} \sin (\sqrt{\omega_o^2 - b^2}\, t + \delta) \qquad (9:6)$$

which assumes the value 0 for $\xi = -h$. The maximum deflection upwards then becomes $(B-h)$, which is obtained if the damping factor b can be ignored. When the movement alters its direction (9:6) again applies, but with a new zero for the sine function as h is now negative. To enable this to be done the amplitude at the moment of turning must be reduced to $B-2h$ and the maximum deflection downwards will be $B-3h$. The damping factor f_2 in (9:3) consequently gives a constant difference between two consecutive readings irrespective of the size of the deflection. In contrast to this the second factor f_1 or b will give a reduced amplitude which is in direct proportion to the amplitude if this is calculated from the zero of the sine function at the actual moment.

Examinations of the natural vibration of the apparatus (fig. 9:3a) show that the amplitude decreases by a constant amount, 0.4 mm for every deflection. It is furthermore evident that the zero of the sine oscillation for the first downward movement lies almost at the same level as the last deflection. The upward movement has its zero at an equally great distance below the inserted middle line. Fig. 9:3a is recorded with the usual gearing of 5:1 for drawing on the celluloid strip. As it is the friction of the needle on the strip which gives the chief damping this can be diminished by reducing the gearing. In drawing with the enlargement of 1:1 a greater number of full vibrations will therefore be obtained before the natural vibration has been damped out (fig. 9:3b). It is quite obvious here that the amplitude practically decreases linearly, which implies that $b \approx 0$, and that the damping is entirely due to a constant frictional force f_2 in (9:2) which counteracts the movement. The decrease here is 0.08 mm for every deflection.

If the weight of a vibrograph is for instance 6 kg, 2 kg of which is in the vibrating system, the apparatus loses contact with the base when the acceleration in the vibrations are greater than 1,5 g. Is the acceleration expected to exceed 1,5 g the vibrograph should be loaded with weights suspended on springs. Fig. 9:4a illustrates vibrograms which have been recorded on a shaking table with a frequency of 50 c/s and the maximum acceleration 3-6g with a loaded (fig. 9:4a) and unloaded instrument (fig. 9:4b) respectively. It is obvious that the vibrogram in the last-mentioned case is very misleading.

The advantage of the loading arrangement is that there is no need to clamp the instrument to the base, which as a rule is difficult.

But the best recording of vibrations with high accelerations is obtained when the vibrograph is fixed on the base with bolts or screws. The recor-

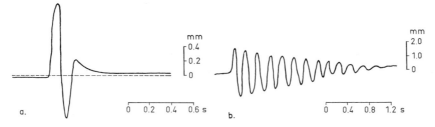

FIG. 9:3. Natural vibrations in a vibrograph whith friction damping a) Enlarged five times; b) without enlargement when recorded (and correspondingly less friction damping).

ding capacity is then limited only by the construction of the vibrograph.

In most cases the instrument is placed on the rock close to, or inside, the building, so that the best possible contact is obtained with the rock. *It is consequently the vibrations of the rock that are first determined*—the different types of vibrations which are obtained in walls and floors are an effect of these ground vibrations and of the special construction of the building. The question of comparing the incoming vibrations with the resulting ones may have to be considered in a second stage of an investigation.

9.2. Reading the vibrograms

The maximum amplitude, frequency, vibration velocity and acceleration are determined from the vibrograms. Fig. 9:5 gives an example of how this determination is made. The vertical distance between two consecutive deflections pointing in opposite directions is determined and stands for the double amplitude, $2A$. If the vibrogram has been magnified the figure obtained is first divided by the magnifying factor, for instance 5 in the primary record, 10 in photographing, 50 in all. The distance horizontally is indicated t_2. During the period of observation it is considered that the vibration can be represented with satisfactory accuracy by a sine oscillation

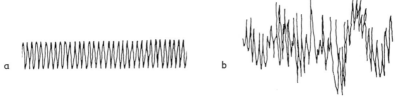

FIG. 9:4. Vibrogram recorded on a vibrating table with the frequency 50 c/s, amplitude 360 μ i.e. 3.6 g by a mechanical instrument. a) the instrument loaded with 40 kg; b) the instrument unloaded.

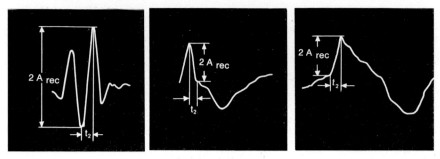

FIG. 9:5. Example of vibrogram reading.

with amplitude A and the period of vibration $2t_2$ for a full period. The frequency f is then obtained by comparison with the timing $t_1 = 0.1$ sec, which gives the relation

Blasting site: Secondary Power Station, surge chamber.

Vibration Report No. I
25 May 1962

Site of vibrograph: Close to relay cabinet T. 7 on the floor of the relay chamber.

Type of instrument: Cambridge Vibrograph.

Measuring done by: B. Kihlström.

Nature and state of building and foundation: Concrete on rock.

Explosive: LF 35

Diameter of drill: 31, 30, 29 mm

Date	Delay No.	Holes			Measuring			Results			
		Charge kg	Number	Burden m	d m	h m	R m	A μ	f p/s	a g	v/c μ/m
12/5 1962	1	1·2	1				38·0	10	150	0·9	1·9
,,	2	1·0	1				38·0	6	150	0·5	1·1
,,	3	1·1	1				38·0	6	150	0·5	1·1
,,	4	0·92	1				38·0	*	*		
,,	5	1·0	1				38·0	*	*		

* Values too small to be read off.

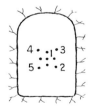

Drilling plan.

Electric detonator: $\frac{1}{2}$ s

Depth of hole: 2 m

FIG. 9:6. Example of vibration report.

264

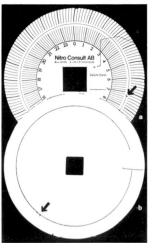

Reference weight	4.3 kg
Natural period of vibration	
during vertical measurement	3.1 cycles/sec
during horizontal measurement	2.7 cycles/sec
Amplification factor	5
Continuous recording time	8 days
Movement of recording paper	0.2 mm/min
Resolution time	10 seconds
External dimensions	width × height × depth
	= 285 × 175 × 70 mm
Total weight	7.5 kg

Rotation speed,	type CV 5 =	5 revolutions/min
high-speed section	type CV 10 =	10 revolutions/min
	type CV 15 =	15 revolutions/min

FIG. 9: 7. Combigraph avd records, a) low speed, b) high speed.

Calculation of amplitude (A) in mm

$$A = \frac{2A_{\text{rec}}}{2 \cdot 5} \text{mm}$$

$2A_{\text{rec}} =$ Double the amplitude registered

The amplitude is registered with an amplification factor of 5 on the combigram.

Calculation of frequency (f) in cycles/sec

$$f = \frac{\pi \cdot d \cdot n}{60 \cdot 2 t_2}$$

d = diameter of zero line circle in mm
n = speed of recording paper in revolutions/min
$2t_2 =$ the registering speed for a cycle in mm

There will then apply for the approximating sine oscillation

$$s = A \sin 2\pi f t \qquad\qquad (9:7)$$

$$v = 2\pi f A \qquad\qquad (9:8)$$

$$a = 4\pi^2 f^2 A \qquad\qquad (9:9)$$

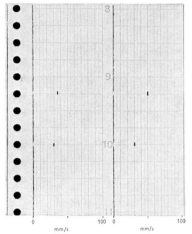

FIG. 9: 8. Vibracorder, geophones and record.

in which v indicates the maximum velocity of vibration and a the maximum acceleration.

An example of a vibration report is given in fig. 9:6. The report should preferably deal with only one round. If it is a question of a delay, and not a short delay round, the different delay numbers can immediately be distinguished and each is then indicated separately if the values are of interest. This is especially the case if a and v assume high values. Even for a single charge several indications may be required because the vibration contains several different frequencies. Every one is then evaluated separately and is given a line in the report if the corresponding a and v (or v/c) values are considered large (c = propagation velocity).

When the amplitude and its corresponding frequency are obtained in the vibrogram the acceleration and the velocity can be calculated with the aid of the nomogram in fig. 9:9. The quantities sought are all on a straight line as shown and can then be read off directly on the vertical scales.

266

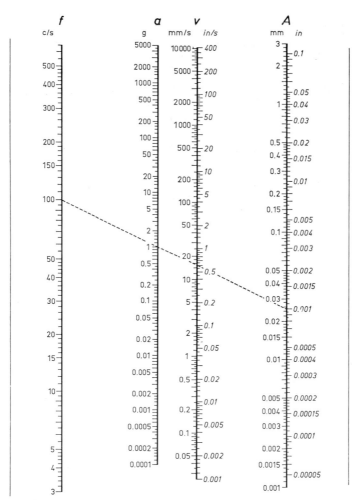

FIG. 9:9. Nomogram for the relationship between frequency (f), acceleration (a), vibration velocity (v) and amplitude (A). Ex: $f = 100$ c/s, $A = 0.025$ mm (0.001 in). The vibration velocity is about $v = 15$ mm/s (0.6 in/s). The acceleration is $a = 1$ g.

9.3. Character of the ground vibrations

The majority of our measurements have been carried out when blasting in hard Swedish rock, with relatively small charges per hole (<100 kg) and at short distances from the charges. This is undoubtedly the cause of the high frequencies which occur in the measurements. For large rounds and longer distances lower frequencies are obtained. Some typical vibrograms are shown in fig. 9:7 and 9:10. The frequencies recorded lie between 5–

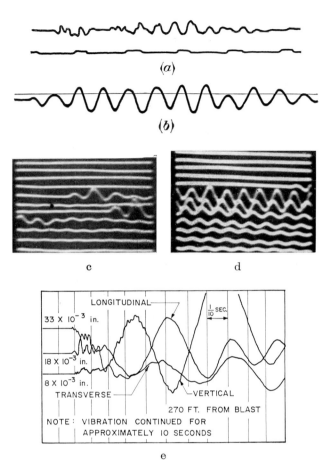

Fig. 9:10. Typical vibrograms recorded by the Cambridge vibrograph (a and b) and by a cathode ray oscillograph and gauge (c and d) where the ray has swept from left to right at a velocity of 20 sweeps a second and (e) a Leet three-component seismograph.

500 c/s, the amplitudes between 5–800 μ (1 μ = 0.001 mm) with the accelerations 0.01–30g.

In simultaneous measurement of the three components of the ground vibrations it has been found that these are generally of the same order of magnitude. In many cases it is quite sufficient to register only the vertical or the longitudinal component, but now and then a check should be made of both of them. According to Northwood the transversal component adds little to the information.

Baule has shown that there is a distinct difference in frequency between vibrations which have been measured in the ground and those measured on

268

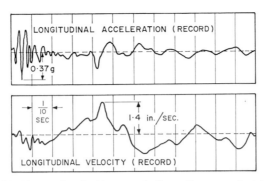

Fig. 9:11. Acceleration and velocity records from the same shot (acc. to Edwards and Northwood).

the ground surface. Blasting below the surface often entails comparatively small quantities of explosives (<50 kg, 100 lb) and measurements of vibrations at close quarters (<100 m, 300 ft), whereas surface operations are often carried out with up to several tons of explosive and measuring stations are hundreds of meters away. In addition a harder rock gives vibrations of a higher frequency than looser rocks.

The records in the figures referred to above show the displacement of the rock face. From the vibrograms the vibration velocity and the acceleration can be calculated by one and two differentiations, but a recording equipment could also be used to give these magnitudes directly. This changes the character of the record from the same observation point at the same blast (fig. 9:11 according to Edwards and Northwood). It is evident from the previous vibrograms, that one has to expect a variety of frequencies in one and the same record, and fig. 9:11 shows how the record is dominated by different frequencies if the displacement, velocity or acceleration is recorded.

9.4. Damage

A wall is deformed in different ways, when the foundations are exposed to vibrations. Decisive factors are, among others, the relation between the natural frequency of the wall, f_0, and the frequency of the imposed vibrations, f. Damage can be caused by elongation, by shearing and bending. The connection between these magnitudes and those of the ground vibrations is first determined with regard to the elastic properties which characterise the wall (fig. 9:12). Parallel with this estimate, however, allowance must be made for the fact that local irregularities and a static state of tension which may exist, can lower the damage limit and thereby finally decide the conditions when there is a borderline case between damage or not.

269

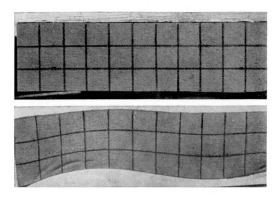

Fig. 9:12. A wave expanding in the ground gives compression, shearing and bending stresses in the wall. The wave-length is $\lambda = c/f$.

The subsequent discussion may be summarised by saying that the damage criteria (S) are all of the type

$$S(\alpha,\beta,\gamma) = \text{constant}\, A f^{\alpha} c^{-\beta} f_0^{-\gamma} \qquad (9:10)$$

in which A is the component in a horizontal or vertical displacement (A_- and A_1 respectively) of the observed ground wave, f the frequency belonging to the amplitude, c a factor that depends on the length along which the deformation is distributed, e.g., the propagation velocity for the ground waves or the propagation velocity of the waves in the structure in question.

When applying the expression to wave propagation in the wall the symbol u is used instead of c. The term f_0 indicates the natural frequency of the wall for forced vibrations of the type which is analysed with the aid of (9:10). When f_0 does not enter as a factor then $\gamma = 0$ and $S(\alpha,\beta,0)$ is indicated as

$$S(\alpha,\beta) = \text{constant}\, A f^{\alpha} c^{-\beta} \qquad (9:10')$$

The object of the investigation is consequently first to determine which values of α and β apply under different conditions, i.e. to demonstrate the connections which apply qualitatively. For practical application it is then necessary to determine by experiment the magnitude of S, corresponding to a definite risk of damage.

In view of the difficulty in determining the propagation velocities c and u in each individual case the same value has been used (2500–3000 m/s), but the errors which arise in doing this are too great to be acceptable, as other factors are now known with considerably greater accuracy. In the

270

absence of wave-velocity determinations we must, however, refrain for the time being from a further experimental discussion of the influence of c. In examining the practical results obtained, a comparison can often only be made between the influence of A and f. If f_0 is not included this means a damage criterion

$$S(\alpha) = \text{constant } A f^\alpha \qquad (9:10'')$$

a. Compression-elongation

For a vertical vibration the frequency of which is slow in relation to the natural vibration of the wall, $f/f_0 < 1$, the wall moves with a constant acceleration along its entire height. For acceleration in a downward direction with the maximum value a the maximum elongation in the wall is

$$\varepsilon = \varrho H a / E \qquad (9:11)$$

in which ϱ is the density of the wall, E its modulus of elasticity, H its height and $a = 2\pi f = 4\pi^2 f^2 A$. Furthermore u is $= \sqrt{E/\varrho}$ and $f_0 = u/(4H)$. Insertion gives

$$\varepsilon = \pi f v (2 f_0 u) = \pi^2 f^2 A / (f_0 u) \qquad (9:12)$$

In this case we have $S(2,1,1)$.

When the frequencies of the incoming vibration are much larger than the natural frequencies of the wall, $f/f_0 > 1$, entirely different conditions apply. The ground then has time to make one or more full vibrations before the compression wave reaches the upper surface of the wall. When the ground moves vertically in the direction dy in the time dt, the former is distributed as an elongation or compression along $u \cdot dt$ of the wall, in which u indicates the wave velocity in the wall. The relative elongation will be

$$\varepsilon = dy/u \, dt = v/u = 2\pi f A / u \qquad (9:13)$$

This corresponds to a damage criterion of the type $S(1,1)$ with $\alpha = 1$, $\beta = 1$ and $\gamma = 0$.

For the relative elongation the damage criterion therefore changes character with the ratio f/f_0. The relation is perhaps best made clear by a representation according to fig. 9:13 in which ε is given as a function of f/f_0. In the region $f/f_0 = 1$ a resonance peak is obtained which, due to the damping conditions in the wall, can give elongation values considerably above the lines indicated in fig. 9:13. These hold good for $v = 50$, 100 and 200 mm/s if $u = 2500$ m/s.

When data are required of the natural frequency of the wall, a diagram for calculating this as a function of the wall height H and the propagation velocity u is made as in fig. 9:14. The horizontal component of the ground

271

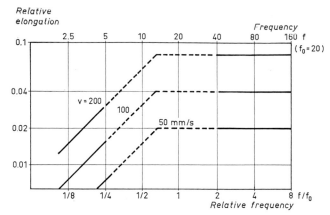

Fig. 9:13. Relative elongation per mille for an oscillating wall as a function of the ratio between frequency f and natural frequency f_0 at vibration velocities 200, 100 and 50 mm/s (8, 4 and 2 in/s).

vibrations also gives the compression-elongation in the wall. For a wave which propagates in the longitudinal direction of the wall at the velocity c, the same discussion applies as for a compression wave which propagates upwards along the wall. A factor $S(1.1)$ is also obtained here.

$$\varepsilon_1 = v/c = 2\pi f A/c \qquad (9:14)$$

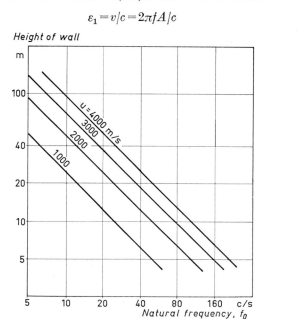

Fig. 9:14. Natural frequency as a function of height of wall and velocity of sound (u) in the wall. The curves are for $u = 1000$, 2000, 3000 and 4000 m/s (3300, 6600, 9800 and 13,100 ft/s).

b. Shearing

A surface wave which propagates at a velocity c along the ground induces vertical vibrations, which, in addition to acceleration effects, cause shearing stresses in minor parts. It is evident from fig. 9:12 that such an effect depends on frequency, amplitude and wave velocity, so that the shearing increases in proportion to A and f and inversely to c. For the shearing angle

$$v = dy/dr = v/c = 2\pi f A_1/c \qquad (9:14')$$

applies with signs according to fig. 9:12, i.e. the same analytical expressions as in preceding cases according to (9:13) and (9:14). It should be pointed out, however, that maximum shearing can be obtained by the slow surface waves while the maximum elongation occurs in another part of the wave system, e.g. the more rapid compression wave. Entirely different values f, A and c are dealt with in (9:14) and (9:14').

c. Bending

Apart from the effects mentioned, the vertical vibrations cause the wall to bend and this gives rise to elongation and compression in the upper part. The maximum value is indicated by

$$\delta = H \, d^2y/dr^2 = Ha/c^2 \qquad (9:15)$$

The horizontal vibrations parallel as well as perpendicular to the wall also give bending vibrations. Without entering into these in detail it can be mentioned that the damage criteria then are of the type $S(2,2)$ as in (9:15).

d. Local effects

The types of deformations discussed here can of course be increased by superposition of stationary local tensions already existing in the wall. This implies that the limit for damage is reduced by a certain amount, but the above-mentioned expression for S can be expected to stand in the sense that, for an actual case, a release of the tension by cracks is obtained if v reaches a given value whether this is attained by high A or f values. The limiting value depends finally on the actual state of the building. Here we must allow for the fact that cracks are normally developed without being influenced by ground vibrations. When a building approaches the limit for such a release even insignificant vibrations can cause cracks.

In taking measurements it is often found that there are considerable variations in the intensity of vibrations in one and the same building even if the measuring is carried out in the foundations. The reason is that the foundations are not in contact with a uniform body of rock, that only part

Fig. 9:15. Cracks released by ground vibrations from blasting. (a) Cracks round a fixing bolt; (b) shear cracks near the foundation of a plastered brick house; (c) cracks close to a bearing wall; (d) vertical cracks due to settlement; (e) chimney damage. (d and e acc. to Edwards and Northwood.)

of the foundations stand on rock, or that holes and fissures in the rock screen off the effect of the vibrations in certain sections. In such cases differences in amplitude or phases can be obtained within a limited section of the building and local deformations may occur which considerably exceed those calculated solely from the wave magnitudes. Such a case has given rise to the pronounced shearing shown in fig. 9:15. Fig. 9:15b shows a case in which existing tensions in a wall have been released by cracks.

A further type due to local effects is illustrated by fig. 9:15c: here the cause of the cracking is to be sought in a locally restricted amplitude or phase difference along a line in the building. Such differences can also arise in a homogeneous foundation when the prerequisites for the wave propagation in two adjoining sectors differ. Another example of this type is the cracking which often occurs close to a chimney, in the joints close to stoves and where different materials meet.

A typical example is the cracking beneath windows and openings in a wall. The shock wave passing vertically throughout the building is reflected at the openings, and adjacent parts of the structure will have different types of movement. For shear waves a concentration of the deformation is obtained in special parts of the building, such as between windows and openings. The damaging effect in these cases should be proportional to the amplitude and the frequency and in inverse proportion to the distance over which the local effect is distributed. The damage may in such cases be related to Af and Af/c and in addition to this to factors depending on the structure.

The first cracks obtained in a building depend on tensions connected with local effects. Such effects occur not only in the structure itself but also in the ground on which it is founded.

e. Vibration energy

It has often been indicated in literature of recent years that the damage may be due to the vibration energy

$$E_{sv} = \tfrac{1}{2} \varrho v^2 \qquad (9:16)$$

This cannot generally be the case. For a vibration with a sufficiently low frequency, the velocity of motion, and consequently the energy, can be high at the same time as the acceleration can be made as small as one chooses. There is then no reason for damage to be caused just as in the stationary case with a constant velocity of translation. The vibration energy cannot be transmitted into any other form of energy.

A possible explanation leading to the same damage criterion $S(1,0)$ is that the effect beyond a certain acceleration is proportional to the acceleration value a and the displacement A.

Another factor that may be tried instead of the energy is the quantity of energy supplied and removed per unit of time. This gives a quantity

$$W = \varrho v a = \text{constant}\, A^2 f^3 \qquad (9:17)$$

of the type $S(3/2,0)$ and of the same character as the magnitude $(1/\pi)va$ suggested by Zeller as a damage factor and for which the unit vibrar (m^2/s^3) is introduced. Proof that W might be related to any type of damage has not, however, been given.

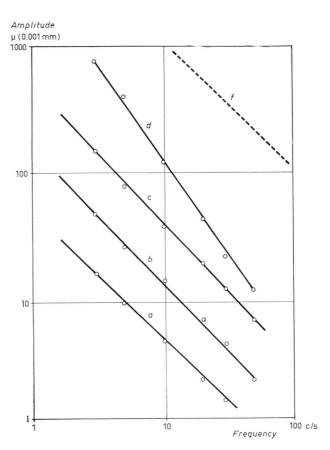

FIG. 9:16. Subjective estimate of ground vibrations (acc. to Reiher and Meister).

f. Subjective estimate

In blasting, consideration should also be paid to how the ground vibrations are subjectively conceived. A great deal of the irritation which arises between the blaster and his environment is due to the fact that a layman is very apt to acquire a wrong conception of the risk of damage. Another aspect is that the blasting is regarded in itself as something disagreeable.

Reiher and Meister have studied human reaction to vibrations. Their results are reproduced in fig. 9:16. The lines indicate when the effect has been considered to be scarcely noticeable (*a*), clearly noticeable (*b*), irritating (*c*) and disagreeable (*d*). This information is of great value in dealing with complaints that the ground vibrations are injurious. The diagram shows that a man normally reacts strongly long before there is reason to apprehend damage to his house. In fact a very large share of common complaints have

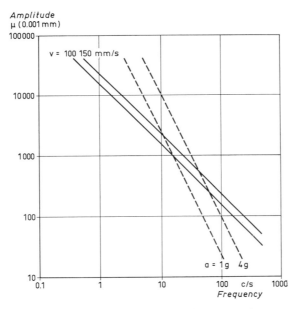

Fɪɢ. 9:17. Connection between amplitude and frequency.

proved to be unwarranted. If people are told that suspicion is aroused long before any damage can possibly be expected this should elucidate matters and reduce the subjective reaction. But even so, there may still be discomfort, which should be avoided as much as possible. The number of blasts can often be reduced, due warning of them given and, above all, blasting at night should be avoided.

In comparison with other causes of discomfort in a modern city, we may well ask whether a blast carried out in a reasonable manner at a reasonable time may not be regarded by a normal human being as a healthy and distinct tone in a blurred cacophony.

g. Discussion

The alternatives which will be discussed are according to (9:12)–(9:17)

$$S(\alpha,\beta) = \text{constant } Af^{\alpha}c^{-\beta} \qquad (9:10')$$

with $\alpha = 1$, $3/2$ and 2 and $\beta = 0$ and $\beta = 1$.

It has previously been shown that at low frequencies ($f < 10$ c/s) there is a risk of damage due to $S(1,1)$ and that the acceleration with $S(2,0)$ or W with $S(3/2,0)$ does not attain a sufficiently high amount for damage until this has already been caused by $S(1,1)$.

The connection between amplitude and frequency for some current values of v and a is indicated in fig. 9:17. It will be seen from the diagram how the

277

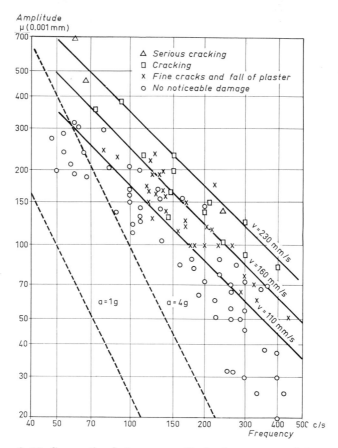

FIG. 9:18. Connection between amplitude, frequency and damage.

relation between the quantity v proportional to $S(1,1)$ and a proportional to $S(2,0)$ changes according to frequency, and demonstrates why $S(1,1)$ dominates at low frequencies. The inverse conclusion that the acceleration $S(2,0)$ is decisive at high frequencies is, however, not correct. This is due to the fact that when $S(2,0)$ has been obtained in estimating the damage, a condition has been low frequencies $f < f_0$. For high f values other conditions, which lead to $S(1,1)$ apply. There remains uncertainty only in the vicinity of $f \simeq f_0$.

Thus we have to consider the damage criteria $S(2,1,1)$, $S(1,1)$ and maybe $S(1,0)$, but not such ones as $S(3/2,0)$. For $S(1,1)$ or $S(1,0)$ this seems to be confirmed by the experience given in fig. 9:18. The full lines give amplitudes at constant vibration velocities $v = 230$, 160 and 110 mm/s (9.2, 6.4 and 4.4 in/s) or relative vibration velocities of $v/c = 46$, 32 and 22 μ/m. The broken lines represent constant acceleration values $a = 1g$ and $4g$. The

TABLE 9:1. Risk of damage in ordinary dwelling houses with varying ground conditions.

TABLE 9:1. Risk of damage in ordinary dwelling houses with varying ground conditions.

	Sand, shingle, clay under ground water level	Moraine, slate, soft limestone	Hard limestone, quartzy sandstone gneeiss, granite, diabase	Type of damage
Wave velocity c m/s	300–1500	2000–3000	4500–6000	
	4–18	35	70	No noticeable cracks
Vibration	6–30	55	110	Insignificant cracking
velocity mm/s				(threshold value)
	8–40	80	160	Cracking
	12–60	115	230	Major cracks

material is obtained in ordinary dwelling houses built on rock where there has probably been no great variation in wave velocity ($c \approx 5000$ m/s). It is evident that the damage factor for these frequencies is not proportional to the acceleration but to the relative vibration velocity. This conclusion is strongly supported by later experience. The risk of damage is characterized in table 9:1.

In the examples shown in fig. 9:15a, b, c the vibration velocity has exceeded 160 mm/s (6.4 in/s).

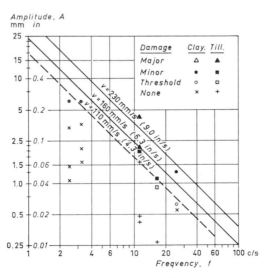

FIG. 9:19. Connection between amplitude, frequency and damage for low frequencies. The experimental points according to Edwards and Northwood.

Investigations by Edwards and Northwood for buildings on sand-clay and till are in agreement with the results of table 9:1 for frequencies down to 2.5 cycles per second (fig. 9:19). In this ground material the propagation velocity (c) is probably about $\frac{2}{3}$ of the velocity in hard rock (where the velocity for the compressional waves is estimated to $c = 5000$ m/s, and for the surface waves to $c = 3000$ m/s). When the factor v/c is decisive for the damage, the vibration velocities given in table 9:1 should be expected to be about $\frac{1}{3}$ lower in sand-clay and till for the damage indicated. We get the value $v = 50$ mm/s (2 in/s) for "no noticeable cracks".

Reports from Germany by Köhler indicate that damage may occur down to 50 mm/s (2 in/s). Even here it is a question of a lower propagation velocity in the ground than in hard bedrock; it is probably about $\frac{2}{3}$ of that value.

The conclusion is that not only the vibration velocity has to be considered but also the propagation velocity of the waves in question. This velocity depends on the ground conditions. The damage criterion is $S(1,1) = v/c$.

Damage factors of the type v/u may also occur. u is the propagation velocity of the waves in the buildings. This factor will give damage for equal v-values in two similar buildings with the same u-value, even if they are founded on rock with different v-value.

In judging the results, it must be kept in mind that one does not always succeed in finding the spot where the vibrations are most pronounced. One may for instance find $v = 35$ mm/s and $v = 18$ mm/s in two records from one round, and in a subsequent round of the same size suddenly receive a value of, say, $v = 100$ mm/s. If results from vibration measurements are used as an argument for increasing the charges in a round, a series of records must be taken from a number of carefully selected spots so as to give an idea of the real magnitude of the severest vibrations in the surroundings and in the foundations of the buildings concerned. When serious practical problems are not entailed we generally recommend keeping the vibration velocity below $v = 70$ mm/s (2.8 in/sec) in rock, and $v = 50$ mm/s (2 in/sec) in ground with a lower propagation velocity. These figures seem to be *definitely below* the actual damage limits.

If it is desired to continue blastings that give vibration velocities above these figures and below $v = 100$ mm/s (4.0 in/sec) attention should be paid to the state of cracks in the buildings concerned and the vibrations from every round should be recorded to make sure that $v = 100$ mm/s (4.0 in/sec) is not exceeded. There is then in ordinary houses a very low, or no risk of cracking.

For higher v-values one has to count on an increasing risk of causing damage. Even so it may well be justified to allow for up to $v = 150$ mm/s (6 in/sec) if one intends to pay the full costs of repairing the damage caused and if there are special reasons for carrying out the blasting in this way. Generally it is advisable when possible to avoid causing any damage. In most cases this can actually be done in keeping with the following sections.

9.5. Reduction of ground vibrations

The ground vibrations which are caused by a single charge acquire a maximum amplitude A only after one or several previous minor deflections. The vibration is ordinarily of relatively short duration, and in most cases only three full vibrations can be expected to have an amplitude greater than $A/2$; all the others can more or less be ignored. This means that at intervals greater than $3T$, ($T = 1/f$ time for a full period) it can be reckoned that there is no collaboration between two different shots.

If the interval is shorter, conditions are more complicated, as there will be co-operation between the different wave systems. How the different factors of time delay, number of holes, and frequency affect the final result has been made clear by Langefors. An interference effect is obtained which implies that, when the delay time τ is as great as the vibration time T, or an integral multiple hereof, co-operation is obtained between the different delays so that the vibration effect adds up. This applies when H is an integer in the ratio

$$\tau = HT \qquad (9:18)$$

On the other hand, when H is an odd number of semi-values the different wave systems extinguish or weaken one another. This takes place almost completely when $H = \frac{1}{2}$; but for higher values of H than $\frac{5}{2}$ the effect, either as regards extinction or amplification, does not play a major role.

An arranged interference can also be obtained in another way. The general condition is that k shall be an integer, but not the ratio between k and n, i.e. $k/n \neq 1, 2, 3, \ldots$

$$n\tau = kT \qquad (9:19)$$

in which n indicates the number of intervals. The implication of this is that the spread of the round in time ($n\tau$) shall be distributed over one or a number of full periods of the vibration concerned. This is an important extension for the practical application. According to the previous simple interference conditions a certain interval is required for every frequency; the interference condition according to (9:19) can be fulfilled within very wide limits.

It has been shown on many occasions in literature that an increase in the size of the round by an increased number of intervals gives the surprising

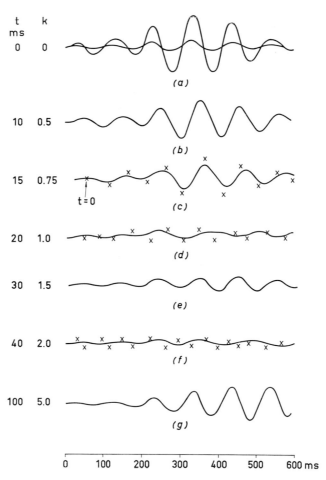

FIG. 9:20. Ground vibrations from five different charges and with delays 0–100 ms. The curves have been calculated with the starting point $\tau = 0$ given by Fish, × experimental values according to Fish.

result of a decrease in the amplitude of the ground vibrations. This is clearly a consequence of the above condition if, by increasing the number of intervals, a k-value is obtained which approaches an integer.

With interval times, which are small in relation to the natural vibration time, the number of intervals is chosen large enough to ensure that n amounts to the time of at least one full period. If it is not possible to adapt the interval times exactly so that k becomes an integer, then it can be calculated that as large a part of the entire round as corresponds to the nearest integral value fulfils the interference condition and that the excess (or missing) part co-operates in practice.

Example 9:1. Ground vibration with $f = 10$ c/s. $T = 100$ ms. $\tau = 30$ ms, $n = 5$. The amplitude from a single hole is $A = 0.020$ mm. This implies that the round in instantaneous ignition can give an amplitude $A = 0.100$ mm. For the short delay round discussed, $n\tau = 150$ ms and $k = 1.5$. The spread of the round in time is 1/3 too long if we count to $k = 1$, and just as much too short if we count to $k = 2$. Two-thirds of the round fulfills the interference condition and are eliminated, the remainder co-operates. An amplitude of approximately 0.030 mm may be expected, i.e. as from 1.5 shots in simultaneous ignition (see fig. 9:20a and e).

With time intervals which are large in relation to the time of natural vibration, we approach the state in which the vibration from each shot has largely died away before the succeeding one manifests its maximum amplitude.

When the interval time is just as large, or almost as large as the vibration time, or double the value, $\tau = T$ or $2T$, it is impossible to avoid the co-operation of the vibrations from different charges. On the other hand, if the rounds in question have a large number of intervals, the shots that lie at a relative time distance of $3T$, or more, can generally be expected to be non-co-operative. For the portion of the round comprising intervals Nos. 0, 3, 6, 9, etc. (or 0, 2, 4, 6, etc. when $\tau = 2T$), the maximum amplitude will be of the same order of magnitude as for a single shot. The resulting vibration is then made up from three (or two) such portions of the round. The whole round thus gives an effect which corresponds to 3 (or 2) co-operating shots. If this reduction of the vibration is not sufficient, all that remains to be done is to leave out one or two intermediate intervals.

These viewpoints can be applied at major stationary blasting places, such as mines and quarries where blasting under similar conditions is carried out for a lengthy period. The character of the vibrations can then be determined, and a suitable interval time and number of intervals calculated.

In practice it is important that the interference condition should also permit unarranged interference to be used. This is obtained if the time spread for a number of charges is large enough to cover at least a full period of the vibrations, even if this spread is distributed quite arbitrarily. The probability that every one of a large number of shots may be in exactly the same phase can then be practically ignored. A ground vibration will be obtained which is reduced to half, or less, according as the spread increases to higher values than the vibration time. In order to calculate the effect of the unarranged interference, the reduction factor γ at different frequency values, the natural vibration period and the values of the scattering $(\pm \Delta \tau)$ are indicated in table 9:2.

The table applies to a number of shots which are detonated at the same interval number with scattering $\pm \Delta \tau$, e.g. defined so that $\frac{5}{6}$ of all

TABLE 9 : 2. *Reduction factor* (γ) *for the total charge* (Q_i) *within one and the same interval in unarranged interference. The vibration effect is assumed not to be lower than the value which is obtained with a single hole.*

f c/s	T ms	$\Delta\tau = \pm 5$	± 10	± 25	± 100	± 200 ms
5	200	1	1	1	1/2	1/3
10	100	1	1	1	1/3	1/6
20	50	1	1	1/2	1/6	1/6
50	20	1	1/2	1/3	1/6	1/12
100	10	1/2	1/3	1/6	1/12	1/25
200	5	1/3	1/6	1/6	1/25	1/50
500	2	1/6	1/6	1/12	1/50	1/100

values of the scattering lie in an interval which is $2\Delta\tau$ ms. If short delay detonators are used with built-in delays the scattering for the two lowest may be ± 5 ms., for the three succeeding numbers ± 10 ms. up to about interval 10, and still greater for higher intervals. For second-interval detonators the scattering is ± 100 ms or more. (See table 6 : 2.)

Example 9 : 2. In a single bench row six charges are initiated, all of them with short delay blasting caps No. 5, $\Delta\tau = \pm 10$ ms. The blasting is carried out in granite and the frequencies which may occur are at least 100 c/s. According to table 9 : 2 the reduction factor γ is then $\leq 1/3$. A vibration which corresponds to two of the charges ignited instantaneously is obtained. On the other hand, if the ground vibrations have a frequency of 20 c/s no essential reduction of the vibration effect compared with instantaneous ignition can be expected.

For different interval numbers it can be assumed that co-operation does not occur when $\tau > 2.5T$. With $\tau = 30$ ms which has often been used, the maximum charge permitted can be used in every single interval after correction with the reduction factor, as long as the frequency is $f = 100$ c/s. This usually applies in blasting in hard bedrock.

In softer kinds of rock or other materials, the frequencies are often between 20 and 50 c/s. For 20 c/s a reduction cannot be counted on between different charges in the same interval, as the distribution will normally be included in a semiperiod. When it is desired to use the maximum charge per hole only one such charge can be used in every interval. In this case the interval time is 30 ms$\approx T/2$ and the vibrations from the consecutive shots may be expected to be almost entirely extinguished.

284

9.6. Planning blasting operations

If ground-vibration problems are likely to arise, an estimate of their effects should be considered in the planning of the work. The risk of damage as a function of the size of the charge and of the distance between the charge and structures has to be determined. Such a relation is given in fig. 9:21, which represents a summary of our experimental experience.

The lines representing the same risk of damage are found to be covered by the relation $Q/R^{3/2} =$ constant down to distances 2–3 times the depth[1] of the charges. Q indicates the charge in one hole in kg, or several simultaneously fired charges at the same distance R in meters. The size of the constant can be used to characterize the charge level. *The diagram is based on the most unfavorable values observed.* If one single case of damage is obtained in a series of, say, 100 blasts at a certain distance, only this observation has been used. Thus the diagram ordinarily gives a much better idea of the risk of damage than some few vibrograms. The diagram covers an enormous amount of material at distances of 2–60 m (6–200 ft) from the blasting site.

It is possible for this reason that the charge levels given here, for example for insignificant cracking, are lower than the corresponding level referred to by other authors. The difference in definition probably cannot be greater than a factor 2. It may be questionable if the lines can be extrapolated to distances of 1000 m (3000 ft) or more, but conclusions of up to some hundreds of meters drawn from extrapolation have yielded satisfactory results.

In principle, the table does not indicate that in blasting, e.g. according to level 0.02, cracking is certain to be avoided, but that in the given conditions it is improbable. Nor can the conclusion be drawn that damage will inevitably be incurred in the region between levels 0.06 and 0.12; the diagram merely indicates that risk of cracking can be expected in this region if the ground conditions are not known, which is generally the case.

A special discussion is required for the region close to the charge. Here a relationship $Q/R^2 =$ constant can be expected to apply for distances less than 2–3 times the depth of the charge. This seems to be in good accordance with the threshold values as well as the major and minor damage points obtained by Edwards and Northwood from experiments with blasting in till and clay. They have got threshold values (level 0.25) for distances above 9 m.

The values for the levels 0.03–1.0 are given in table 9:3 referring to single charges.

Of these charge levels 0.06 is indicated in fig. 9:21 as "threshold in granite" when blasting close to ordinary houses founded on the same rock. For concrete structures the corresponding value is 0.12–0.25.

---

[1] Under free surface.

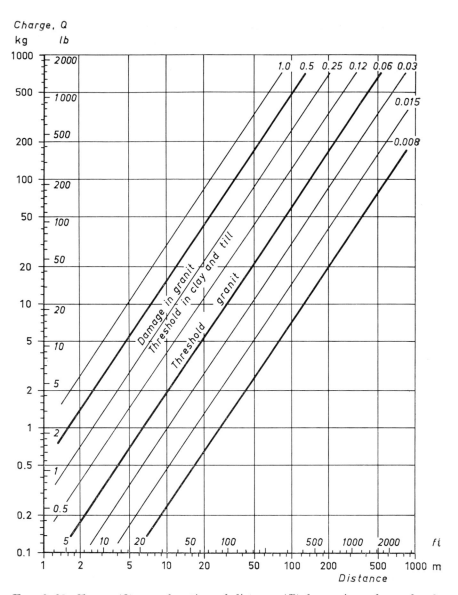

Fig. 9:21. Charge (Q) as a function of distance (R) for various charge levels $Q/R^{3/2} = 1.0 - 0.008$. Relations between charge level and damage are given in table 9:4.

The course of the calculation is as follows: Taking into consideration the nature of the buildings likely to be affected, the risk level is first chosen so that the probability of damage is estimated to be practically negligible. By making use of the possibilities which we have already enumerated, the

286

TABLE 9:3. *Amount of charge for different charge levels* $Q/R^{3/2}$.

Distance R		Q kg					
m	ft	level 0.03	0.06	0.12	0.25	0.50	1.0
0.4					0.07	0.14	0.28
1.0	3	0.03	0.06	0.1	0.24	0.5	1.0
1.4	5	0.05	0.10	0.2	0.4	0.8	1.6
2.0	7	0.09	0.2	0.4	0.7	1.4	2.8
2.8	9	0.15	0.3	0.6	1.2	2.4	4.8
4.0	13	0.25	0.5	1.0	2.0	4.0	8.0
6.0	20	0.5	0.9	1.8	3.7	7.5	15
10.0	30	1.0	2	4	8	16	32
14	45	1.7	3.4	7	14	28	56
20	70	2.8	5.6	11	22	44	88
28	90	5	10	20	40	80	160
40	130	8.5	17	35	70	140	280
60	200	15	30	60	120	240	480
100	300	32	64	130	260	520	
140	450	50	100	210	420	840	
200	700	90	180	360	720	1440	
280	900	150	300	600	1200	2400	
400	1,300	250	500	1,000	2000	4000	
600	2,000	500	900	1,800	3700		
1000	3,000	1000	2000	4,000	8000		
1400	4,500	1700	3400	7,000			
2000	7,000	2880	5600	11,000			
2800	9,000	5000					
4000	13,000	8500					

To get the values in lbs multiply by 2 and add 10 %.

problem can often be solved without reducing the size of the individual charges in the holes, or the total quantity of charge used in the round. In other cases the planning must be changed so that the charges and the size of the round are reduced.

If this is not economical, blasting at a higher level may be accepted, but only if the cost of compensation for damage that may occur is taken into account.

If the charge allowed (fig. 9:21) is indicated by Q, the charge quantity in the same interval by Q_i, the reduced charge by Q_{red}, then acc. to table 9:2

$$Q_{red} = \gamma Q_i \qquad (9:20)$$

287

TABLE 9:4. *Relation between the charge level $Q/R^{3/2}$ and the vibration velocity v in hard rock at smaller distances (in softer material a certain v/c-value requires 2–4 times larger charges).*

$QR^{-3/2}$ kg m$^{-3/2}$	v mm/s	v/c μ/m	Description of damage in normal houses on granit
0.008	30	6	Fall of plaster. No cracking
0.015	50	10	No evidence of cracking
0.03	70	14	No noticeable cracks
0.06	100	20	Insignificant cracking (threshold value)
0.12	150	30	Cracking
0.25	225	45	Major cracks
0.50	300	60	(Fall of stones in galleries and tunnels)
1.00	—		(Cracks in rock)

will apply, the first condition being that this charge quantity will be smaller than Q.

$$Q_{\text{red}} \leqslant Q \qquad (9\!:\!21)$$

This condition is also sufficient if the interval time (the mean value) exceeds $3T$.

At slow natural vibrations in the ground when $\tau < 3T$, co-operation may take place between the different groups, each one of which is represented by Q_{red}. This co-operation does not occur between the interval numbers which have a time distance of $3T$, or more. If $\tau = T$, for example, then intervals 1, 2 and 3 will co-operate, but not 1 and 4 or higher numbers, nor will 2 and 5, etc. Here the condition will be

$$3Q_{\text{red}} \leqslant Q$$

If the interval times are $\tau < T$, then generally $\Delta\tau < T/2$ and the reduction factor is $\gamma = 1$. The total charge in an interval, Q_i, co-operates, but a reduction can be effected between the different intervals by a regular interference in keeping with the condition (9:19) so that the effect does not exceed that from an individual interval.

The relation between the charge levels and vibration velocities are not conclusively determined but are about the size given in table 9:4.

9.7. Practical directions

To make clear how the calculations are made in practice some types of blasts which have been carried out in recent years will be discussed. These examples have been chosen as typical cases. Unless otherwise indicated, a

Interval No	0, 1, 2, 3, 4, 5 ... 15
Total charge quantity, kg	2, 4, 4, 4, 6, 6 ... 6

reduction factor γ according to table 9:2 is used, with $f = 100$ c/s, and with level 0.06 in fig. 9:21 indicating the largest co-operating charge (Q) allowed, this being given for different distances in table 9:3.

Example 9:3. Blasting of 60,000m³ of a foundation close to flats. Frequency of the ground vibrations $f \geqslant 100$ c/s. The foundation is to be blasted to a depth of 20 m. The bench height can either be 5 m with 4 shelves or 6–7 m with 3 shelves. If the first alternative is chosen the depth of hole will be about 5.6 m, and the drill hole diameter at the bottom 28 mm. This makes a maximum burden of 1.3 m at the bottom. Allowing for normal inaccuracies in drilling (5 cm/m) the burden must be reduced to $V = 1.0$ m. The corresponding distance of hole is $E = 1.25$ m. A drill hole will then suffice for the blasting of an area $VE = 1.25$ m. This applies even if the burden and spacing are changed, so that the area per hole remains constant, for instance $V = E = \sqrt{1.25}$ m. This makes 6.25 m³ per hole. The desired fragmentation determines the final amount of charge in kg/m³ (the specific charge) to be $q = 0.30$–0.35 kg/m³. The total size of charge per hole will be $Q_t = 2$ kg. According to table 9:3 this charge can be used for distances $\geqslant 10$ m from neighboring houses. Down to this distance the placing of holes need not be changed with regard to the ground vibrations, provided that the charges do not co-operate. In the present case the interval time is $T = 30$ ms and the natural vibration time of the ground wave $T \simeq 10$ ms, i.e. the condition $\tau \geqslant 2.5\ T$, has been fulfilled. The charges in different intervals do not then co-operate and 2 kg can be blasted in every interval. The round can amount to 32 kg with intervals Nos. 0–15. The blasting can be done according to Alt. I in fig. 9:22 b.

A reduction effect can also be counted on within the same interval with short delay caps, as will be seen from table 9:2. For interval No. 0 $\Delta\tau = 0$ applies, for Nos. 1, 2 and 3 $\Delta\tau = \pm 5$ ms and for higher numbers $\Delta\tau = \pm 10$ ms or greater. The reduction factor then is $\gamma = 1$, $\frac{1}{2}$ or $\frac{1}{3}$ respectively, i.e. the size of the charge in every interval will be as indicated in table 9:5.

The total charge of the round can then amount to 84 kg. This is an enormous increase in the charge quantity, but with the ignition pattern mentioned the only risk incurred is that two charges within the same interval may just happen to co-operate by chance in a vibration effect corresponding to 4 kg. Utilizing the reduction factor γ implies that a relatively small probability is counted on for co-operation taking place, not that it is excluded. If we insist that the charge quantity according to level 0.06 (table 9:3) must in no circumstances be exceeded, the total charge quantity is limited to 32 kg with only one charge in every interval.

In fig. 9:22 a vibrogram is shown of a blast according to Alt. with 60 kg which has been correctly distributed for interval Nos. 1, 2 and 3, and also 7, 8 and 9, having regard to the reduction factor $\gamma = \frac{1}{2}$ or $\frac{1}{3}$. For the other

Fig. 9:22a. A foundation for a new building excavated in bedrock to a depth of 20 m (67 ft) below street level and close up to old buildings in the city of Stockholm, Sweden.

interval Nos. 4, 5 and 6 the total charge quantity/interval has been 4.2, 6.2 and 6.2 and the reduced charge quantities 2.7, 4 and 4 kg, i.e. double the desired charge quantity. The vibrogram shows also that the greatest vibration effect was obtained in these intervals 4, 5 and 6. This could easily have been avoided if the number of intervals had been increased. The vibrogram also clearly shows that a further increase in the charge quantity to 84 kg by utilizing, according to table 9:5, all intervals 0–15, would not in any way augment the vibration effect.

Some further facts may be pointed out in conjunction with the vibrogram. The two holes with No. 1 are more tightly wedged than the two which follow as No. 2. This is directly shown on the vibrogram by stronger

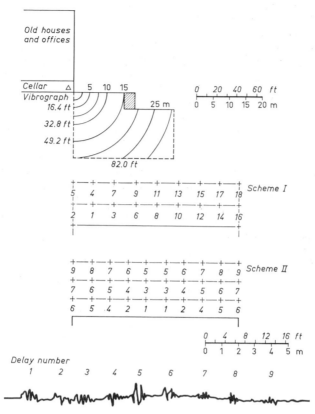

FIG. 9:22b. Blasting of foundation close to dwelling houses. Upper picture is a profile; the two lower diagrams are two alternatives for the drilling and ingition patterns for rounds. Size of rounds c. 150 m³ (5,300 cu.ft). Foundation 60,000 m³ (2,100,000 cu.ft). A vibrogram from blasting acc. to Scheme II.

deflections for No. 1 than for No. 2. If we compare Nos. 1 and 4 it will be seen that No. 4 has given a greater velocity than No. 1, but the difference is insignificant. It seems as if Q_{red} for No. 4 has been 35 per cent larger than for No. 1 (2.7 kg compared with 2 kg) has been compensated by freer breakage conditions for No. 4. Another fact which is worth pointing out is that the vibration effect diminishes the higher the interval numbers; compare, for instance, Nos. 4 and 7. The reason for this, of course, is the increasing scattering in ignition time. As will be seen from Table 9:2 this diminishes the reduction factor.

Example 9:4. Blasting of a canal below a road bridge (fig. 9:23). The bridge was not built when the work was being planned, but it was considered important for various reasons to complete the bridge before the blasting was

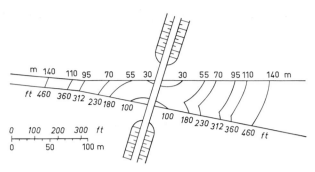

FIG. 9:23. Blasting of a canal beneath a bridge. Bench height 11 m (36 ft). Figures give distance in m (ft) from nearest bridge foundation.

started. The bench height enabled large diameter holes to be drilled. Those of immediate interest had diameters = 100, 88 and 51 mm. Fully loaded these holes required 108, 70 and 25 kg of explosive per hole.

Taking these values as a starting point, a plan can be made showing the connection between the distances from the bridge abutments and hole diameters. Taking values according to level 0.06 in fig. 9:21 as the limitation for the charge in each individual hole, 140 m, 110 m and 55 m are then obtained as the minimum distances with fully charged holes. This means that blasting at distances greater than 140 m can be done by taking into consideration the ground vibrations only as far as a suitable division of intervals is concerned. For shorter distances, down to 95 m, it is also possible to maintain the diameter of holes at 100 mm if the charge in a hole is divided into two parts of approximately 60 kg each. The two part charges are then ignited at different intervals. If so, it assumes that to separate the charges dry sand is inserted between them to a length of 2 m.

Dividing a charge into more than two units in one 100 mm hole is not advisable in view of the relationship between the burden (about 4 m) and the length of the charges.

The diameter of the holes must be reduced for distances under 95 m. Even with 80 mm holes it is necessary to divide up charges into 2 or 3 units, which can be ignited independently of one another. This division of the charges can be used at distances down to 55 m. At lesser distances diameter of holes must be still further reduced. Table 9:6 gives a compilation of the possibilities which are offered. The rock affected by the restriction due to ground vibrations will be about 100,000 m³.

As has been mentioned, however, the portion of this rock which lies between a distance of 140–95 m can be blasted without changing the diameter of holes, if the charge in the hole is divided in two units. The increase in cost for this procedure will in all probability be insignificant. At distances of less than 95 m the measures for reducing ground vibrations must appreciably increase the blasting costs. With 30,000–40,000 m³ of solid rock involved the extra cost of the blasting, even with a moderate increase in percentage cost, may be considerable. When the extra cost has been calculated it can be decided whether the blasting shall be carried out as originally intended or whether the following alternatives should be adopted:

I. That the blasting be carried out in accordance with a level above 0 06 in fig. 9:21, so that all distance values are halved for the same charge quantity.

This will reduce the extra cost by more than half, but the risk of damage to the bridge will of course be increased. If this risk implies cracks which can be tolerated, or repaired at a moderate cost, this course may be chosen. The final decision depends on the extent to which the bridge must not be exposed to risks. The great difference which is inherent between a house and a solid concrete structure should be emphasized here, and it should also be pointed out that fig. 9:21 refers to houses.

II. Unless the advantages of building the bridge before the blasting has been carried out are deemed to outweigh the extra cost of the precautions needed, its erection should be postponed until the blasting has progressed 140 m beyond the site. The latter alternative was adopted.

TABLE 9:6. *Distribution of charges according to data in ex. 9:4.*

Distance from bridge abutment m	Diameter of holes mm	kg/hole	Inter- vals/hole	kg/interval
140	100	108	1	108
140–110	88	70	1	70
110–95	100	108	2	60
95–70	88	70	2	35
70–55	88	70	3	25
70–55	51	25	1	25
55–30	51	25	3	10

Example 9:5. Blasting of a tunnel close to a power station. The section area is 18 m². At the spot where the tunnel is nearest to the power station the crown lies 6 m below the lowest part of the concrete foundations of the power station. The length of the round is 4.8 m and the blasting is directed towards a central uncharged hole having a diameter of 110 mm.

Fig. 9:24 shows the distances $R = 8$, 10, 13 and 16 m from the structure. The corresponding maximum charges are 1.5, 2, 3 and 4 kg. The distance lines are drawn with regard to the difference in level that is 6 m. The total distance will therefore be $R = \sqrt{d^2 + 6^2}$, d being the horizontal distance. The connection between advance and charge per hole is determined by the depth of hole being approximately equal to the advance and the charge in the drill hole being $1 = 1.0$ kg/m. The last half meter of the stope holes is left uncharged. Hence it follows that the advance will be $l = Q + 0.5$ m on the assumption that, as in previous examples, an ignition pattern is used which makes it unnecessary to count on the co-operation of two or more holes.

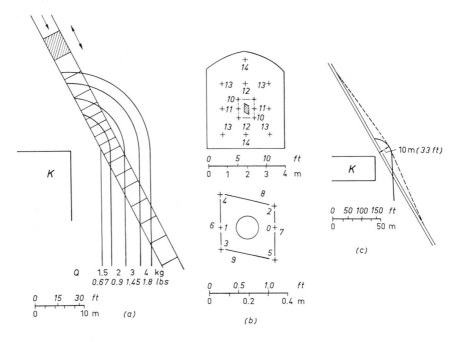

Fig. 9:24. Blasting of 18 m² (195 sq.ft) tunnel close to a power station K. (a) Intended direction of the tunnel close to K. The length of the round 4.8 m (15.5 ft) has to be successively reduced to 2 m (6.6 ft). (b) Drilling diagram and ignition pattern for the central part of the round. Parallel arrangement of holes round one large hole with a diameter of 110 mm (4.3 in). The figures indicate the sequence of short delay detonators. (c) If the distance between K and the tunnel is increased by 10 m (32.8 ft), the tunnel is extended by 1 m (3.28 ft). All rounds can then be made with full advance.

This condition is fulfilled by the ignition pattern shown at the top of fig. 9:24b. For the central part, with the ten first intervals, the drilling and ignition pattern has been drawn on a larger scale. With the given conditions the length of the rounds must be reduced in stages from 4.8 to 2.0 m, and then again increased to 4.8 m. In this manner a stretch of 43 m of the tunnel length will be blasted with a reduced advance, the number of rounds being increased from 9 to 14.

As an alternative the possibility of increasing the distance between the tunnel and the power station may be considered. If the tunnel is moved 10 m from the nearest point of the power station the length will be increased as shown in fig. 9:24c. The distance from the power station to two points where the location of the tunnel was determined for other reasons was about 100 m in either direction. This will only add 1 m to the total length of the tunnel. It is clear, therefore, that the extra cost of the cautious blasting, with 14 rounds instead of 9, will be considerably higher than for this insignificant addition to the length.

9.8. Problems

1. A vibrograph with the unsuspended mass 4.1 kg and the spring-suspended mass 2.4 kg is intended to record the vertical component of the vibration. a) For what accelerations will it loose contact with the ground? b) Calculate the relation between the suspended mass and the unsuspended one if vibrations with up to $a = 8$ g are to be accurately recorded.

2. Calculate the amplitude and frequency in that part of the vibrogram in fig. 9:7 where half a period of a vibration (approximated as sinusoidal) is indicated. a) What are the vibration velocity and acceleration in this part? b) Are there in other parts of the record higher velocities with the same type of evaluation? If so, how high?

3. A house with a height of $H = 7.5$ m (25 ft) has an estimated wave velocity of $u = 2.250$ m/s (8500 ft/s) in the walls which corresponds to a proper frequency of $f_0 = 75$ c/s. Suppose that a crack due to pure elongation occurs for a relative elongation of 0.04%. Calculate according to fig. 9:13 the vibration velocity and acceleration corresponding to this elongation for a) high frequencies ($f \geqslant 2 f_0$), b) low frequencies ($f \leqslant \frac{1}{4} f_0$).

4. When blasting with single holes in the vicinity of a house it is desirable not to exceed the charge level $Q/R^{3/2} = 0.06$ in fig. 9:21. Calculate the maximum charge for $R = 10$, 15 and 20 m (33, 50 and 67 ft).

5. A row of holes with 10 charges of 3 kg (6.6 lb) each is ignited with MS-detonators of one and the same delay, with a scattering in delay time $\Delta t = \pm 5$ ms. Determine the reduction factor (table 9:2) for a frequency of $f = 100$ c/s. How big is the corresponding reduced charge?

6. Calculate the closest distance for blasting the row in problem 5 if the charge level 0.12 in fig. 9:21 is not to be exceeded.

7. At this distance the number of charges can be increased if they are ignited by detonators with the time-spread $\Delta \tau = \pm 25$ ms. How much?

8. The frequency of the ground vibrations that is to be reduced is $f = 10$ c/s for a blasting job carried out at a distance of $R = 40$ m from a building. The MS-detonators used have a time-spread of less than $\Delta \tau = \pm 25$ ms, thus $\gamma = 1$ and all the charges in one interval cooperate. It is found that a "charge level" (fig. 9:21) of 0.25 can be accepted with the existing ground conditions. Decide a) the largest total charge in an interval, b) in every separate interval if 2 delays $\tau = 25$ are used.

9. Discuss the situation when 8 delays with intervals of 25 ms are used. Take a vibrogram from a single charge and construct by superposing the resultant vibration.

10. Calculate for one of the short delay detonator series of table 6:2 the biggest bench round that can be blasted at a distance of 200 m (670 ft) from ordinary houses if the vibrations from the round should not exceed the level 0.06 in fig. 9:21 and the frequency in the ground vibrations is $f = 20$ c/s.

10. SMOOTH BLASTING AND PRESPLITTING

As a monument to the work done the final contours of the walls in open cuts, of walls and ceilings in underground work remain long after the blasted portion has been removed and the technical and economic problems have been forgotten. In the case of water power tunnels, the effect of badly blasted contours will be felt continuously in greater resistance in the flow of water which reduces the energy output; in mining, in store-rooms and air-raid shelters through increased maintenance costs and decreased strength.

Apart from mines, most underground projects constructed in rock are of a permanent character. A number of them include important buildings such as work-shops, power stations, staff premises, oil stores, etc. Inconvenience and expense are generally incurred in such undertakings if blocks of stone have to be removed after completion of the work and when dressing of the roof is needed and further safety measures have to be taken. There are also other, chiefly constructional reasons, for taking every care in blasting sections for rock premises. When supporting concrete vaults are used, they are carried up so close to the roof that it cannot be inspected once the interior of the building has been completed. In this form of construction it is highly desirable to blast the roof of the rock as close to the theoretical section as possible so that the height of fall will be as small as possible should any blocks become loose. The demands for accurate blasting apply especially where cavities caused by surplus rock have to be filled up with concrete at very high costs, and to all types of blasting carried out in rock of low strength, where it is essential for an uninterrupted rhythm of the job to get a contour with the best curvature and to avoid or delay slides in the rock.

In conventional blasting the explosive has in fact been used in such a way as to destroy the quality of the remaining rock. The result is rough uneven contours, systems of cracks created by the explosive penetrating far into the rock and a great overbreak. This has been accepted, partly as an inevitable consequence of using modern high explosives, partly in the belief that the final result was due to the rock and not to the method of blasting.

With the introduction of methods for accurate perimeter blasting as has

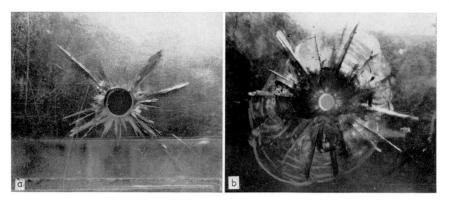

FIG. 10:1. More cracks are obtained with the same charge when it entirely fills the hole. In (a) the volume of the hole was four times larger than in (b), but the charges were the same.

been done in Sweden in the early 50's in work by Hagthorpe, Dahlborg, Kihlström, Lundborg and Langefors and with the development in the U.S. by Holmes it has proved possible to make the final contour appear almost as if it was cut out of the rock with a knife and to leave the remaining rock practically undamaged. These methods have been called "smooth blasting", "perimeter blasting", "presplitting" or simply "smoothing".

10.1. Formation of cracks

Some headlines concerning the formation of cracks are already given in the introductory chapter. Some further details of importance in this connection shall be given and illustrated in model-scale blasting. Fig. 10:1 shows how a drill hole filled with explosive gives heavy cracking when detonated and how the cracks can be almost suppressed with the same charge if the pressure on the walls of the hole is reduced by increasing the diameter. Then only some few cracks of almost the same length remain. The problem is then to guide these cracks in the desired direction.

The dependence of the cracks on the size of the charge is shown in fig. 10:2. a) and b) have the same burden but a) has double as large a diameter of the hole and the charges and the cracks are also about twice as large as in b). It is therefore important to avoid heavy overcharging.

When an elastic material with an empty circular hole is in a state of tensional stress, for instance from a detonating charge in an adjacent hole, it can be shown by calculations that there is a three-fold increase of the stress at the two points of the empty hole that are closest to and farthest

Fig. 10:2. For minimum stress on remaining rock it is essential to avoid over-charging. In this scale model the same burden has been blasted with varying sizes of charge. In (a) a charge has been used four times larger than the correct one in (b).

from the loaded one. This gives an effect causing cracks at this empty hole if it is close enough to the charged one. These cracks tend to connect the holes as shown in fig. 10:3. The effect can be used to direct the cracks in a precise direction quite independent of the influence of a free face in front of the row. An example of such cracking in rock is given in fig. 10:4. Fig. 10:5 shows the influence of the distance to the charged hole. The hole closest to this one receives the dominating effect.

Fig. 10:3. At a low concentration of charge in the holes and a spacing close enough, a line of cracks is created through the holes almost without cracks in the other directions.

a

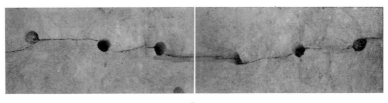

b

FIG. 10:4. Creating cracks from hole to hole in hard rock in the same way as shown in fig. 10:3. Spacing $E = 0.2$ m (0.7 ft), diameter $d = 30$ mm (1.25 in). (Widmark o. Platzer AB.)

Another way of guiding the cracks is to make a primary indication in the hole as in fig. 10:6 The radial cracks from the charge created immediately at the moment of detonation compete, however, and have to be suppressed if the desired result is to be obtained. Fig. 10:7 gives an excellent example of this effect.

Primary cracks can also be created with a special form and ignition of the charge, but the influence of such shaped charges may be entirely changed in an unfavorable direction by irregularities in the rock. Hence it is found that in ordinary application in rock the charge should be symmetric in relation to its axis, which simply means cylindrical.

299

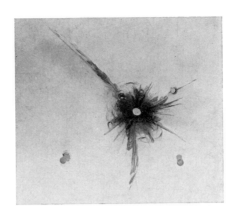

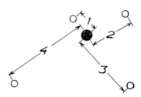

FIG. 10:5. Effect of guide holes at different distances (1, 2, 3 and 4 cm) from the charged hole.

FIG. 10:6. Effect of notches for directing cracks in a hole filled with explosive.

Fig. 10:7. Effect of notches for directing the cracks in a hole with the same charge as in fig. 10:6 and a diameter twice as large.

10.2. Reduction of cracking

To ensure a good rock face the cracks outside the blasting contour, and especially those which orginate from the contour holes, must as far as possible be kept within reasonable bounds. Several factors must be considered to achieve this.

In the first place, the section of rock which is to be blasted should not be too tightly wedged, for in such a case an unnecessarily large quantity of explosive will be required, resulting in strong vibrations and considerable cracks in the surrounding rock. The placing of holes and the detonating sequence must be chosen also for the main round so that the rock is blasted away successively with breakage towards free surfaces (fig. 10:8). When splitting the contour before the blasting of the rest of the round, the remaining rock is to some degree protected from the cracking round the heavily loaded holes. As shown above it is important to use the correct charge with a low concentration compared to the volume of the hole. One tenth of the

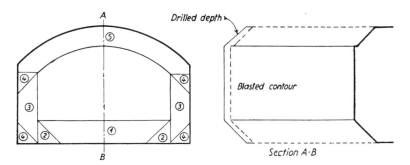

Fig. 10:8. Appropriate detonating sequence for blasting roofs and walls in tunnels with minimum damage to the remaining rock.

concentration in a fully loaded hole ($P = 1.5$) can be generally recommended. This is quite sufficient even in hard rock to create a contour crack by tension straight from one center of a hole to the adjacent ones when fired simultaneously and when the spacing is 10–20 times the diameter of the holes. But the pressure at the walls of the hole is below the compressive strength of the rock and not enough for the creation of tension cracks around the hole, thus leaving the remaining rock in an almost undamaged condition.

10.3. Burden and spacing

How should the burden and the spacing of the holes be adjusted to give the best results for smooth blasting?

Scale-model blasting, as well as practical blasting experience, show that the quality of the remaining rock face is largely dependent on the relation between the distance of the holes E and the burden V. If a good result is desired E/V should be $\leqslant 0.8$, i.e. the burden should not be too small compared with the spacing.

The upper picture in fig. 10:9 shows a model scale experiment, in which $E/V = 2$. This high value for E/V has resulted in a very uneven final contour and, in addition, has left a wall which is seriously damaged by large cracks. A wall face with such cracks would obviously not be very strong.

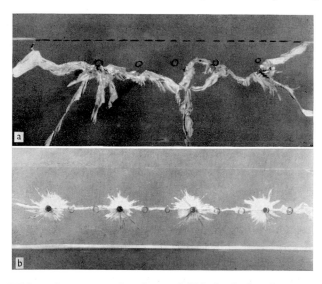

FIG. 10:9. With a given mass of rock per drill hole the burden must not be too small. In (a) the burden was $V = 1$ and the distance of the holes $E = 2$. In (b) the corresponding values were $V = 2$, $E = 1$. In (b) can be seen the influence of guide holes in blasting homogeneous material.

In the lower picture in fig. 10:9, $E/V = 0.5$, the quantity of "rock" per drilled meter is the same as in fig. 10:9a, but as only every third hole was loaded the size of the charge per unit of volume is only one-third. In fig. 10:9b the easiest course for the cracks has been through the holes, the uncharged holes also favoring cracking along the intended final contour which proved to be very even. If all the holes had been charged, i.e. if the charge per cu.m had been the same as in fig. 10:9a, there would have been a larger number of crack roses but a considerably better final contour than in fig. 10:9a.

In full-scale rock blasting the effect of uncharged guide holes is greatest if the guide holes are placed from 0.1–0.2 m from the charged holes when their diameter is $d = 30$ mm (1.2 in). The distance between the charged holes can then be 0.6 to 0.8 m. Uncharged guide holes are not required except in special cases, e.g. when it is desired to obtain contours with a small radius of curvature (connection between wall face and roof, or something similar).

10.4. Ignition

There should be as little time-spread as possible in the ignition of a row of holes for smoothing. In those cases where the row can be blasted separately after the rest of the round, so-called *trimming*, the ignition can be made with instantaneous caps or with detonating fuse. This gives the best result. When using instantaneous ignition the tearing in cracks around the hole is reduced if the charge is ignited from the top of the hole.

In most cases the contour holes are part of a large round and have to be detonated by electric caps of a high interval number, short delays or half second delays. If it can be arranged, short delays are definitely to be preferred. Equal or even better results are obtained in those cases where a detonating fuse can be used for the contour without risk of being torn off by previous charges in the same round. This is often possible in open cuts, but cannot be recommended in tunneling. The new LEDC ignition system may solve this problem.

The influence of the time-spread is shown in fig. 10:10 where the upper row is detonated shot by shot, the middle one by instantaneous blasting, the lower one with short delays. The contour created in these cases is best with instantaneous and short delay blasting. Delay times of more than 0.1 secs (100 ms) mean that every charge has to work separately as in blasting shot by shot. This effect is to be counted on with most half-second delay detonators where the time-spread in one and the same delay is greater than ± 0.1 secs.

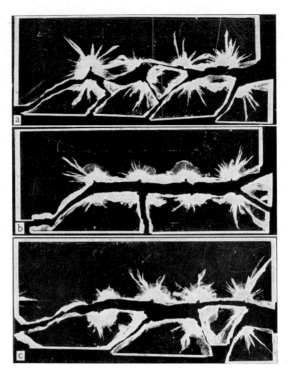

FIG. 10:10. Blasting of a row of holes (experiment in plexiglass): (a) shot by shot; (b) instantaneous blasting; (c) short-delay blasting.

It is also important from another point of view to keep the spread in ignition time below 100 ms. In fissured rock there is a tendency for charges to be ejected by the detonation gases in neighboring holes.

Fig. 10:11 shows a close-up of a rock face which has been blasted with gurit pipe charges for smooth blasting, using short delay detonators of the same interval number (No. 1). The drill holes remain unbroken up to where the bottom charge has been placed, marking the boundary line between two consecutive rounds.

10.5. Presplitting

Another application of the general principles given in section 10.1 has been found by D. K. Holmes at the Lewiston Power Plant, later applied in the Niagara Power Project and reported by S. Paine, D. K. Holmes and H. E. Clark. This is a genuine and important contribution to the technique of smooth blasting that offers new possibilities for applying instantaneous ignition and to reduce overbreak and ground vibrations.

304

FIG. 10:11. Close-up of rock face blasted with gurit pipe charges for smooth blasting. Instantaneous ignition or short delay detonators of the same interval number give the best results. Note that even a small bottom charge destroys the drill hole contour.

In presplitting, cracks for the final contour are created by blasting prior to the drilling of the rest of the holes for the blast pattern. Instantaneous electric ignition or ignition with detonating fuse with no or a minimum of time scatter can then be used. Once the crack is made it screens off the surroundings to some extent from ground vibrations in the main round. The charging of the holes should be the same as in ordinary smoothing (see following section).

Presplitting works in loose as well as in hard rock, but in hard granite the crack created for the contour is rather tight. No cracks are formed at the bottom of the holes perpendicular to them. In hard rock this means that the holes for the breakage of the rock must be placed rather close to the presplit row or that the preslit holes have to be charged once more with bottom charges and ignited in the main round. This is a drawback for the method compared with ordinary smoothing and so is the fact that in presplitting more holes are required.

An example of drilling pattern and final result is shown in fig. 10:12 and 10:13. Note the excellent drilling precision, equally important in all types of precision contour blasting. *The result cannot be better than the drilling, which is often forgotten.* The pictures from the Niagara Power Project are extraordinary also from this point of view.

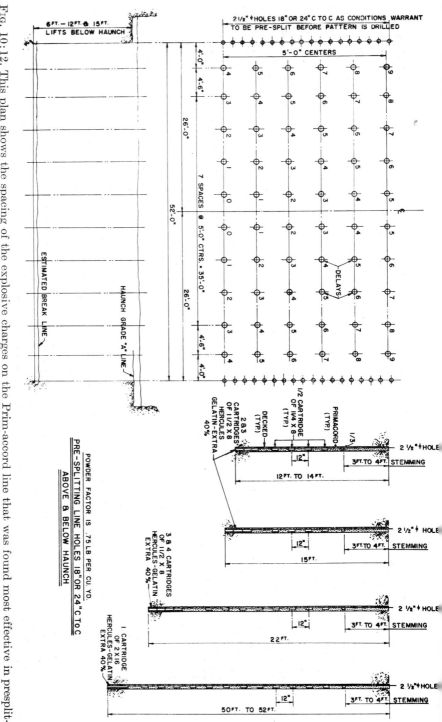

FIG. 10:12. This plan shows the spacing of the explosive charges on the Prim-accord line that was found most effective in presplitting line holes both above and below the haunch (Courtesy Hercules Powder Company).

Fig. 10:13. This nicely cut conduit wall from the Niagara Project is the result of the presplitting technique described by Paine, Holmes and Clark. There will be a minimum amount of scaling, and less concrete (Courtesy Hercules Powder Company).

10.6. Practical directions

The results from investigations and practical results in Europe and USA are condensed in table 10:1 for diameters (d) between 30 mm ($1\frac{1}{4}$ in) and 200 mm (8 in) for ordinary smooth blasting and presplitting.

The concentration (l) of the charge in kg dynamite per m is given together with the definition of the suitable charge for the diameter and spacing (E) in question. The spacing is given as the greatest value for a perfect result, as it is ordinarily desirable to use a minimum number of holes. If the number of holes is increased and the spacing reduced the burden (V) can be increased in such a proportion that the surface VE is left unchanged. If the spacing is reduced to half the value the burden can be doubled.

When the E-values are reduced in proportion 4:5 and the V-values increased in proportion 5:4 the burden is twice as big as the spacing (EV-value unchanged). The cracks from one hole to another can then in the case of simultaneous ignition be expected to be already formed when the reflected shock waves return from the free face. A further increase of the burden cannot reduce this cracking and it does not matter whether there is a free face as in ordinary smooth blasting or not, as in presplitting. This is the reason for the upper E-values in table 10:1 for presplitting. The lower values have been mostly used, and give a somewhat better result.

The gurit charges specially designed for smoothing (fig. 10:14) are extensively used for small diameter holes with excellent results. The bottom of the holes should then preferably have a charge of 0.1–0.2 kg (0.2–0.4 lbs). Bottom charges greater than 0.2 kg (0.4 lbs) in small holes increase the fragmentation and the risk of blowing through cracks into adjacent holes.

In order to reduce overbreak the charges should be loaded up to the top of the hole, or as far as can be permitted while having regard to the scattering and throw of small pieces of rock from the free face. Excellent results especially for the reduction of backbreak are reported with charges reaching up to the top and with covering material to prevent the scattering of stones. This is of importance when blasting very close to buildings. Further, if the empty space in the holes around the charge is filled with dry sand the flow of gas into cracks in the surrounding rock is prevented and the tearing in these cracks is reduced.

For drill hole diameters of 62 mm ($2\frac{1}{2}$ in) and larger ones, similar charges have been developed in Sweden with diameters from 22 mm up to 50 mm, the latter for diameters of the drill holes of 150 mm (6 in). They are charged with so-called Nabit, a powder ammonium explosive, or with LFB, a semi-gelatine NGL-explosive (35 % ngl). The bottom charge in these holes can be up to 0.5 kg (1 lb) for a 62 mm hole and up to 2 kg (4 lbs) for a 150 mm

308

Fig. 10:14. Loading with the NAB gurit charges for presplitting in holes of a diameter $d = 50$ mm (2 in).

hole. If the special charges are not available the concentration of charge recommended in table 10:1 can be arranged by taping explosive cartridges on a detonating fuse as in fig. 10:15. This way of loading has been used with Primaccord in presplitting in the Niagara conduit.

In cases with extreme demands for smooth contours, or when it is a question of following a curved contour exactly one or two uncharged guide holes between the charged ones may be used. The spacing is then reduced to one half or one third of the value. If drilled in order to reduce overbreak they need not be drilled to a greater depth than three times the burden for the smooth blasting row. With one guide hole between two charged holes the result may be improved if the guide hole is asymmetrically placed at a distance of about $5 \times d$ from one of the charged holes. This should only be done if the drilling is done with great accuracy. The burden should be greater in smooth blasting than the distance between the holes. If desired, the burden can be increased to twice the value given in the table if the spacing is halved.

TABLE 10:1. *Smooth blasting and presplit blasting.*

Drill hole diameter d		Concentration of charge l		Charge units[a]	Smooth blasting		Presplitting E_1	
					E_1	V_1		
mm	(in)	kg/m	(lb/ft)		m		m	(ft)
30	$(1\frac{1}{2})$			Gurit	0.5	0.7	0.25–0.	$(1–1\frac{1}{2})$
37	$(1\frac{1}{2})$	0.12	(0.08)	Gurit	0.6	0.9	0.30–0.5	$(1–1\frac{1}{2})$
44	$(1\frac{3}{4})$	0.17	(0.11)	Gurit	0.6	0.9	0.30–0.5	$(1–1\frac{1}{2})$
50	(2)	0.25	(0.17)	Gurit	0.8	1.1	0.45–0.70	$(1\frac{1}{2}–2)$
62	$(2\frac{1}{2})$	0.35	(0.23)	Nabit 22 mm	1.0	1.3	0.55–0.80	$(2–2\frac{1}{2})$
75	(3)	0.5	(0.34)	Nabit 25 mm	1.2	1.6	0.6 –0.9	(2–3)
87	$(3\frac{1}{2})$	0.7	(0.5)	Dynamite 25 mm	1.4	1.9	0.7 –1.0	(2–3)
100	(4)	0.9	(0.6)	Dynamite 29 mm	1.6	2.1	0.8 –1.2	(3–4)
125	(5)	1.4	(0.9)	Nabit 40 mm	2.0	2.7	1.0 –1.5	(3–5)
150	(6)	2.0	(1.3)	Nabit 50 mm	2.4	3.2	1.2 –1.8	(4–6)
200	(8)	3.0	(2.0)	Dynamite 52 mm	3.0	4.0	1.5 –2.1	(5–7)

[a] If no special charges are available dynamite taped on detonating cord to a concentration l kg/m (lb/ft) can be used.

Example 10:1. A smooth blasting is to be performed in the final part of an open cut with 75 mm (3 in) holes. These should be spaced $E = 1.2$ m (4 ft) at a distance of $V = 1.6$ m (5 ft) from the row in front. The concentration of the charge should correspond to 0.5 kg dynamite per m (0.33 lb/ft). With a burden of $V = 3.0$ m (10 ft) the spacing should be about $E = 0.7$ m (2 ft).

The last column in table 10:1 gives the spacing recommended for presplitting for the diameter of the drill holes in question. The figures apply in soft as well as in hard rock. The charging should be the same as is the practice in smooth blasting. In application the presplit row can be blasted in the same round as the main round and ignited by instantaneous detonators. This presupposes that the quality and character of the rock are good enough to prevent misfires in the rest of the round by tearing from the presplit row.

On the other hand in presplitting, especially in soft rock where the splitting causes a comparatively wide crack in the upper part of the hole, the crack shields the material behind from the effect of the explosion from the heavy charges in the main round. This serves to reduce the overbreak in a way that differs from ordinary smoothing where the contour cracks are created only after the main part of the round has detonated.

In open cuts, where an overbreak is caused also by the reflection of the shock waves at the upper free face (horizontal) a correct smooth blasting presupposes that the holes closest to the contour row are loaded with a

FIG. 10:15. Loading in holes with $d = 50$–100 mm (2–4 in) for smoothing with detonating fuse and dynamite cartridges. The hole is then filled with minus $\frac{3}{8}$ clean stone chips. (Courtesy Ensign-Bickford Company.)

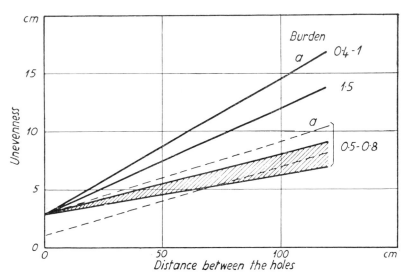

FIG. 10:16. Average unevenness of the rock contour as a function of the hole distance, the burden and the size of charge. Unbroken lines refer to site A; broken lines to site B which, contrary to A, has exceptionally homogeneous rock. Curves a, large overcharge; other curves, correct charge.

Fig. 10:17. Tunnel roof in unhomogeneous rock, free from large cracks and crevices (AB Skånska Cementgjuteriet).

concentration of not more than 3–4 times the values indicated in table 10:1 in an area closer than 3–4 V from the free upper face.

Example 10:2. With $d = 75$ mm (3 in) $E = 1.2$ m (4 ft), $V = 1.6$ m (5.3 ft). As in example 1 the holes at the distance 1.6 m (5.3 ft) from the contour row should have a concentration of the charge not exceeding 1.5–2.0 kg/m (1.0–1.4 lb/ft) in 5–6 m (17–20 ft) from the top of the hole.

10.7. Results from smooth blasting and presplitting

Results from tests with smoth blasting are given in fig. 10:16. These investigations were carried out in Sweden in 1953 in a joint program in which the Royal Fortifications Service, the Nitroglycerin AB and Atlas Copco AB took part. The diagram shows the average unevenness of the rock face in relation to the distance between the holes, the burden and the size of the charge.

The diagram can be used as a guide in choosing the burden and distance of the holes. The aim of the investigation was to arrive at average values of 5 cm (2 in) or less for the unevenness of the rock. The practical recommendations in the previous section (10.6) correspond to this unevenness, when using small diameter drill holes (below 40 mm). With greater dia-

FIG. 10:18. Rock from the right hand side of the valley at Valdecanas, Spain. The photo gives a good idea of the nature of the rock after ordinary excavation.

meters and spacing according to table 10:1 the unevenness increases in proportion if given in cm (in), but remains constant if given as angles of deviation from the intended line.

The lines marked a show where overcharging has occurred and confirm the importance of avoiding overcharges. It is interesting to note that this is especially true in unhomogeneous rock (unbroken lines). The trials were not carried out in bad rock as has later been much practised; it would have given even steeper lines for overcharging in figure 10:16 showing that smooth blasting is even more important in bad rock than in rock of good quality. This is an important practical experience.

The diagram explains the reason for the deep-rooted and false idea that careful blasting of the contour requires line-drilling. With heavy overcharges, as has always been the practice before, it is necessary to have very short distances in order to reduce the unevenness of the rock (upper unbroken line in fig. 10:16). This may give an even contour but will destroy the rock behind due to the great concentration of charge per m² (sq.ft). With correct loading according to table 10:1 (shaded section in fig. 10:16) no line-drilling is required and the very low concentration of charge per m² (sq.ft) does not cause cracking inside the remaining rock. With $d=63$ mm ($2\frac{1}{2}$ in), for example, the spacing is $E=1.0$ m (3.3 ft). It is essential to keep in mind that smoothing does not call for line-drilling and therefore entails only a small—if any—extra cost for drilling.

The effect of smoothing in unhomogeneous rock without great cracks and crevices is shown in fig. 10:17 (Sweden). With a special drilling technique

313

FIG. 10:19. Results of smooth blasting in the right hand abutment, Valdecanas, Spain. Soc. Hydroel. Española and Widmark and Platzer AB, Stockholm.

314

Fig. 10:20 Two examples of presplitting in hard igneous rock (The Royal Swedish Fortifications Administration and Nitro Nobel AB, Sweden).

Fɪɢ. 10:21. Presplitting soft rock in Torrejon (Soc. Hydroel. Española and Widmark and Platzer AB, Spain). Spacing 0.2 m (0.7 ft), diameter 44 mm (1.7 in).

it has been possible to avoid "steps" between the rounds. By using small bottom charges of about 0.1 kg (0.25 lbs) the crushing effect in fig. 10:11 has been avoided and the course of the drill holes can be seen in almost every inch of its length, 21 m (70 ft).

Quite a different type of rock with irregularities and cracks is shown as seen after ordinary blasting methods in fig. 10:18 (Spain). Even in this rock the smooth blasting works perfectly (fig. 10:19), which seems almost more impressive than the pictures from more homogeneous rock.

In igneous rock presplitting can be carried out with excellent results as shown in fig 10:20 and in fig 10:23 for three various kinds of application.

From another working site (Spain) the following picture (fig. 10:21) in softer rock shows that perfect results can be obtained with presplitting, when the rock has no open cracks. The same is shown in the original pictures from the Niagara project, U.S.A. (fig. 10:22, 10:13) where not only the blasting result but also the drilling precision are to be admired.

Another example of blasting an open cut with presplitting is given in the photos of fig. 10:23, the upper one showing the holes of the round and

FIG. 10:22. Presplitting in the Niagara Project reported by Paine, Holmes and Clark. A nice clean smooth face. Holes drilled from haunch elevation to invert, 15 m (45 ft). Tough area at bottom, which is caused by loads not reaching bottom, is removed by light explosive charges loaded in alternate holes (Courtesy Hercules Powder Company).

of the presplit row charged with gurit charges and detonating fuse in holes with a diameter of $d = 51$ mm. These holes were ignited in the main round with instantaneous detonators. The bottom photo in fig. 10:23 shows the resulting wall left after the blasting.

Finally some classical figures from the Stornorrfors Power Station, Sweden (fig. 10:24, 10:25), give an example of how the very delicate contours can be cut out even in extremely hard rock.

FIG. 10:23. Presplit row with gurit charges in benching with drill-hole diameter $d = 51$ mm and spacing $E = 0.4$ m. Upper picture before, lower picture after blasting (Widmark and Platzer AB, Stockholm).

FIG. 10:24. A fine example of how the final contours required can be cut out of the actual rock by smooth blasting (Stornorrfors Power Station, Swedish State Power Board).

10.8. Problems

1. Decide the burden and spacing for the final smooth blasting row in an open cut where the diameter $d = 50$ mm (2 in) is used.

2. Calculate the charge and discuss how to arrange the loading and to ignite the holes in problem 1. (Bottom charge has to be calculated according to Chapter 3.) Bench height 7.5 m (25 ft).

3. Calculate the column charge in the row closest to, and in front of the smooth blasting row if a very careful blasting is to be done. Discuss according to figures given in Chapter 3 how the burden for the row in question must be reduced.

4. Calculate the burden and spacing for the perimeter holes in a tunnel with a diameter of $d = 38$ mm (1.5 mm) a) if smooth blasting is to be performed, b) with presplitting.

5. Discuss the special problems in using presplit blasting in tunnel perimeters arising from the presplit row not being able to create cracks perpendicular to the holes at the bottom. Should the edge holes be ignited as presplit holes, or later?

6. Calculate the spacing of the holes and how to charge them in an open cut with drill hole diameter $d = 100$ mm (4 in).

7. In the final blasting of the rock in problem 6 some of the holes in the presplit row have to be loaded a second time to break out the rock in front of the presplit row.

FIG. 10:25. Smooth blasting of roof and wall faces in Stornorrfors Power Station.

11. UNDERWATER BLASTING AND BLASTING THROUGH OVERBURDEN

Underwater blasting, or more generally, blasting where the rock surface lies beneath water, clay mixed with water, or beneath a layer of soil or till, entails so many problems that this branch of blasting technique deserves to be specially analysed.

The costs of drilling and loading have been very high when divers have had to do the work, and at the same time it has been considered necessary to have a larger number of holes and from three to six times as much explosive as in ordinary bench blasting. There has also been a large percentage of holes which could not be loaded after being drilled (2–10 %), and a still higher occurrence of faults in occasional detonators and their wires (5–20 %). A fundamental fact which has not been realized is that conventional underwater blasting has been carried out in such a way that there has been a great risk of flash-over between different holes. From the point of view of safety alone this is very unsatisfactory; with respect to the blasting result it means that it is no longer possible to control and decide the sequence of breakage in multiple-row short delay blasting. Fragmentation will be unsatisfactory, and there will be a risk of incomplete tearing at the bottom.

In view of the complicated phases of the work which proper underwater blasting entails it has been important to consider various possibilities for simplifying the work and the technical factors determining the final result.

11.1. Calculation of the charge

For rational underwater blasting it is essential to achieve the desired result with a minimum of drill holes. The calculation of the charges will be discussed here on this basis.

The cost of the blasting is very closely connected with the number of holes; on the other hand, the footage drilled plays an entirely subordinate part. An increasing diameter of holes, of course, increases the actual drilling cost per hole, but other factors are comparatively little affected. This applies to collaring the holes, identification, placing the charge and ignition media, and subsequent checking. In addition, divers' work in connecting, checking and searching for defects is facilitated by a wider spacing of holes.

322

FIG. 11:1. Diver descending with loader (Stockholm City Street Board).

As a matter of fact, a connected round of 200 holes, for example, at a hole distance and with a burden of 1×1 m is an impenetrable jungle of wires for a diver if each one is drawn up to the surface separately. In a round with the same number of holes, but with a distance between the holes and the rows of 2.0×1.5 m^2 it will be possible to get at an individual hole inside the round in an entirely different manner if it should be necessary to change a detonator or the electric wires. This last round is consequently simpler in spite of the fact that it affects 600 m^2 instead of 200 m^2 as in the first example.

a. Breakage

The energy required for detaching the rock lies chiefly in the comparatively slow tearing which follows when the shock wave has given an indication of radial cracks round the hole. In the case of a covered rock face the movement of the rock is restrained by a static counterpressure $p = 0.1$ D kp/cm^2, in which D indicates the depth in meters. If the movement of the rock face is assumed to be less than 0.1 m when the loosening is completed, the additional energy required on account of the covering will be less than $10^4 \times 0.1 p$ kp m/m^2. At a depth of 10 m the value of the additional energy $= 1$ ton m/m^2, which is only 0.2 % of the chemical energy per kg of the explosive. For other reasons drilling below the bottom to be blasted will be recommended later; this gives an additional charge of at least 0.25 kg/m^2 compared with blasting above water. An efficiency of 0.8 % is consequently all that is

323

Table 11:1. *Density and sound velocity for various materials.*

Material	p g/cm³	c m/s	m
Water	1	1500	
Granite	2.7	3950	7
Gneiss	2.1	5000	7
Limestone	2.6	3700	6.4

required to give full compensation for the effect of the load of water. With other covering materials the additional energy is increased by a factor ϱ so that the value will be ϱ ton/m² at a depth of 10 m.

In ordinary blasting, the shock wave contributes less than 20 % of the energy required in loosening the rock. If the rock face lies beneath covering material part of the energy of the shock wave is lost to the covering. This loss, F, is greatest for a plane wave with its front parallel to the rock face. In this case

$$F = 1 - \left(\frac{m-1}{m+1}\right)^2$$

where $m = \varrho_0 c_0/\varrho c$ in which ϱ_0 and c_0 indicate the density of the rock and the velocity of the sound, ϱ and c the corresponding values for the covering material. The densities and sound velocities for various materials are given in table 11:1.

If we take $m = 7$ we get a loss of $F = 0.44$, i.e. at the most 44 % of the shock wave energy is lost, corresponding to less than 8 % of the total energy required for loosening the rock.

This can be compensated for by the higher weight strength ($s = 1.07$–1.15) for the explosives and the tables of Chapter 3 can be directly applied when $E = 1.25 V_1$, $P = 1.27$ and the unloaded part of the drill hole is of a length less than V.

In underwater blasting, however, it is recommended to have a square net pattern, $E = V$, and to use the pneumatic cartridge loader giving $P \geqslant 1.35$, which means that the burden can be increased to values corresponding to a 15 % bigger diameter of the drill hole in ordinary benching. In addition to this for bench heights below $K = 2V$ the burden can be increased even more in %, allowing the bottom charge to rise higher than is presupposed in the tables of Chapter 3. If the charge in underwater blasting is permitted to rise to $\frac{2}{3}V$ from the top of the hole we get the curve (b) instead of (a) in fig. 11:2. The relation between bench height and burden is indicated in

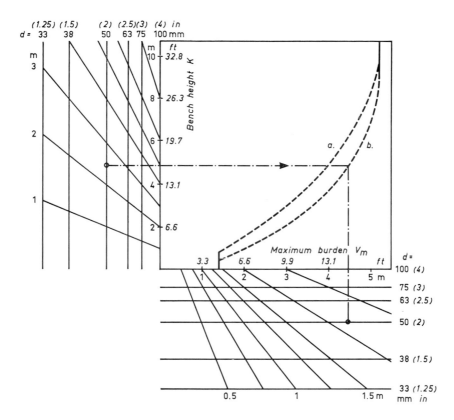

FIG. 11:2. Relation between bench height and maximum burden at diameters between 33–100 mm (1.25–4 in) for bottom charge, which for a rises to the length V_m, for b to the length $\frac{2}{3} V_m$ from the surface of the rock. The example inserted proceeds from a bottom diameter of 50 mm (2 in), and a bench height of 2.5 m (8.2 ft) and gives $V_m = 2.2$ m (7.2 ft).

fig. 11:2 for a bottom diameter $d = 50$ mm (2 in) and is transformed into other d-values according to the relation.

$$V_m = df(K/d) \qquad (11:1)$$

This can be done with the aid of the inserted side scales to a diameter between 30–100 mm as is proved by the example given. With a bottom diameter $d = 50$ mm (2 in), and a bench height $K = 2.5$ m (8.3 ft) a maximum burden of $V_m = 2.2$ m (7.3 ft) is obtained.

With vertical holes the burden and spacing have to be 5–7 % less than the values in fig. 11:2.

b. Deviation in drilling

The burdens and hole distances of the drilling pattern must be adjusted so that they give a reliable break at the bottom. This means that if a drilled hole can deviate R m from the value intended because of wrong alignment and possible deviation inside the rock, the burden must be corrected to $V = V_m - R$. The correction values concerned must be estimated at every working site, as the conditions vary considerably.

When a diver does the drilling it may be necessary to count on faulty collaring of 0.2 m (0.7 ft) and errors in alignment of 5 cm per m (5 %) or more. In drilling with rigid platforms the faults in collaring depend on the depth to the rock face, and the mistake in alignment, in its turn, on where the drill happens to strike the rock surface. In practice the total deviation from the fixing point of the drilling machines to the completed bottom must be observed. It involves at best values of 3 cm per (3 %) or 0.36 m (1.2 ft) at a depth of 12 m (40 ft) for a drilling device with good fixation, but usually the values are about 6 cm per m (6 %). Values considerably in excess of this should not be accepted. The last-mentioned figure implies that at $d = 40$ mm (1.6 in) the burden and the spacing must be reduced from $V_m = 2.1$ m (7 ft) to $V_1 = E_1 = 1.4$ m (4.1 ft) at a depth of 12 m (40 ft).

The effect of faulty drilling diminishes with an increasing diameter.

c. Swelling

Without cover. In ordinary multiple-row blasting a certain part of the charge over and above that needed for loosening the rock is required to give the round sufficient swelling possibilities. The overcharge required increases linearly with the height of the bench. In hand-held drilling the overcharge preponderates from the very beginning on account of faulty drilling so that it has not been necessary here to allow especially for the swelling.

In drilling with fixed platforms and larger diameters of hole it has been possible to reduce faulty drilling considerably, but the overcharge required for the bulging remains unaltered and becomes a dominating factor. It is of importance to realise these connections seeing that they decide what measures are to be taken to improve the result of the blasting. When it comes to the swelling, conditions change considerably with the slope of the hole. A dip that gives a direction of throw of 1:2 obliquely forward and upwards needs only half the overcharge required at 1:3 to provide the same lifting energy. The maximum burden for an inclination of the holes of 2:1 and one row of holes is given in fig. 11:2 and fig. 11:3.

For breakage of the bottom 0.36 kg/m³ (0.47 lb/cu.yd) is needed, and an additional charge to give sufficient swelling, proportional to the bench height.

326

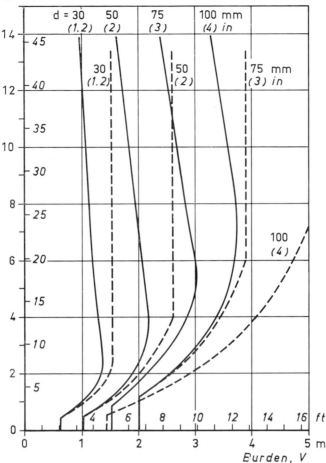

FIG. 11:3. Maximum burden for one row of holes $---$ (V_m) and for multiple-row rounds $——$ with regard to the conditions for swelling (V_1). Slope of the holes $2:1–3:1$, $P \cdot s = 1.5$; $E = V$.

For the total specific charge at the bottom we have the equation from Chapter 2 (2:37):

$$q_b = 0.36 + 0.04 K \qquad (11:2 \text{ M})$$

K indicating the bench height in meters and M that the formula is given in metric units (as the original formula). The charge which is required for breakage alone is indicated by q_0.

If a certain burden for a given bench height and diameter of hole is obtained according to formula (11:1) and fig. 11:3, the same result can be used, if we allow for the swelling, by increasing the diameter of the hole to $d_1 = d/\varkappa$ in which $\varkappa = \sqrt{q_0/q_b} < 1$. We obtain from (11:2)

$$V = \varkappa d_1 f(K/\varkappa d_1) \tag{11:3}$$

In fig. 11:3 the burden is indicated as a function of the bench height for multiple-row rounds. With a drilling diameter of 50 mm (2 in) the reading can be done directly on the inserted main scales. The broken curve indicates the maximum burden V_m in blasting a single row of holes which does not require an additional charge for swelling. The same curve indicates the maximum burden for single holes at the bottom in multiple-row rounds. A deviation in placing and drilling alignment, which together do not give larger values than according to the broken curve, can be tolerated when the average value in the row is V_1.

Effect of covering. Pioneering work in blasting without uncovering the rock has been made by AB Skånska Cementgjuteriet at the construction of the Lindo canal, where the method of overburden drilling and blasting, the so-called ODB- or OD-method, was first applied. Let us observe a multiple-row round with the bench height $K \geqslant 2$ m (7 ft) and the interval period 25 ms between the rows. The first row will obviously be retarded in its forward movement by the surrounding material. Assuming an initial velocity of 20 m/s (67 ft/s) (which corresponds to an energy of 50 tonm/m³) the first row will at the most have moved forward half a meter when the charges in the next row detonate. The rock in this row is, however, retarded by the water, whereas the process for the second row sets in without any effect from the water, as it is not then between the rows. After moving less then 150 ms and 3 m (10 ft) the back row collides with the one in front. The difference in velocity is then only 3 m/s (10 ft/s). After the collision both the rows continue at a greater velocity than that of the first row before the collision. Their velocity will moreover be retarded by only half the force per m³ since the mass has been doubled. The time between the collisions when rows 3, 4, etc. strike the one in front will become longer and longer, finally culminating, for a heavily charged round with a sufficient number of rows, in an equilibrium which implies that water is forcing its way between the rows, and collisions are not necessary between two rows during the phase when both are in motion.

If, on the other hand, the round is weakly charged so that the initial velocity is, say, only 5 m/s (3 ton m/m³), row No. 1 will only have moved 0.12 m when row No. 2 begins to move. If no regard is had to the fact that the rock demands more room after breakage than in its compact state, rows 1 and 2 will collide after 80 ms and a length of 0.25 m. In relation to the distance between the masses of rock from rows 1 and 2, which is at the most 0.12 m at the beginning, the increase in the volume of the rock cannot of course be ignored. In reality the two different sections of rock make contact after a much shorter time than 80 ms and thereafter combine in removing the covering. The following rows contribute in the same way when their charges detonate.

In the case of a round with vertical holes one of the conditions for swelling is the removal of a volume of water about as big as that of the compact rock. This requires energy equal to pV, where p is the average pressure on the moving front. Supposing the energy per kg of the explosive to be 600 tonm and the efficiency of the overcharge to be 16 %, then an overcharge of about 0.1 $p\varrho$ kg/m³ is required.

For a round with sloping holes the same overcharge is needed only in a part of the front section, and a smaller overcharge for the remainder.

It is consequently sufficient in all instances to overcharge 0.1 kg/m³ at a depth of water of 10 m ($p=1$, $\varrho=1$). At a bench height of $K=1$ m, and the reduction of the spacing by 20 % proposed under "Breakage", an increase in the specific charge from 0.4 to 0.5 kg/m³, i.e. 0.1 kg/m³, is obtained, half of this being intended to compensate for the loss of shock-wave energy. As, in addition to this overcharge, the drilling extends below the bottom (see 11.2 Drilling pattern) which increases the bottom charge about 25 %, the proposed reduction of the distance between the holes for underwater rounds should be quite sufficient to compensate for the effect of the water. This applies to small K values (or in the bottom part of a high bench), and therefore also in the pipe part where there is more room for overcharging than in the bottom part. In addition the value of p diminishes higher up in the pipe. The formula (11:2 M) which indicates the specific charge in the bottom will be for underwater blasting

$$q_0 = 0.45 + 0.05\,K \qquad (11:2'\text{ M})$$

With a covering material of greater density than 1, and at a greater depth of water than 10 m the charge must be increased.

The above description applies if the covering material is a liquid. For solid materials, or such as have a comparatively high viscosity, it may be necessary to insert charges in the covering also, if this is needed to move the rock.

d. Fragmentation

With a correctly adjusted distribution of charge and sequence of ignition the same type of relation applies between fragmentation and specific charge as is given for short delay blasting.

Specific charge, kg/m³	0.24	0.30	0.40	0.50	0.60	0.70	0.85	1.0
Fragmentation, m³	1	$^1/_2$	$(^1/_2)^3$	$(^1/_{2.5})^3$	$(^1/_3)^3$	$(^1/_4)^3$	$(^1/_5)^3$	$(^1/_6)^3$

A more uniform fragmentation can even be counted on in underwater blasting since the charge here can be distributed farther up in the pipe and the counterpressure of the water increases the collision effect when using heavy overcharges. On the other hand the strength of the explosive is reduced when the dynamite is "stored" in the holes under water some weeks before the entire round is loaded. In table 11:2 the reduction of the effective energy is assumed to be 20 %.

The largest boulders usually derive from the part nearest the rock face. The relation mentioned gives a simplified description of fragmentation achievable.

In underwater blasting $q \geqslant 0.60$ kg/m³ (0.8 lb/cu.yd) applies for the bottom charge at bench heights over 3 m (10 ft). If the specific charge can be increased with a given drilling pattern, without increasing the number of holes, this should be done in all instances where improved fragmentation facilitates removal. The drill holes should be fully utilized with the important proviso, however, that the risk of a flashover must be reduced to a minimum. Should a q-value which gives sufficient fragmentation according to table 11:2 not be achieved, then and only then must the more expensive procedure of increasing the number of holes be resorted to.

When the fragmentation is less satisfactory than expected the reason probably lies in an unfavorable distribution of cracks and crevices in the rock, or a wrong sequence of ignition due to flashover.

e. Effect of faulty holes

It is important to avoid stumps at the bottom. An additional margin in the practical burden has generally been allowed to compensate for the fact that a certain number of the holes cannot be blasted as intended. The reason for this may be that the holes have become choked in charging, that the electric wires have been destroyed, or that the insulation has been damaged. Projections of rock cannot be tolerated to a corresponding extent, and in this case overcharging must be adopted.

It is here a question of the bottom charge only, as the column charge has no influence on the breakage at the bottom.

Example 11:1. A portion of the round, according to fig. 11:4a, has 10 % faulty holes, one of which has been indicated as No. 1. $E_1 = 1.3 V_1$. The burden at the bottom in front of this hole cannot be torn away by holes 2 and 3 even

```
                    2   I   3
          X    X    X   ⊗   X   X   X
               7    5   4   6   8
               X    X   X   X   X
```

(a)

```
               6  4  2  I  3  5  7
               X  X  X  ⊗  X  X  X  X
```

(b)

```
               X  X  X  X  X  X  X     X
```

Fɪɢ. 11:4. Arrangement of holes with the same volume of rock per drill hole gives different probability for bottom rock projection close to a "faulty hole" (No. 1).

with a strong overcharge. Holes Nos. 5, 6 and 4 in the back row, however, can give breakage if they have been driven $0.6V$ under a completed bottom, and if they have an overcharge of 50%. It is assumed in this connection that these three holes are ignited with the same interval number so that they collaborate. There is still a chance of a projection if, at the same time, one of the holes 5, 6 or 4 is faulty. For about 3% of the holes there are therefore projections in spite of the 50% overcharge. If this is increased to 100%, breakage can be counted on even if either hole 5 or 7 fails. For this example we get the relation between the risk of projections and overcharge according to table 11:3, in which the relation for 15 and 5% faulty holes has also been included.

Example 11:2. With the same mass of rock and charge per hole as in the preceding example, but with double the burden, as in fig. 11:4b, the contribution from any given hole for the loosening of the burden in front is only 25%,

Tᴀʙʟᴇ 11:3. *Percentage risk of projections for various burdens and percentage of faulty holes (fig. 11:4 a).*

Faults %	Burden		
	V_m	$0.8\ V_m$	$0.7\ V_m$
15	15	6	1.6
10	10	3	0.8
5	5	1	0.2

Percentage risk of projections for various burdens and numbers of faulty holes (fig. 11:4 b).

Faults %	Burden V			
	V_m	$0.97\ V_m$	$0.87\ V_m$	$0.7\ V_m$
10	50	30	3.8	0.2
5	30	15	0.6	

the other 75 % being given by the six adjacent holes. This implies that an overcharge of 33 % suffices for breakage. The relation between extra charge and risk of projections is given in table 11:4.

The example indicates that a pattern having the holes in a row wide apart gives the lowest frequency of projection at the bottom when the maximum burden V_m is approached (the broken curve in fig. 11:3). With burdens for less than $0.8\ V_m$ on the other hand, the smallest number of remaining projections is obtained with the holes closer in the row but with greater burdens. The comparison has been made with the same mass of rock per hole.

With regard to the deviations in drilling the burden is usually not more than $0.7\ V_m$ and if a good fragmentation is not required a drilling pattern with great burdens and small distances between the holes could be used.

When it is of importance to obtain good fragmentation the spacing should be greater than the burden as in table 11:3 and it is important to reduce the percentage of faults. It is cheaper to reduce this percentage than to increase the overcharge values to larger amounts than bulging and faulty drilling demand. In one case in table 11:4 there is a risk of projections of 3.8 % at a fault percentage of 10 % and at an overcharge of 33 %. To reduce this value to 0.2 % the overcharge can be increased to 100 %, e.g. by increasing the hole diameter from 40 to 50 mm with the same drilling pattern. But if the number of faulty holes can be reduced from 10 % to about 3 % the same result as regards projection risk is achieved. The fault percentage is due to the fact that a number of holes cannot be loaded as they have been choked with stone, etc. This may be 10 %, but is on an average about 5 % of all drilled holes of 30 mm, and 2 % of 50 mm diameter. The wires of the ordinary electric detonators are damaged in 5–20 % of all cases. The total fault percentage should be reduced to below 5 % (see following section) and, if possible, be eliminated. This presupposes loading the holes with the pneumatic loader (see Chapter 4). If loading, connecting and checking are carried out with the help of divers, it is possible to inspect an occasional hole if the rows are more than 1.8 m apart. Faulty holes can then be practically eliminated. With drilling diameters under 50 mm (2 in) this spacing can be achieved by reducing that of the holes and increasing the burden (or inversely) instead of using quadratic placing of the holes.

11.2. Drilling pattern

After drilling, every hole must be easily found for loading, connecting and checking. This means that the arrangement of the holes must be of the simplest geometrical pattern even if this means the drilling of additional holes. A reduced number of holes facilitates control and supervision.

The first row, close to the line on the rock indicating the stipulated bottom level, can with vertical drill holes loosen the rock down to about $K = 8d$ below that level when the spacing is $E_1 = 20d$, the burden $V_1 = 20d$ and the sublevel drilling $H = 30d$. Note that the blasted level is ordinarily not the same as, but lower than the stipulated one. With sloping holes $3:1-2:1$ an additional $7-10d$ gives a vertical bench height of at least $15d-18d$ for the second row. In both cases the first row breaks the rock through cratering (compare fig. 8:14).

Example 11:3. With $d = 50$ mm (2 in) and vertical holes we have for $K = 0$ $E_1 = V_1 = 1.0$ m (3.3 ft); $H = 1.5$ m (5 ft) and the rock is broken $K = 0.4$ m (1.3 ft) below the grade.

Example 11:4. With $d = 100$ mm (4 in) the corresponding values are $E_1 = V_1 = 2.0$ m (6.7 ft), $H = 3.0$ m (10 ft) and $K = 0.8$ m (2.7 ft).

Example 11:5. With the holes sloping $2:1$ the K-value increases to $K = 18d$ or 1.8 m (6 ft) below the grade. The bench height for the next row measured along the slope is 10% more, or $K = 20d$.

For the *subsequent rows* the rock is broken as in benching. The burden can be calculated with the aid of fig. 11:2 for small bench heights and fig. 11:3 for greater bench heights. These two diagrams presuppose $E = V$, $P \cdot s = 1.5$ and holes sloping $3:1-2:1$. For vertical holes and large multiple-row bench rounds the drill hole diameter required is about 4% per m of the bench height from $5-7\%$ valid for a bench height of 2 m or less.

Example 11:6. At a bench height of $K = 10.5$ m (35 ft) and vertical holes the drill hole diameter should be 40% larger than at a slope $2:1$ and the same drilling pattern.

In any case the holes are driven about $25d$ below the intended bottom of breakage.

A much greater number of holes per m² has to be used at the lowest bench heights. In order to perform the job with a minimum of holes these are

TABLE 11:5. *Downward slope of the bottom blasted in relation to the direction of the hole (V_m = maximum value for loosening).*

Burden	V_m	0.7 V_m	0.5 V_m	0.35 V_m	0.25 V_m
Slope	1:∞	1:15	1:7	1:3	1:1.5

drilled to such a depth that the bench height reaches $K = 50d$ as soon as possible even when this calls for blasting far more rock than required. In fig. 11:6b the row with $K = 2.2$ m (7.3 ft) has a burden $V = 2.0$ m (6.7 ft), whereas a burden of only $V = 1.0$ m (3.3 ft) could be used if the blasted bottom coincides with the grade level. By blasting down to a lower level in this way not only the number of holes, which is the important point in this connection, but paradoxically also the drilled footage in the round can be reduced.

For the drawing of the drilling pattern it is necessary to know the bottom that will be created by a row of holes. The breakage at the bottom is in principle perpendicular to the holes in the row, but for burdens smaller than $0.7 V_m$ a gradual extra downward slope is obtained according to table 11:5.

The greatest practical burdens (V_1) including conditions for the swell are given for diameters of the drill holes of $d = 30$–100 mm in table 11:6. The values of the table can be used when the total deviations of the drill holes at the intended bottom is less than ($V_m - V$).

To illustrate the placing of the holes an example is given in fig. 11:5 of an underwater area which is to be broken to a depth of 10 m (bottom stipulated). A profile is drawn through A_1–A_2 and the position is indicated for vertical holes and for sloping ones in fig. 11:6. Finally, a plan of the position of the holes at the stipulated bottom is shown for vertical and sloping holes (fig. 11:7) according to the profiles of fig. 11:6 for $d = 50$ mm.

TABLE 11:6. *Greatest burden (V) for a slope of the holes of 2:1.*

d						V					
mm	in	$K = 0.3$	0.4	0.5	0.6	0.8	1.0	1.4	1.8	2.2–6.5 m	V_m
30	1.2	0.5	0.6	0.7	0.8	0.9	1.0	1.2	1.3	1.3	1.5
40	1.6	0.6	0.7	0.8	0.9	1.0	1.2	1.4	1.5	1.6	2.0
50	2	0.7	0.9	1.0	1.1	1.2	1.4	1.6	1.8	2.0	2.5
75	3	1.0	1.3	1.5		1.8	2.1	2.4	2.7	3.0	3.7
100	4	1.4	1.8	2.0	2.2	2.4	2.8	3.2	3.6	4.0	5.0
150	6	2.1	2.7	3.0	3.3	3.6	4.2	4.8	5.4	6.0	7.5

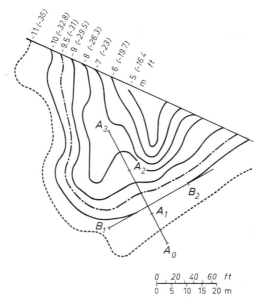

FIG. 11:5. Plan of an underwater shoulder of rock to be blasted off to a level of 10 m (32.8 ft) below the water surface.

It is striking for the lowest as well as for the highest bench heights how the number of holes can be reduced when using sloping holes instead of vertical ones. For $d = 50$ mm (2 in) in fig. 11:6 only about half as many holes are then required for a given width of the bench. On the other hand the drilling may be somewhat easier with vertical holes.

A further and essential advantage with sloping holes is that they give the broken bottom a tendency to work its way down automatically in the event of one or several holes for some reason not breaking the rock satisfactorily. The practical drilling pattern is made with the diagrams of fig. 11:3 as a basis, but the following adjustments must be made:

1. Correction for faulty position of the holes

This has not been done in the profiles of fig. 11:6 and the plan of fig. 11:7, as this factor varies very much according to the conditions. Full breakage can be expected when the burden at the bottom is less than V_m, which for greater bench heights may leave a sufficient margin $(V_m - V_1)$ for the deviation in the drilling. For $K = 10$ m (33 ft) and $d = 50$ mm (2 in) this margin is as much as 0.75 m (2.5 ft) and is proportional to the diameter of the drill hole, thus leaving a greater margin for larger diameters than for small ones.

335

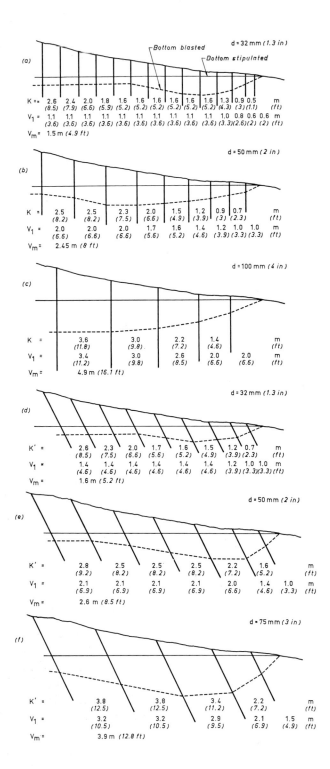

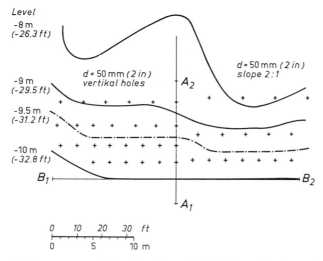

Level
-8 m
(-26.3 ft)

-9 m
(-29.5 ft)

d = 50 mm (2 in)
vertikal holes

A_2

d = 50 mm (2 in)
slope 2:1

-9.5 m
(-31.2 ft)

-10 m
(-32.8 ft)

B_1

B_2

A_1

0   10   20   30 ft
0   5   10 m

FIG. 11:7. Drilling pattern indicating intended placing of holes at the level for the requested bottom. Left of center line, holes arranged according to (a) right of centerline according to (e) in fig. 11:6.

For lower bench heights the difference $(V_m - V_1)$ approaches zero and the burden and spacing $V = E = 20d$. If the deviations in the position of the holes in the first row don't exceed $6d$ the drilling pattern can be left unchanged as the holes are so close to one another that they cooperate, thus compensating for small deviations in the drilling.

As an easy practical rule deviations at the bottom of up to $(6d + 0.05 K)$ can be accepted with burdens according to fig. 11:3. If the deviations exceed this value the burden and spacing will have to be reduced.

Example 11:7. According to 11.1 b a diver makes a faulty collaring of 0.2 m and a faulty alignment of 0.05 m/m. The total deviation at the bottom is then about $(0.2 + 0.05 K)$, which can be accepted without changing the pattern if $d = 34$ mm (1.3 in). If the collaring faults are 0.4 m the burden and spacing have to be reduced by 0.2 m each.

2. Drilling below the grade

To facilitate removal and to simplify the indications for the depth of holes, it is generally recommended that the bottom be blasted and laid 1 m (3 ft) under the stipulated one. At the lowest heights of rock the bottom blasted should be laid still lower according to the example in fig. 11:6 c.

FIG. 11:6. Profiles A_1–A_2 from the rock in fig. 11:5 with vertical holes (a–c) and holes with a slope of 2:1 (d–f). The holes are indicated at their greatest distance for full breakage $E = V$. The burdens have to be reduced due to practical deviation in the position of the holes.

3. Fragmentation

In order to improve the fragmentation it may be necessary to reduce the burden and spacing in such cases where the rock is removed with a dredger or where the diameter of the drill holes is larger than 75 mm (3 in). A reduction of the greatest burdens (V) by 20 or at most 25 % can in these cases be recommended. If the boulders obtained are still too big this is often due to faulty sequence of ignition through flash-over and there is no reason for a further reduction of the distance between the holes.

4. Simplifications of the drilling pattern

The idea of this is to improve the control of the round. The simpler the drilling pattern the less correction is generally required due to faulty drilling. The first step is to use as big a diameter as possible and load with the pneumatic cartridge loader. Then the first and subsequent rows are drilled with the same and smallest burden and spacing, until attaining the bench height at which the burden for the rest of the round can be used.

Example 11:8. A round is to be drilled with inclined 50 mm (2 in) holes. The drilling pattern is made with $V = E = 1.0$ m (3.3 ft) for bench heights below $K = 1.8$ m (6 ft) and the rest of the round is blasted with $V = E = 1.8$ m (6 ft) in a square pattern.

The simplest pattern can be obtained in the following way. The profile with the drill holes of diameter d_1 is drawn as in fig. 11:6. The burden (V) obtained from fig. 11:3 for the greatest bench heights may be used with or without a reduction of 0–20 % according to other conditions and is denoted V_1. The whole shoulder of rock is then drilled in a square pattern where $E_1 = V_1$ but with the first row of holes drilled with a diameter = (d_2) big enough for this first row to lower the bottom level sufficiently to use the burden V_1 in the following row.

Example 11:9. The diameter $d_1 = 50$ mm (2 in) is to be used and the bench height reaches $K = 12$ m (40 ft). We obtain the value $V = 1.7$ m (5.7 ft) and intend to use it without any reduction. $V = V_1$. This burden requires a bench height of $K \geqslant 1.5$ m (5 ft). In order to create this bench height with the first row (sloping 2:1) and with the same spacing $E_1 = 1.7$ m we can here use drill holes with $d_2 = 83$ mm (3.3 in).

By not filling out the field forward to the level -10 m in the left part of fig. 11:7 we have indicated that the pattern is inserted from behind in the example shown in fig. 11:7.

338

TABLE 11:7. *Number of days for explosive to deteriorate under water.*

Dynamite	35	50	60	93 % ngl
Number of days until				more than
flash-over capacity = 0	2	7	21	360
Weight strength s	1.0	1.07	1.15	1.27–1.30

11.3. Explosive

An explosive must be used that will give a full detonation even when stored underwater in the holes for the time they have to remain charged until the round is blasted. Regard must also be taken here to the fact that unpredicted circumstances more often than not may add some days, or even weeks, to the time expected when the work was planned.

Table 11:7 gives for some different explosives the number of days which a cartridge of 25 mm diameter can be suspended in water before its flash-over capacity = 0. Blasting gelatine, which is given in the last column, can lie in water for an almost unlimited time without being destroyed. This is hardly desirable for ordinary underwater blasting. For a dynamite with 35 % nitroglycerin, the upsetting test value after 2 days diminished from 25 to 23, after 4 days to 21 and after 7 days to 0. This may give some idea of how the strength decreases.

For a charge which is tightly rammed into a drill hole the time will be considerably longer as the surface exposed to the water is comparatively small and the water in the holes soon becomes saturated with the salts of the explosive which are soluble in water. When keeping them there longer it may be advisable to place the electric detonator in a cartridge of Dynamite 60 (60 % ngl) even if the rest of the drill hole should be charged with Dynamite 35 or Dynamite 50 (50 % NGL.). Rounds with Dynamite 35 blasted three weeks after loading have detonated satisfactorily, provided that the water had not been able to flow through cracks and crevices. Blasting gelatine has often been used, but Dynamite 50 and Dynamite 60 and sometimes Dynamite 35 are more suitable. For reasons of safety it is preferable to have an explosive that is destroyed after a certain time.

Another point that has not been taken into consideration hitherto is that the blasting gelatine entails great risks of a flashover between different holes. One of the most important measures from the point of view of safety should be to prevent the spreading of such unintentional ignition of the charge from an individual hole to other holes in the round. Table 11:8 gives a comparison between some kinds of dynamite.

TABLE 11:8. *Distance in cm for a flash-over/non flash-over at a high and a low velocity of detonation in the primary charge.*

Detonation velocity in the primary charge	Dynamite with a percentage of ngl. of			
	35	50	60	93
High	60/65	65/70	65/70	85/90
Low	45/50	50/55	60/65	—

The values have been received for two parallel 25 mm cartridges of the same explosive suspended in water. When loaded with a pneumatic loader, holes of 30 mm diameter achieve a high detonation velocity. In full-scale tests occasional cases of flash-over have been obtained with 3 kg/m (2 lb/ft) at a distance of 2 m (6.7 ft) between the holes. Some further values are given in fig. 11:8 in which a curve has been drawn according to the law of conformity. It will be seen in comparison with fig. 11:3 that burden V_1 for $d=50$ mm (2 in) and $l=3$ kg/m (2 lb/ft) is already near the limit for a flash-over between the holes (about 1.8 m or 6 ft). If V_1 is corrected to lower values the risk of a flash-over increases. This means that at present it appears to be difficult to carry out underwater blasting entirely without risk of flash-over. To improve the conditions only alternate holes need be charged in a first operation so that during this part of the job the distance to the nearest hole is about $\sqrt{2}\,V_1$. There is then a very low probability of a flash-over occurring here. In this way more than half the time for the loading

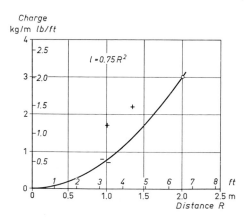

FIG. 11:8. Relation between weight of charge per meter and flash-over distance in water. + indicates flash-over, ○ limit, − no flash-over. Inserted curve indicates charges which do not give flash-over at distances less than 2 m (6.6 ft).

has passed when the turn comes for the second part of the loading. In unbroken rock there can be no flash-over; fig. 11:8 applies to flash-over in water, i.e., if there are big crevices between the holes. When this is the case the water helps to reduce the sensitivity to flash-over. The risk of flash-over in the event of unintentional ignition is consequently diminished by loading alternate holes during the first half of the loading operation, as well as during the second and the consequent blasting of the rock. In deep water blasting when the explosive must lie a long time under high pressure a special explosive is needed.

11.4. Loading of holes

The NAB pneumatic and pressure water loaders are used for loading. Further details are contained in .2 and .3 of Chapter 4. They play such a decisive part in rational underwater blasting that after having tried out this loading method there is no reason to mention any other type. The job is done rapidly and smoothly with a high degree of packing and less risk of the holes becoming choked and the detonators damaged. There are two types of the loader, S and R, the one with a continuous polythene tube, the other with jointed 1.6 m metal tubes.

With the polythene loader the diameters of the drill hole at the bottom must be at least 37 and 40 mm for loading with 22 and 25 mm cartridges. The corresponding values for the metal-tube loader are 30 and 33 mm. As the polythene loader offers many advantages a larger diameter than 40 mm should be selected for the holes in underwater blasting, if this is possible with regard to the ground vibrations.

There is no need for jointing, as in the metal-tube loader; there is less strain on the electric detonators, their wires and insulation. Bännmark states that the percentage of damaged detonators in the work at Stenungsund and Lindö was 20 % with the metal loader, but only 7–5 % with the polythene one. Later on the figure has been reduced to below 2 % through the so-called OD (Lindö) detonators, developed to meet the special requirements for blasting in salt water and through overburden.

When there is loose material at the bottom the transparancy of the water is improved for the diver when using the pressure water loader where water is used as a driving medium instead of compressed air. The loader is inserted in the hole either by a diver, or, if possible, direct from the working level above the water. A charge quantity of 2.5–3.0 kg/m (1.7–2 lb/ft) in a 50 mm (2 in) hole is achieved. Fig. 11:9 shows the pressure water loader in use.

The charging is done by two men. The explosive is inserted in the loader

FIG. 11:9. Loading with pneumatic tube loader in underwater rounds. The explosive is inserted in the tube through the breech. (Stockholm City Street Board.)

by one man. The other is not visible because he is a diver who inserts the loader and detonator in the drill hole and afterwards carries out the rest of the loading. While the diver is at work the man at the upper level has only to check the pressure in the apparatus on a manometer. When the apparatus has been emptied the pressure falls from 3 Atm to a level corresponding to the depth of the water. The loader must then be refilled with just as much explosive as is intended for the next drill hole.

The division of labour is similar with the pneumatic loader for a round drilled and blasted without removing the overburden. The work which was done in the earlier instance by a diver is now done by the man on the right in fig. 11:10.

In charging it is important to see that only the intended quantity of explosive is inserted in the drill hole. If we allow for 2.5–3.0 kg/m (1.7–2.0 lb/ft) this quantity should not be appreciably exceeded. In certain cases this may happen if the loader presses out the explosive into large crevices and fissures in the rock. When this occurs it is impossible to avoid a flash-over between the rows, thus increasing the risk of an unsatisfactory breakage at the bottom.

When using pneumatic loaders it will easily be noticed if there is any crevice connecting the hole being charged with adjacent ones. A note of this should be made and marked on the drilling pattern. The holes affected should be charged with an explosive which is less sensitive to a flash-over.

342

FIG. 11:10. Loading of a round through over-burden of clay and till. (AB Skånska Cementgjuteriet.)

11.5. Electric ignition and control

Ordinary electric detonators can nowadays stand exposure to such water pressure as occurs in underwater blasting. The insulation of the wires is often not strong enough to resist mechanical stress; some damage occurs during the loading. Detonators with nylon insulation or a reinforced insulation reduce these faults considerably. It is found, however, that even when the loading is performed without any earth faults, they nevertheless may occur in some detonators after some days and the earth resistance is lowered continuously day by day. For the OD-detonators the influence of this factor is eliminated and with them the number of faulty detonators after some days or weeks is reduced to 1–3 % instead of 5–20 % (or even more) in the case of ordinary ones and 2–10 % for such with stronger insulation.

For depths greater than 20 m (70 ft) special deep water detonators are required.

There should be no joints below the surface of the water. Should such joints have to be made they must be insulated by dipping in plastic of the type which is now being used to protect tools kept in storage against corrosion (hot metal plastic). It must be plastic without oil, for example cellulose acetobutyrate, which is heated to 160°–180°C (320°–356°F).

343

Every charge should be provided with two or more electric detonators, especially if they cannot be easily placed there after something has gone wrong with the first one. With the OD-detonators one or two are quite sufficient.

The wires from each detonator should be drawn up separately above the surface of the water. When the demand for certain ignition is especially great the detonators in the same drill hole can be connected up to different series.

Short delay blasting caps are used. Ignition patterns are made up in suitable parts and the blasting is carried out in keeping with instructions which apply to short delay blasting.

Underwater rounds make heavy demands on the ignition machine, which must be ample for a larger number of detonators. This implies condenser-igniting machines preferably with a voltage of 1000 V or less. The ignition system is checked not only with an ordinary resistance meter but also with a so-called earth-fault tester. When blasting in salt water it is recommended to have an instrument for testing the insulation as well.

The *earth-fault tester* (fig. 6:5) has been constructed by Enoksson and is based on the principle that the explosive, subsoil water and such like in contact with a metallic conductor form a galvanic element which can generate a very weak, but measurable electric current. This can be determined with a sufficiently sensitive instrument (μA meter). The measuring should be carried out when the electric detonator is being pushed down the hole, or as soon as possible after the charging, because after some time the instrument may record complete deflection even if the insulation resistance is great enough. In so doing the detonator, or a series of detonators, is connected to the one pole of the instrument, while the other pole is appropriately earthed.

The earth pole must not be connected to any object, such as rails and pipelines, which can transmit power current. There are two instruments for measuring insulation faults. They are described in Chapter 6.1 c. An ordinary *insulation meter* cannot be used at the round as it gives too high electric tension. The so called *earth-fault meter* is recommended.

It is not necessary to check the insulation if the OD-detonators and the earth-fault tester are used and faulty detonators immediately replaced.

The earth resistance for every series should be at least twice the resistance of the series itself. For underwater rounds that have been examined, a large number of the detonators have had an insulation resistance between 5–100 kohm.

Fig. 11:11 shows a connecting schedule for an underwater round with 162 holes. The insulation resistance has been measured and indicated hole by

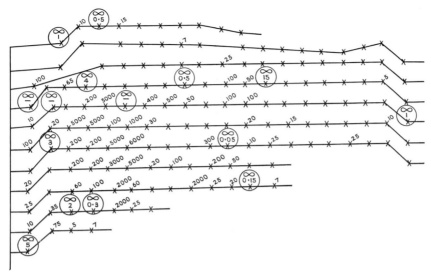

Fɪɢ. 11:11. Result of insulation measurement in kohm for a round with depth of hole between 1 and 9.6 m (3.3 and 31.5 ft). The resistance has been infinity where no figures are indicated.

hole with the detonating caps in every hole. In cases where the resistance has been lower than 5 kΩ the detonator is disconnected and a new one substituted, the insulation resistance of which is indicated above the original one. For the holes which have had infinite insulation resistance from the beginning no numeral is given in the figure.

In order to reduce the risk of incomplete ignition as a result of earth faults it is advisable to limit the number of detonators per series to 100 or less. With large rounds the number of parallel series will be large and require a thick ignition cable.

11.6. Ignition through flash-over

Until now rock blasting under water has not been performed without detonators in the holes. It seems possible, however, to systematize the occurrence of flash-over, that in common underwater blasting is only an embarrassment, and to adopt this system in order to simplify the ignition in a similar manner as in trenching in marshy grounds.

In such trenching a single row of holes can detonate from flash-over several tenths or hundreds of meters if the spacing of the holes is small enough (fig. 11:12a). A high-percentage NGL-explosive (≥93 %) is used.

The width of the trench obtained is 2–3 times the depth of the holes. If

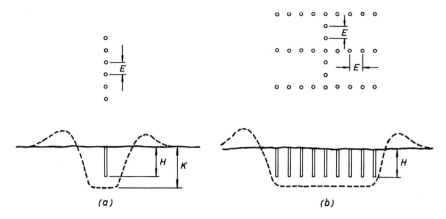

Fig. 11:12. Trenching in marshy ground.

the required width of the trench or channel is greater every second or third of the holes in the line (according to table 11:9) can ignite a row of holes perpendicular to the line, as in fig. 11:12 (b). This forms a multiple-row round with delays between the rows of about 0.7–1 ms per m.

If the same principle is to be applied in rock blasting the charging of the holes is arranged so that one special part of the charge consisting, for instance, of a tube of blasting gelatine is above the rock in the water and the spacing close enough for the shock wave to set off an adjacent hole in the row at the same time as the distance between the rows is too big for such a flash-over. An example showing the principle is given in fig. 11:13. This way of ignition gives greater shock pressures in the water than in ordinary blasting but not necessarily greater ground vibrations. The shock pressure can be reduced by an air bubble curtain (see 11.8).

TABLE 11:9. *Charges for trenching in watery sand and till.*

Charge per hole Q		Depth of hole H		Spacing (Burden) E	
kg	lb	m	ft	m	ft
0.25	0.55	0.6	2	0.4	1.3
0.5	1.1				
1	2.2	1.2	4	0.8	2.7
2	4.4				
5	11	1.7	9	1.5	5

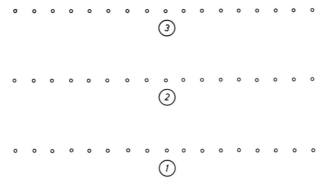

Fɪɢ. 11:13. Multiple-row underwater blasting in rock with flash-over ignition in the rows. One hole in every row is ignited by MS-detonators.

11.7. Ground vibrations and water shock waves

General bases for dealing with problems in connection with the ground vibrations are given in Chapter 9. The characterisation of underwater rounds as opposed to ordinary blasting is, first, the risk of an unintentional flash-over and, second, the larger quantity of charge. This makes estimates not only more complicated but also more uncertain since there is no assurance how a flash-over will occur. Apart from the fact that a flashover causes a larger total charge to be ignited within a certain time, there is the fact that faulty ignition sequence causes greater fixation and also for this reason stronger ground vibrations.

When the round is very extended "distance delays" must also be taken into account. This effect is simply illustrated in the blasting of a whole round with instantaneous detonators with interval time $\tau = 0$. If the distance to the drill holes from a measuring place for registration of the ground waves is 90–150 m (300–500 ft), there will be a difference in the time of arrival corresponding to 60 m (200 ft) between the ground wave from the nearest holes and those which are farthest away. In time this implies 20 ms if the ground wave travels at a velocity of 3000 m/s (10,000 ft/s). The ground vibrations are thus affected so that the frequencies higher than 100 c/s are practically eliminated through interference.

As the highest frequencies also give the greatest stresses in the rock this implies that the distance-interference in the case of widely extended rounds is of great practical significance. If the rock is covered with large layers of earth, clay or such like which extend into the surroundings, and there comprise the foundation for dwelling houses, conditions will be different since the waves which then have the greatest damaging effect have low

347

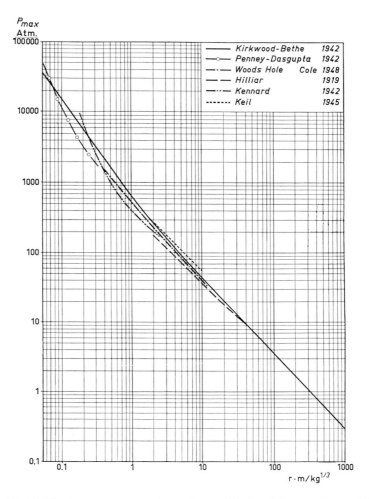

FIG. 11:14. Maximum pressure p in at (1 at $= 1$ kp/cm²) from a charge Q hanging freely in the water at a distance $R = rQ^{1/3}$ (according to Enhamre).

requencies, $f = 10$ c/s, or lower. Ordinarily the distance-interference is then of no importance. Reduction of the low-frequency oscillations here is achieved exclusively with the aid of delay ignition.

Another problem which is sometimes met with in underwater blasting is the effect of shock waves in the water on intakes, dam gates and other constructional details near the blasting site. For spherical charges suspended freely in the water questions falling under this category have been dealt with by E. Enhamre. Fig. 11:14 gives the connection between the maximum pressure p_{max} of the water shock wave and the scaled distance r (m/kg$^{1/3}$).

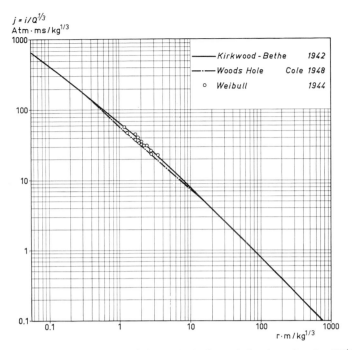

$j = i/Q^{1/3}$
Atm·ms/kg$^{1/3}$

——— Kirkwood - Bethe   1942

—·— Woods Hole   Cole 1948

○   Weibull   1944

r·m/kg$^{1/3}$

FIG. 11:15. Reduced impulse j for calculation of the impulse $i = jQ^{1/3}$ from a charge hanging freely in water (according to Enhamre). The impulse is given in Atm.milliseconds.

With a charge $Q = 1$ kg the scaled distance equals the distance R to the charge. If Q is larger than unity, r is obtained from the relation

$$r = R/Q^{1/3}$$

The duration of the water shock wave is so short that it has dropped from its peak value to half the pressure in a fraction of a millisecond. This implies in practice that co-operation does not occur between charges which are ignited at different intervals when the difference in time amounts to 12–50 ms, nor even for different charges which are ignited with the same interval number in which the spread in interval time is ±5 ms or greater. In ignition without a flash-over a maximum pressure of the same order of magnitude as from individual holes can be counted on.

If there is a flash-over the "interval time" is equal to that for the pressure wave to be transmitted in water from one hole to the next. This time is about 0.7 ms per meter. If such flash-overs affect a large section of the round, cooperation of the water shock waves can be obtained between several different holes.

349

11.8. Reduction of pressure and impulse in the water shock waves

The above discussion deals with a single charge freely suspended in water. The maximum pressure in the water is considerably reduced when the charge is placed in rock below the water level. According to investigations by A. T. Edwards the pressure was reduced to only 10–14 % when the consumption of explosive was about 0.7 kg/m³ (1.25 lb/cu.yd) of the broken rock. For the impulse we allow for a reduction to about 10 % for a charge in rock.

a. Reduction of pressure through the choice and placing of the explosive

This great difference between a buried and a free charge indicates that the first step in order to reduce the water shock wave from a blast is to avoid part of the charges to detonate while more or less uncovered in the water. One kg of the charge detonating in the water may give the same peak value as 200 kg well packed in the holes of the round and ignited simultaneously. One has to avoid

the top of the charges being too close to the top of the hole
risk of flash-over between separate holes in the round
several charges detonating simultaneously
part of the charges being exposed by tearing from adjacent holes.

This means that at least $15d$ of the top of the hole has to be left unloaded or loaded with a lower concentration of the charge, that an explosive with a low flash-over value should be used (see 11.3) and that short delay detonators should be used in a pattern (see Chapter 6) which ensures free breakage.

The water shock wave from a charge in rock can be regarded as a positive half-period of a vibration cycle (fig. 11:16b). Its duration is denoted $T/2$. This value is at a scaled distance $(R/Q^{1/3})$ proportional to the linear dimension of the charge $(Q^{1/3})$. For charges less than 10 kg (20 lbs) and distances below 20 m (70 ft) the duration is $T/2 < 2$ ms. In a round with several charges in rock the superposition of the pressure takes place in the same way as for ground vibrations. Charges less than 100 kg (200 lbs) in separate delays do not cooperate at all. In one and the same delay two or more of the compression waves add up depending on the relation between $T/2$ and the scattering in delay time $\pm \Delta \tau$.

The reduction factor γ given in table 9:2 can be used in the sense that it indicates how many of the charges in a delay may cooperate out of the total number of charges.

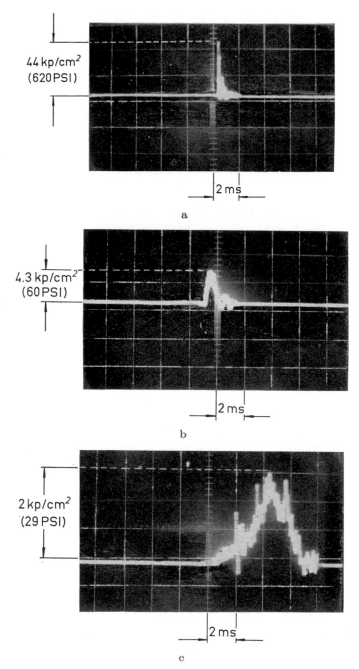

FIG. 11:16. Water shock waves from a) 1.35 kg (3 lb) (nitrone) in water at 14 m (47 ft) distance, b) 0.9 kg (2 lb) explosive in 0.8 m³ (29 cu.ft) of rock at a distance of 14 m (47 ft) and c) 0.45 kg (1 lb) explosive in water through an air curtain with 0.004 m³/s and per m (0.004 cu.ft per ft) at a distance of 10 m (33 ft). (Courtesy of A. T. Edwards, Hydro-Electric Power Commission of Ontario.)

Example 11:10. A round with 2400 kg (5300 lb) distributed in 240 holes is ignited in 10 delays of $\tau = 30$ ms between each. $T = 4$ ms. The scattering in delay time is $\Delta\tau = \pm 10$ ms. The reduction factor is $\gamma = 1/6$ which means that at most 4 of the 24 holes in an interval may cooperate, or at most 40 kg of the whole round.

Example 11:11. Calculate the maximum pressure at a distance of 22 m from the above round if the pressure from one single charge in the rock is supposed to be 14% of the pressure from the same charge in water.

Calculation: 10 kg in water at a distance of 22 m gives the same maximum pressure as 1 kg in water at a distance of $22/10^{1/3} = 10$ m. Fig. 11:14 gives here $p \simeq 40$ kp/cm². The pressure from 10 kg in rock at a distance of 22 m is then $p \simeq 5.6$ kg/cm². Four cooperating charges give at most 22 kg/cm² which is the maximum pressure that can be expected from the round as a whole.

b. Reduction of pressure by an air-bubble curtain

If excessive maximum pressure is feared even with the precautions mentioned above, the structures affected may be screened off by an air bubble curtain. This method is suggested by A. Lapraire, C.I.L., Canada, and the effect is described by R. C. Jacobsen, A. T. Edwards and others.

When the pressure wave arrives at the air bubble curtain, part of it is absorbed in the bubbles. These are compressed during some ms or some fractions of a ms and emit during some ms new compression waves (with lower peak values) in all directions. The part of the shock wave that passes between the bubbles gets a more or less reduced peak and a more rapid attenuation than the original undisturbed shock wave. The more air-bubbles the greater is the reduction in peak value at the same time as the duration of the pressure in the water increases at a distance from the curtain. According to investigations by Edwards at the Hydro-electric Power Commission of Ontario, the peak pressure from 1 lb of explosive at a distance of 10 m was reduced by a factor of 10 with an air flow of 0.004 m³/s per m length of the pipe. With an air flow of 0.008 m³/s the reduction factor was 70. (The corresponding factors at a distance of 50 m were 8 and 40 respectively.)

The pipe in these experiments had a diameter of 75 mm (3 in) and two 1.5 mm ($\frac{1}{16}$ in) diameter holes diametrically opposite each other spaced about 20 mm apart (0.75 in) for one pipe and 40 mm (1.5 in) for another. For ordinary charges of the size 10 kg, or even more, it ought to be possible to achieve the same reduction factors but at a cost of more air per second.

These results are important when structures have to be shielded from charges detonated in the water and in such underwater rock blasting where the flash-over effect is used for the ignition.

In other ordinary rock blasting under water when the above mentioned general precautions are taken there is by no means the same effect with an air-bubble curtain, as the pressure is then already reduced before it reaches the curtain. When this reduction is obtained experience has hitherto indicated that if there is still a risk of damage this is often due to the impulse being too high and not the maximum pressure. In such a case more effective means can be used than an air curtain which in principle distributes the impulse over a longer time period, but does not reduce its total value.

c. Reduction of the impulse

The impulse that hits a structural element causes this to vibrate with its natural frequency (f_0) and possibly also with overtones. If the deflection is too great, the element in question will be damaged, even if the compression in the shock wave is—as ordinarily—far below the compressive strength of the material. The deflection is determined by the pressure-time curve, or with a simplification approximately by the total impulse delivered during $T_0/2$, where T_0 is the time for a cycle of the natural frequency. The duration of the water shock wave, or the main part of it, is denoted $T/2$. This time is at a distance of 10 m of the order 0.15 ms for 1 kg explosive in water, 1 ms for 1 kg in rock below the water and 4 ms with an air-curtain according to Edwards (fig. 11:16a, b, c). The corresponding T-values in an actual case may be denoted T_w, T_r an T_a.

If $T_0 \geqslant T_a$, which with the above figures means $f_0 \leqslant 125$ c/s, there is no essential reduction in the effective impulse and consequently not in the risk of damage through an air-bubble curtain either. In such a case it is quite misleading to use an air-curtain and rely upon the reduction in peak pressure.

The measures to be taken are then, *first*, those mentioned above (11.8a), which means that the impulse from every individual charge, Q, is reduced to less than that from $0.1Q$ free in water.

Second the impulse can be reduced through a suitable ignition pattern. The reduction factor γ is calculated in the same way as in table 9:2, if the T_0-value is used instead of T.

Example 11:12. A dam gate with $T_0 = 50$ ms is at a distance of 20 m from the round in Example 8. The reduction factor is $\gamma = 1$. All the 24 charges in one delay cooperate.

Example 11:13. Calculate the total impulse on the gate during $T_0/2 = 25$ ms.
Calculation: A charge of 10 kg in the rock gives a maximum impulse corresponding to, at most, 1 kg in water. With this charge $j = i$ and $R = r$ in fig. 11:15 and we get for $r = 20$ m $j = 4$ Atm. ms. 24 charges can at most give an impulse of 96 Atm. ms. It is the same impulse that a static pressure $p = 3.7$ Atm. exerts during the time $T_0/2 = 25$ ms. The individual delays do not cooperate.

Example 11:14. Reduce the impulse to 40 Atm.ms during $T_0/2$. This is done by using 24 delays with 10 holes in each. They cannot together give a higher impulse than 40 Atm.ms. (the corresponding static pressure during $T_0/2$ is $p = 1.6$ atm).

Third. If the holes closest to the structure to be shielded are ignited first and with a successively increased number of holes in the following intervals the structure will receive a lower impulse from the first delay. The impulse from the following delays will be partly screened off by the broken rock and the explosion gases from the holes in front of them.

Fourth. If there is still a risk that the impulse values acting on a structural element will be too high they can be further reduced by some kind of support which transmits the impulse from, say, a particularly sensitive intake door to adjacent, less sensitive elements, or if possible to the ground; or the door can be loaded from behind with water pressure or a mechanical support to which part of the impulse will be transmitted.

Only when the T_0-values are so small that $T_0 < T_a$ does the air-bubble curtain reduce the effective impulse from a single charge to that part of the time pressure curve that can be included in the time $T_0/2$. The smaller the T_0-value and the higher the proper frequencies f_0 the higher degree of reduction is obtained.

For rounds with several holes the reduction that can be obtained through the delay effect (of τ and $\Delta\tau$) is more effective.

Example 11:15. $T_a = 10$ ms, $T_0 = 5$ ms. Calculate the total impulse on the gate in Example 10 during $T_0/2 = 2.5$ ms.

Calculation: The reduction factor according to table 9:2 is $\gamma = 1/6$. Only 4 of the charges in one delay cooperate. The greatest possible impulse is 1/6 of that in Example 4, or 16 Atm.ms.

The conclusion is that the air-bubble curtain is effective for charges detonating in water and for single charges in rock when the frequencies of structural element are higher than 100 c/s. In ordinary rock blasting under water the air curtain is either ineffective or not needed when the desired short delay detonators are available and the round is loaded and ignited according to the recommendations above (11.8a).

FIG. 11:17. Working platform for blasting an underwater round in open water (a and b). The plastic tubes which "extend" the drill holes up above the water surface are seen in the lower picture c and d show two stages of the round during the blasting (AB Skånska Cementgjuteriet).

Fig. 11:18. Checking of ignition system in underwater round drilled from ice-covered water surface (AB Skånska Cementgjuteriet).

11.9. Various types of objects

Underwater blasting covers a very wide range of objects with most varying conditions. This is perhaps best illustrated by the fact that, apart from blasting small and ordinary rounds, the world's largest rounds with more than 1 Mkg (2 Mlb) in a round must also be included in this group. Then there are in addition the external circumstances, which vary considerably, and the fact that in working in open waters one is at the mercy of the weather and the vagaries of nature to an extent practically unknown in ordinary bench blasting.

It is characteristic of all general planning of the work that endeavours are made to reduce or eliminate work with divers. A platform is built from which the work can be carried out, as will be seen from fig. 11:17. If a platform is used which can be transported to the spot with the help of pontoons or such like, it should be firmly anchored to the bottom, and to avoid the effects of swell and water waves the pontoons should be placed beneath the surface of the sea if they are to have a bearing function. Work from a raft in open waters entails great difficulties, not least in drilling. Then the drill holes have to be driven vertically and a greater number of holes is needed.

Fig. 11:17 c and d show the movement and rise of the water when blasting a round of 5 tons (11,000 lb) of explosive. One photograph was taken some tenths of a second after ignition, the other after some seconds.

The problems with a working platform are eliminated if the ice-cover of the water can be used for drilling and charging. It must not be forgotten,

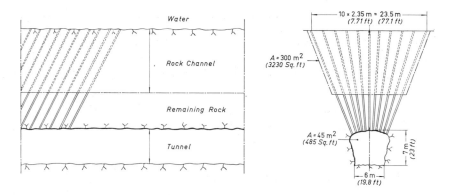

Fig. 11:19. Proposal for blasting rock channel under water. Work carried out from a working tunnel blasted out beneath intended channel. The lower parts of the drill holes in the "Remaining rock" have cement injected after the loading.

Hole diameter 50 mm (2 in). Channel to be blasted to depth of about 50 m (164 ft) below water surface.

on the other hand, that variations in the level of the water which alter the level of the ice crust may hazard the entire round if the extension tubes of the drill holes are withdrawn. Fig. 11:18 shows an example of a wellorganized plan for underwater blasting from ice.

Another possibility of avoiding many of the practical difficulties entailed in deep underwater blasting is illustrated in fig. 11:19.[1] This applies to a channel which is to be blasted in underwater rock to a depth of about 50 m (165 ft). With a working tunnel under the channel, as will be seen from the figure, the drilling and blasting for the channel can be done as in simple stoping. The problems to be solved here are rendering the drill holes and joints between two consecutive rounds leak-proof if the drilling is done in advance. Here too, as is usually the case with underwater rounds, it will be desirable to blast as large rounds as possible.

Fig. 11:20 shows a round of 15 tons (33,000 lb) which is blasted under an overburden of clay and earth. The photograph was taken 1000 ms after the current was switched on. A more detailed account of this and the previous type of blasting is given by Brännfors.

---

[1] The method, which is based on a proposal by the Nitroglycerin Company in 1957, has a number of features in common with the so-called Janol method, but it should be pointed out that the latter was evolved without a knowledge of the former.

11.10. Practical directions

As a result of the investigations the following general directions can be given.

(1) The aim should be to have the smallest possible number of holes that will give a full guarantee for breakage and satisfactory fragmentation.

(2) This implies, among other things, that large drill holes should be used.

(3) The holes should be arranged in a very simple pattern, even if some additional ones are needed at occasional points in the round.

(4) It is essential to avoid a flashover between the holes. A flash-over causes inferior fragmentation and risk of incomplete loosening of the rock at the bottom. This implies that the result can be impaired by increasing the quantity of the charge in a round. Moreover, the charge must not rise too close to the top face of the rock. At least a length $= V/2$ should be left uncharged possibly less, with a greatly reduced charge per meter.

(5) The risk of a flash-over can be reduced by charging every alternate hole with an explosive which is less sensitive to a flash-over.

(6) Sloping drill holes give a better guarantee for bottom breakage and, at small bench heights, a considerably reduced number of holes.

(7) By blasting 1 to 2 m (3–7 ft) below the bottom at low bench heights a considerable reduction in the number of holes is obtained. The drill holes should be driven at least 0.6 of the maximum burden under the intended bottom.

(8) Charging should be done with the NAB pneumatic or water-pressure loader.

(9) Only alternate holes should be charged in the first relay.

(10) Every hole should be provided with one or two OD-detonators or two or more ordinary (high quality) electric detonators.

(11) Increased inspection of the ignition system is needed. Every detonator should be tested directly with an earth-fault tester in the course of charging, as well as before being finally connected to the round.

(12) When connecting up large rounds with two bottom detonators per hole in different series it is advisable to do this in teams of two men. Then the one man will advance slightly before the other and connect the one series, marking it, for instance, with yellow plastic sheaths over the joints; the second connecting the remaining detonators marks the joints, for instance, with black sheaths.

FIG. 11:20. A phase in a round of 15 tons (33,000 lb) and 800 holes blasted without removing layer of clay (AB Skånska Cementgjuteriet). Picture taken 1.000 ms (1 sec) after the ignition of the round.

(13) The explosive should be able to stand storage in water for the time required, but not for an indefinite period.

(14) In many instances work by a diver under water can be avoided by using special drilling platforms and polythene tubes which will extend the holes to above the surface of the water or by drilling from the icecovered surface if there is a possibility to do the work during the winter.

(15) A rock face which has a cover of material other than water need not be laid bare in order to achieve free breakage in blasting.

(16) In dealing with the problem of ground vibrations special attention must be devoted to the risk of changes in the ignition sequence and fixation conditions, if the possibility of a flashover between different rows of the round must be taken into account. For large and scattered rounds regard must be had both to reduction by the delay ignition and by the difference propagation times for waves coming from different parts of the round.

11.11. Questions put by a contractor before the planning of the job

Questions

1. Is there any danger to the explosives and detonators in the day to day store on board the drilling-pontoon, due to vibration from compressors or drilling?

2. Is there any danger of detonators exploding through galvanic action from steel barge on spuds in *salt* water and copper wires of detonators?
Static electricity?

3. Which drilling pattern seems to be indicated when $d = 51$ mm?
a) for the layer thickness of 3.6 m + overdepth
b) for the maximum layer thickness of 1.8 m + overdepth
c) for the maximum layer thickness of 0.30 m + overdepth.
Assuming a fixed distance between the rows of drillholes of 1.5 m.

4. What will be the required overdepth in case 3 *a*, in case 3 *b*, in case 3 *c*?

5. Which type of explosive is indicated?
a) % of Nitroglycerin?
b) % of nitrate of ammonia?
c) Speed of propagation?
First in the case of using the pneumatic loader, second when loading by hand.

6. What is the probable quantity of explosive per m³ required, in case 3 *a*, in case 3 *b*, in case 3 *c*? In average?
Assuming that a fragmentation giving cubes of *20 cm-sides* is required:

7. Is a special explosive-cartridge required to guarantee an explosion better?
If so, which type?

8. Which type of detonator is best suited for 10 m under water, in salt water, and staying 3 days underwater?

9. Are the ordinary detonators for underwater blasting sufficient in this case?

10. What are the advantages of the hydraulic loader, of the pneumatic and of both in respect to hand-loading?

Connections

11. Why are the series-connections mostly used in Sweden? What are the advantages?

12. Do they not require a very high ohm-resistance to the earth for, say, 160 to 200 shots fired at once?

13. Is it not safer to used parallel connections requiring less ohmic resistance to the earth?

14. Which test should be made for control? At what stage of the charging and connecting

a) to the detonators?

b) to the circuit?

15. How can connections be made waterproof (very salt water)?

Which substance is suitable to apply cold?

How is the insulation applied *hot*?

Is the insulation effective immediately, so that the joint can be dropped in the water at once?

16. Which protection or insulation would be suitable if connections are kept above water, say hanging on a line 0.50 m above water?

17. Which wire thickness is recommended for the detonators?

Are the wires to be protected by an extra plastic-tube?

Up to what height?

a) just at the bottom of a hole (loader-end)?

b) up to just above rock level?

c) up to above water level?

18. Is it preferable or necessary to use

a) 1 detonator only in each hole?

b) 1 detonator and one special cartridge?

c) 2 detonators?

19. In using millisecond detonators, how many milliseconds should the intervals be?

20. Why are the 0-detonators not recommended? What is the danger of using them?

21. Is a special bottom explosive required?

22. Can it be applied with the pneumatic loader?

23. Up to what level must the holes be charged with explosives? (How far below the rock surface.)

a) in 3.60 m rock-layer + overdepth under 4 m water?

b) 1.80 m and 6 m water?

c) 0.20 m and 9 m water?

24. When the connections are in parallell, say 160 to 200 shots to be fired at once which exploder should be used?

Can the dynamo of a welding set be used?

Which characteristics are required?

Voltage?

Amperage?

25. What is the influence of high velocity explosives on flash-over? Can an explosive with 45% nitroglycerine and 7000 m/sec speed be used?

26. Which diameter of the firing wire is to be used, say, for a length of 100 m?

27. Which test or control apparatuses are to be used?

28. Can a cartridge of $7/8'' = 22.2$ mm be used in the

a) pneumatic loader of $\varnothing$ 25?

b) hydraulic loader of $\varnothing$ 25?

29. Can a pneumatic or hydraulic loader be quickly made available for $1\frac{1}{8}''$ or $1\frac{1}{4}''$ cartridges?

Answers to the questions

1. Not for ordinary civil explosives nor for good quality detonators (see below).

2. No. No.

3. Depends on deviations between the pattern and the practical drill hole position at bottom level and the inclination of the holes. The following suggestions are made for a)–c):

Vertical holes: First row $E = 0.67$ m (the distance $V = 1.5$ m from the contour line), Second row $E = 0.67$ m, Third row $E = 0.85$ m. Fourth row $E = 1.2$ m. Subsequent rows $E = 1.5$–2.0 m.

Sloping holes 2:1. First row $E = 0.67$ m. Second row $E = 1.0$ m. Subsequent rows $E = 1.5$–2.0 m.

4. See fig. 11:6.

5. A dynamite with 45–50% ngl is to be preferred. (This is *not* the same as what Americans call 45–50% strength.)

If it has to be in the holes for an extremely long time (more than half a month) a 50–60% ngl explosive can be used. This increases slightly the risk of a flash-over. A higher % of ngl should not be used.

6. For the row Nos 1–4 the consumption of explosive in kg/m³ (q) is roughly

Row No	vertical holes	sloping holes
1	$q \approx 13$	$q \approx 9$
2	$q \approx 5$	$q \approx 3$
3	$q \approx 4$	$q \approx 1.5$
4	$q \approx 2.5$	$q \approx 1.5$

Note that the figures are calculated for the broken rock (not for the rock above the grade).

The charges are in themselves sufficient for the fragmentation desired, 0.2 m. If it is not acquired it is due to some fault in the ignition pattern, or to sympathetic detonation that destroys the ignition sequence intended. The fragmentation is better with inclined holes.

7. No.

8. All ordinary good detonators can be stored for 3 days at a depth of 10 m. What is also decisive in underwater-blasting is how all the details of the cap, including the leg wires and the insulation, are affected during the loading process. In this respect, only the NAB Lindö detonators seem to fill all demands, and maybe also the Du Pont and other detonators with nylon insulations.

362

9. See above. With ordinary detonators at least 2 detonators should be used per hole.

10. The hydraulic loader is not so dependent on the accuracy of the diameter of the cartridges. It is easier to overcome a jam in the tube as pressure is higher in this loader. The pneumatic loader is usually satisfactory. They both have a much better loading capacity than hand-loading.

Compared with loading with tamping poles the loaders mentioned give 15–40% higher density, and there is less risk of damage to the detonators. When the loader with the detonator is inserted in the hole it is loaded before the tube (hose) is taken out of the hole for loading of the next hole.

11. It is easier to connect the detonators in series. With a given igniting cable and blasting machine, you can have a greater number of detonators in the round. The loss of energy in the cable is much higher in parallel connection. In a 200 shot round with 5 ohm per shot and requiring, say, 0.6 amp each you need 120 amp from the blasting machine, the resistance in the round as a whole is 0.025 ohm. With a cable-resistance of only 1 ohm (1.025 total) 40 times as much energy is consumed in the cable as in the round.

12. A round of 200 shots is, e.g., connected in 4 series with 50 detonators in each; you need a resistance to earth of more than twice the resistance in one series. This may be about $0.5 \, k$ ohm (500 ohm). This is in fact a very low value and if you have it, there is definitely a fault in the insulation and it is in any case advisable to exclude the detonator in question.

13. If you have a bad system with a lot of earth-faults the loss of energy may be less with a parallel connection. But don't forget that, e.g., as in 11. 97.5% of the igniting energy is lost already in the cable.

a) It is most important that testing with the earth-fault tester shall be done while loading the hole. It reveals if any serious damage occurs during that operation. In such a case the detonator is to be changed immediately.

b) With an ohm-meter (specially designed for blasting) and supervision.

15. See 11.5 first paragraph.

Apply some kind of insulation tube 3–4 cm in length. Connect up and then dip the free metal part of the joint in the molten plastic. Remove and draw the tube over the joint. The plastic is kept hot by placing it, e.g., on an electric plate.

Then the insulation becomes quite effective after some seconds.

16. In principle no insulation is needed if there is no risk of water reaching the connections.

17. Special demands on the mechanical strength of the insulation are: firstly just close to the cap, secondly for the part of the wire in the rock and in the overburden.

A good submarine detonator does not need any extra protection. See answer 8.

18. a), c) With Swedish submarine detonators and an earth-fault check (see 14 a) one or two are needed in the hole (practical fault percentage up to 2% each). With other detonators you should have 2–4 because of the higher percentage of faults (up to 10% and sometimes more depending on conditions). Note that a single fault in a round doesn't necessarily mean that this very hole will be a misfire. If there is an electric current leak it implies a danger to the rest of a round or part of it.

b) If the charges are to be for a long time in water the detonator can be used with a 60% NGL cartridge when the rest of the charges is with 50% NGL.

19. Between 5 and 100 ms. The best fragmentation is received with 3–10 ms interval between the holes in a row (ordinarily spread in delay time) and 10–30 ms between the rows.

20. No O may detonate before all the others have obtained their igniting energy.

21. Ordinarily not.

22. Should be, as the packing at the bottom is most important.

23. With 50 mm holes up to 0.6 m from the rock surface.
With 100 mm holes 1.2 m. This is restricted by the risk of a flash-over.

24. The NAB blasting machines for 250, 600 or for 2400 shots are by far the best ones. The two biggest give an even better margin, as their electric energy is much higher. The machines mentioned give less tension than 700 V at the round.

No.

The characteristics can be given in simple terms of voltage and amperage only if you have DC-current, and by DC I mean in this case that the ampere-time current is a steady one. Then in a series of detonators everyone should have a current 1–2 amp depending on the type of detonator.

25. The flash-over distance increases (table 11:8). Yes, see answer 5.

26. The resistance of the wire should be less than 5 ohms in all.

27. The earth-fault tester and an ohm-meter (compare answer 14 b).

28. a) It may be used, but we don't recommend it. If so, however, the jet should have a diameter of 22.3 mm (according to Ljungberg).

b) Yes, here too the jet must be diminished.

29. Yes, there are such apparatuses for immediate delivery.

12. NON ELECTRIC BLASTING

The possibility of firing large rounds with short intervals as described in previous chapters caused something of a revolution in the field of rock blasting. It is now possible to calculate and master in an entirely different manner the problems of fragmentation, scattering and ground vibrations. In multiple-row blasting, drilling and loading can be carried out on a larger scale, which with the proper mechanization substantially increases the capacity per unit of time. Since the technique of calculating the charges has also been refined to a high degree of precision, very large blasting operations can now be carried out in accordance with predetermined blasting patterns. Rock blasting has come to be an industry.

Short-interval blasting has hitherto been possible only with electrical ignition. Great progress has been made when it comes to precision and safety, especially with modern detonators with increased safety, but absolute safety is unlikely to be achieved as long as electricity remains in the picture. Electrical ignition, moreover, is not really convenient—hooking up is a tricky business and initiation requires blasting machines which often have extremely powerful ignition impulses.

12.1. The Nonel system

The new non-electric firing system is designed for the completing of the *Controlled Blasting*[1] technique adding new safety and reliability to the work. It is easier to train new personnel in Nonel firing—control and calculation of resistances and earth faults are not required.

The possibility of faulty connections has been practically eliminated with the Nonel GT system. The percentage of misfires is reduced to a level below that in electric ignition, which is also an essential advantage.

12.2. The Nonel GT detonator

The Nonel GT detonator consists of a Nonel tube with the free end sealed, a (delay) period marking tape, a packing tape and a delay detonator. The periods available are given in table 12:1.

[1] Sometimes incorrectly referred to as "Swedish Blasting Technique". It is in fact completely international.

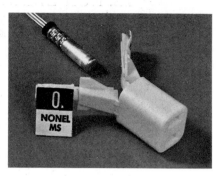

FIG. 12:1. A Nonel GT detonator.　　　FIG. 12:2. A Nonel GT connecting block.

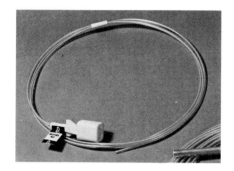

FIG. 12:3. A Nonel GT 1 connector with *one* connecting block.

TABLE 12:1. *Periods and delays for Nonel GT detonators.*

Period number	Number of periods	Delay ms	Interval ms
3–20	18	75– 500	25
24, 28, 32 36, 40, 44	6	600–1100	100
50, 56, 62 68, 74, 80	6	1 250–2 000	150

366

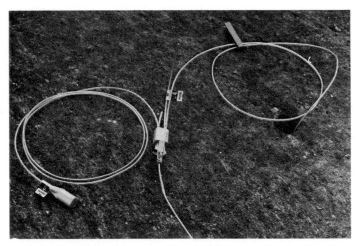

Fig. 12:4. Connector, starter and detonator (in the hole).

It is to be noted that the range of the Nonel GT timing does not only offer short delays 1–18 and half delays in between as described later, but also the application of deci-second blasting (DS-blasting) with 1–17 delays of 0.1 s = 100 ms and 0.15 s = 150 ms, replacing the older second or half-second delays. This will meet another requirement of the controlled blasting technique.

The systems with MS-blasting and with DS-blasting complete each other in effective rock blasting.

12.3. The connectors and starters

A connecting block includes a transmitter cap, see fig. 12:2. The strength of a transmitter cap corresponds to about 1/7 of a high strength detonator and it has been adapted merely to initiate Nonel tube. Because of the

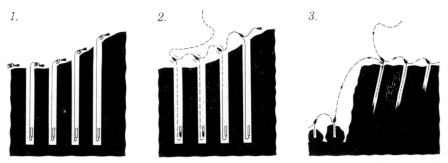

FIG. 12:5. A bench round before connecting (1), some ms after initiation (2) and a connection of a multiple-row round combined with boulder blasting (3).

367

Fig. 12:6. Connecting a Nonel detonator to a connector. The detonator tube is folded double and inserted in one of the openings in the connecting block.

control obtained through the design of the connecting block, the Nonel tube connected up is in contact with the transmitter cap, which gives each Nonel connection a standard quality which does not change under the effect of either external conditions or varying degrees of professional skill.

For connection and joining up *connectors* are used. A connector GT 1 consists of a Nonel tube with the free end sealed, a packing tape and one yellow connecting block with an instantaneous transmitter cap.

For tunnelling and other applications with closely spaced holes a connector is available with two connecting blocs, GT 2.

The starter is a connector with a greater length of the tube as seen in the table.

Fig. 12:4 shows a starter connected to a detonator (to the right) placed in a hole and to a connector (to the left) which is open for connecting of three other Nonel GT detonators.

TABLE 12:2. *Nonel GT connectors and starters.*

Connecting block	Delay ms	Standard length of tube m				Description
1 Yellow	0	1.8,	3.0,	4.8,	6.0	GT 1
2 Yellow	0		2.4			GT 2
1 Yellow	0	30,	50,	100,		Starter (GT 1)

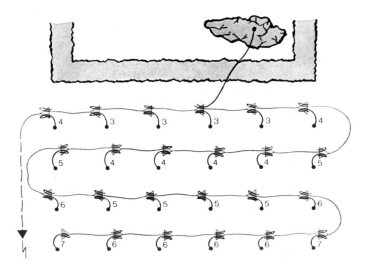

Fig. 12:7. Nonel GT being used in a bench round with small extension. In the simplest case, one connector is used per hole. The holes in the rock are symbolized as ●.

12.4. The connecting of a round

a. Initiating

The Nonel rounds can be initiated by 1) a starter and a starting pistol, 2) by safety fuse and a fuse detonator or 3) by an electric detonator in a starting block, which is joined to the round through a connector.

b. The ignition pattern

The ignition pattern is made according to the general rules already given in Chapters 6–11. A difference to be noted is that there is a certain time for transmission of the impulse through the Nonel tube from one hole to another. This causes an *additional delay* time of 1 ms for every second meter of the tube (1 ms/2 m). Between two adjacent holes in one and the same row this is ordinarily less than the spread in the delay time in a given delay number. Take for instance a row with delay numbers No. 3 and two meters between the holes. The delay time here is 125 ms±10 ms. The average time to be expected for the first hole along the initiation sequence is then 125±10 ms, the next hole 126±10 ms and so on. But for the following row, if initiated through the last hole in the row in front of it (as in the example in fig. 12:7) there may be a significant addition to the time between the rows if the extensions of the row is more than 10 meters. The

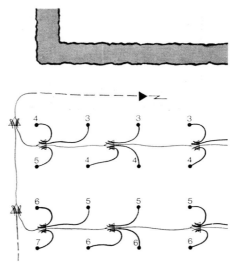

Fig. 12:8. Detail of a bench round with fully utilized connectors, which gives the best safety and economy.

additional delay in ms caused by the Nonel tube will then be to the order of the bench width given in meters for the holes last initiated in the second row.

With bench width W = 10 m or more it is recommended to arrange the gnition as in figs. 12:8 or 12:9.

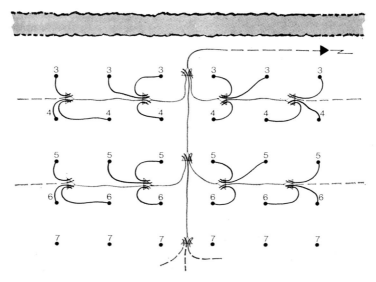

Fig. 12:9. A symmetrical central ignition of a multiple-row round.

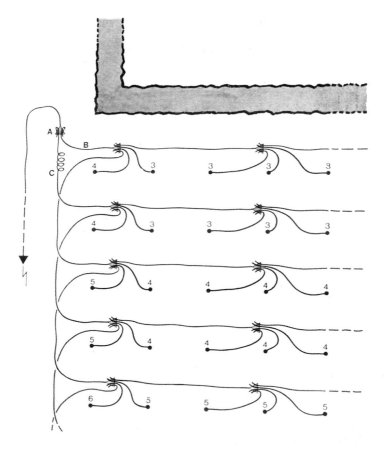

Fig. 12:10. A standard method for using half delays with the aid of a starter (C) which delays the ignition 12 ms for every second row.

c. Half delays

The system offers a special advantage in the possibility of arranging an ignition pattern with delays of 12.5 ms as an extension of the ordinary delays of 25 ms. At the same time the number of effective delays can be doubled, which can be of great importance in controlled blasting. This is done connecting a starter (C) with a Nonel tube length of 30 meters (15 ms delay) instead of a connector GT 1 (B) with 4.8 meter length (2.4 ms delay) and derived from this starter ignite every second row as in fig. 12:10. The shorter delays can reduce the ground vibrations and so also the throw in open air blasting, the greater number of delays can permit greater rounds and better accuracy when correctly used in controlled blasting.

371

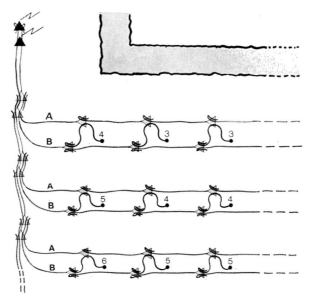

FIG. 12:11. Two parallel connecting lines with GT 1 connectors.

d. Duplicated connection

In certain cases, incomplete blasting can imply considerable risks, for example when demolishing constructions, blasting out residual sections in mines, such as pillars, ore headings, etc.

It is possible to take precautions in such situations by reinforcing the connecting system of a round.

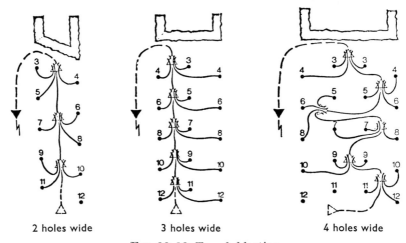

2 holes wide   3 holes wide   4 holes wide

FIG. 12:12. Trench blasting.

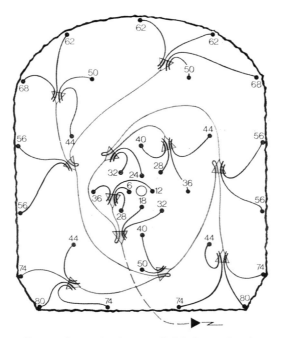

FIG. 12:13. A small tunnel round with parallel holes and twin connectors GT 2.

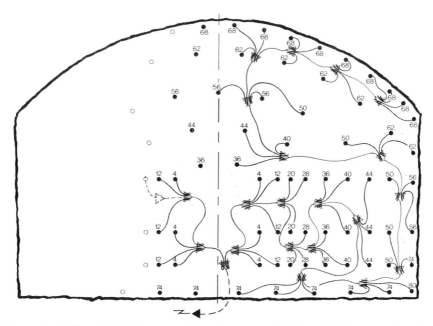

FIG. 12:14. Nonel GT being used in a large tunnel. Since hole location is relatively well spaced in large areas the mono connector of GT 1 type is used.

This is also recommended when conditions require two detonators per hole. The higher degree of initiation dependability needed when using double detonators is only obtained when the connecting system has been reinforced accordingly.

e. Trench blasting

The best ordinarily used drilling patterns are given in Chapter 6.8. How the connection for Nonel blasting is to be done is shown in fig. 12:12. Even here the Nonel possibilities to use half (and generally split delays) can be taken advantage of. It can contribute to less overbreak and better swelling of the broken rock.

f. Tunnel blasting

Some simple examples of the application of Nonel firing in tunneling are given in figs. 12:13 and 12:14. For small tunnels the twin connectors GT 2 are used, and for larger tunnels the mono connector GT 1 can be used.

13. THE USE OF UNDERGROUND SPACE

As the technology of dealing with rock has advanced it has also opened new opportunities in using the underground space for various new purposes. One or two decades ago the general approach in planning and architecture was to try to avoid the rock if such was in the picture. Now the situation is often reversed. The rock offers conditions with unparallelled stability for the foundation of buildings and for air-fields. It means even temperature conditions in various forms of underground facilities. It is today well-known how to handle the rock and how to plan the constructional work. The cost as related to other alternatives is in many cases essentially reduced. At a conference on engineering sciences in a big city partly founded on rock the host from the city addressed the conference with the statement that the current planning for the city had in its actual and effective form been made possible by the modern technique of rock blasting.

In cases where the underground space is needed or wanted and today's technique makes an economic use of it possible the decision-maker has to use an ETI-analysis (economy–time–intangibles) where economy and time can be clearly quantified. The uncertainty and the intangibles are related to such fields as: throw, vibrations, overbreak. These factors are with today's knowledge possible to quantify to a great extent with methods described in other chapters and thus reducing the uncertainty. Other remaining intangibles of various kind have to be labelled for each individual project, quantified (even if roughly) when possible and when not ranked according to importance and whether positive or negative in favour of the project.

In a study of the subject of this chapter the "Underground Construction Research Committé" (UCRC) have indicated some areas for improvements with the greatest probability of short-term payoffs:

Better contact form
Materials handling
Standardization of tunnel areas
Geological exploration ahead of face
Analysis of operation as a system
Pre-construction geological evaluation
Information dissemination

As an illustration to the use of underground space a series of figures, diagrams and tables are given.

13.1. Geological areas of the earth

13.2. Technical objects

a. Transportation
b. Energy
c. Water
d. Sewage—drainage
e. Underground extension of city buildings
f. Underground industry
g. Storage
h. Shelters
i. Defence facilities
j. Nuclear waste

13.3. Economy

Fig. 13:1. Geological areas of the earth (Courtesy AB Skånska Cement-gjuteriet).

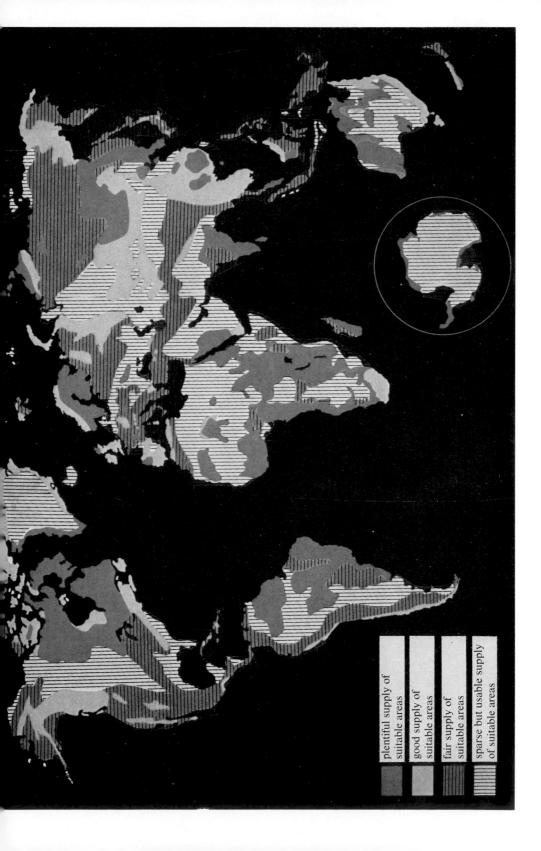

plentiful supply of
suitable areas

good supply of
suitable areas

fair supply of
suitable areas

sparse but usable supply
of suitable areas

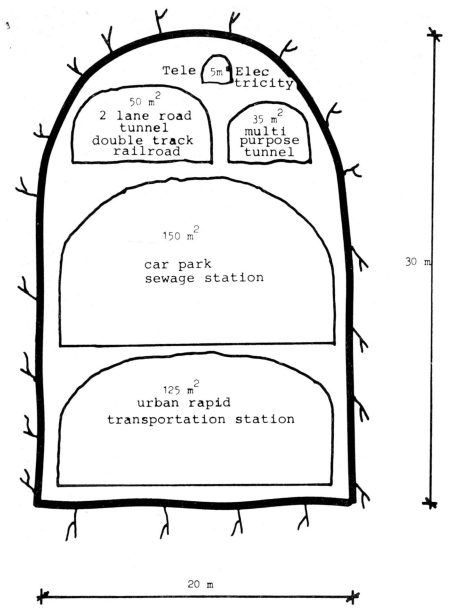

Tele | 5m | Electricity

50 m^2
2 lane road tunnel
double track railroad

35 m^2
multi purpose tunnel

150 m^2

car park
sewage station

125 m^2
urban rapid
transportation station

30 m

20 m

FIG. 13:2. Sections indicating relative proportions for some UC objects. The surrounding contour shows the size of a room for oil storage.

Fig. 13:3. Top and bottom gallery excavation for a large cavern (Courtesy Rock Tank Group). Photo: Manne Lind.

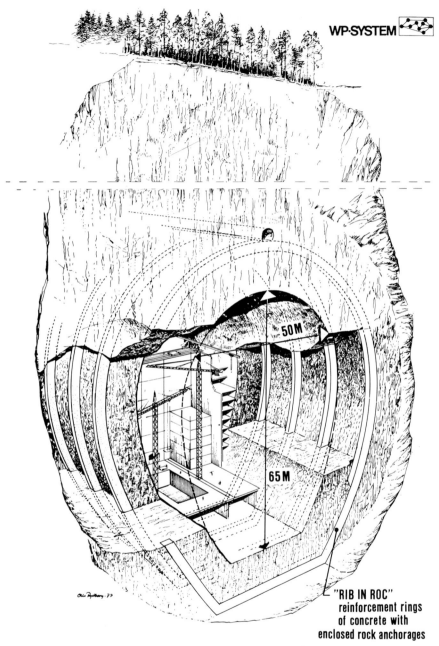

WP·SYSTEM

50 M

65 M

"RIB IN ROC"
reinforcement rings
of concrete with
enclosed rock anchorages

FIG. 13:4. The WP "rib-bone" method for excavating with very big spans
(Courtesy WP System AB).

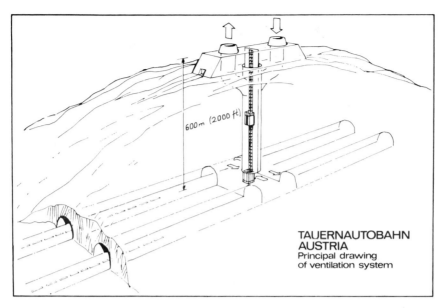

FIG 13:5. Road construction in Austria (Courtesy Linden-Alimak AB).

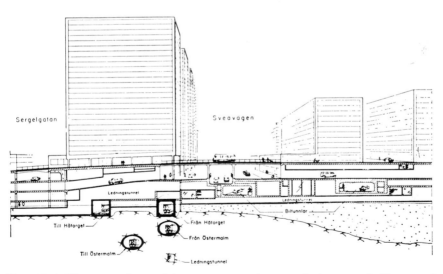

FIG. 13:6. The great demand for more space in central city areas is illustrated by this section of the metro-system of Stockholm.

381

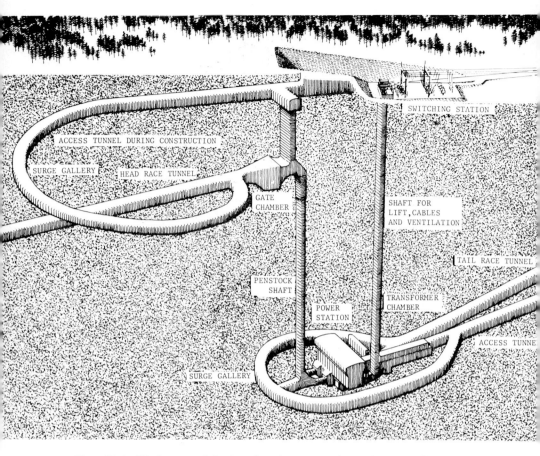

ACCESS TUNNEL DURING CONSTRUCTION

SURGE GALLERY   HEAD RACE TUNNEL

SWITCHING STATION

GATE
CHAMBER

SHAFT FOR
LIFT, CABLES
AND VENTILATION

TAIL RACE TUNNEL

PENSTOCK
SHAFT

TRANSFORMER
CHAMBER

POWER
STATION

ACCESS TUNNE

SURGE GALLERY

FIG. 13:7. Underground hydro-electric power plant (Courtesy VBB, Sweden).

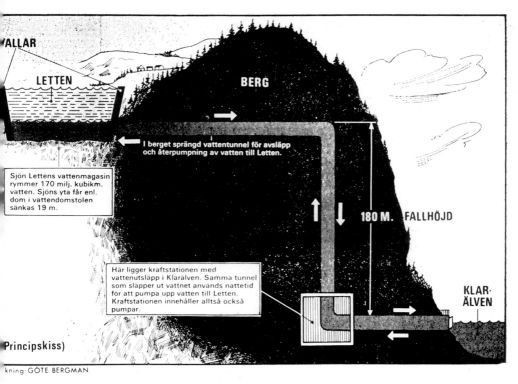

Labels within figure:
- ÄLLAR
- LETTEN
- BERG
- I berget sprängd vattentunnel för avsläpp och återpumpning av vatten till Letten.
- Sjön Lettens vattenmagasin rymmer 170 milj. kubikm. vatten. Sjöns yta får enl. dom i vattendomstolen sänkas 19 m.
- 180 M. FALLHÖJD
- Här ligger kraftstationen med vattenutsläpp i Klarälven. Samma tunnel som släpper ut vattnet används nattetid för att pumpa upp vatten till Letten. Kraftstationen innehåller alltså också pumpar.
- KLAR·ÄLVEN
- Principskiss)
- kning: GÖTE BERGMAN

FIG 13:8. Pump-power station. Pumping between two magazines at different height with the lower one being replaced by an underground room. This gives additional possibilities in the choice of place.

383

FIG. 13:9. Tunnel for fresh water and distance heating in a multi purpose tunnel, Stockholm. Photo: Sture S:son Lindvall.

FIG. 13:10. Sewage station in Stockholm area (Käppala). Safe-guarded against high and low temperatures. Photo: Manne Lind (top); Lars Hallberg (bottom).

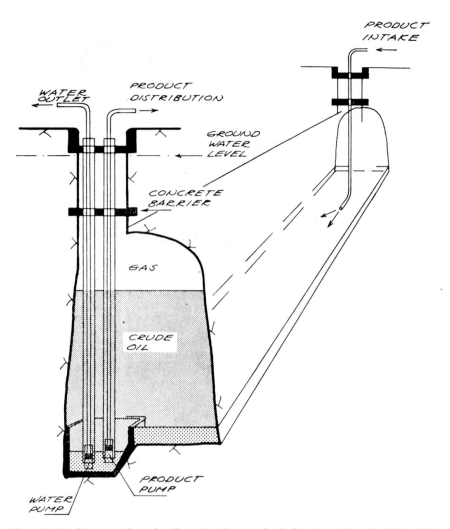

Fig. 13:11. Storage of crude oil on fixed water bed (Courtesy Svenska Väg AB).

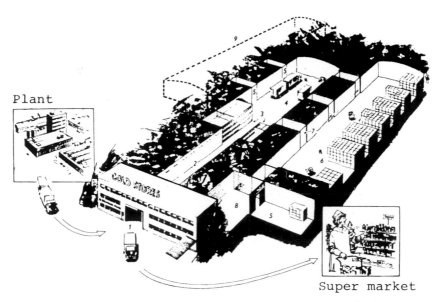

Plant

Super market

Fig. 13:12. Cold store for food in the city of Stockholm (Årsta).

387

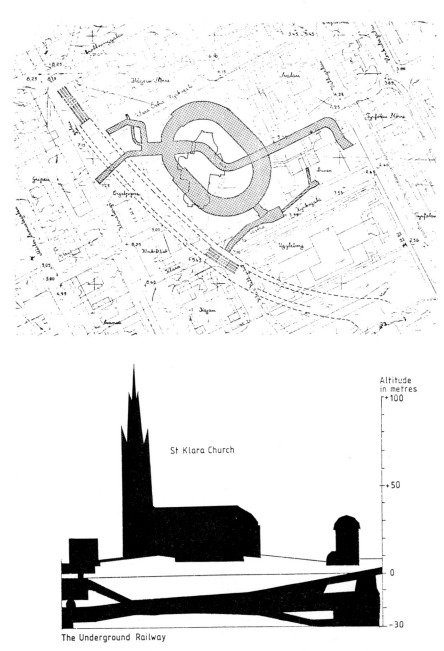

St Klara Church

Altitude
in metres
+100

+50

0

−30

The Underground Railway

Fɪɢ 13:13. Underground shelter and subway (Courtesy Scandiaconsult International AB).

FIG. 13:14. The underground shelter in the naval base is of impressive dimensions.

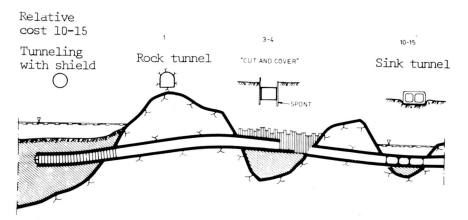

FIG. 13:15. Tunnel in various material.

Cost for blasting
(K = cost in $/m acc. to the diagram)

ect	Length (m)	Price/m $/m	Total $
in tunnel incl.			
ecut		K	
se		$(2-3) \times K$	
k		$(3-4) \times K$	
nsport tunnel		$1.2 \times K$	
		Total A	

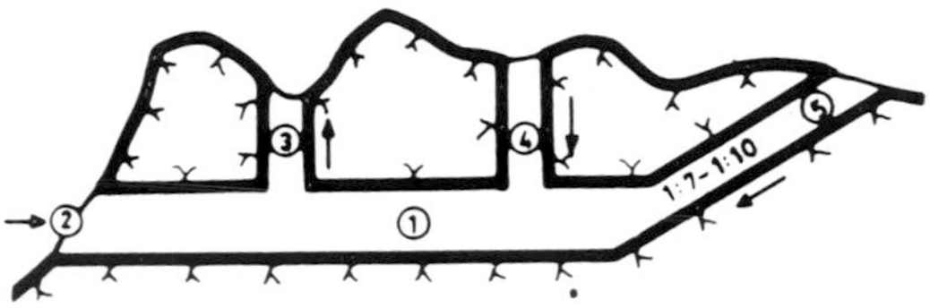

① Main tunnel
② Precut
③ Raise
④ Sink shaft
⑤ Transport tunnel

Cost for the treatment of the rock excavated

ect	Volume (m³)	Price $/m³	Total $
ck transport		0.4/km	
p fee		0–2	
sses sold (income)		0–4	
		Total B	

Cost for reinforcement and injection

ect	Length (m)	Price $/m	Total $
ayed concrete			
lting			
injection			
r-injection			
ng			
		Total C	

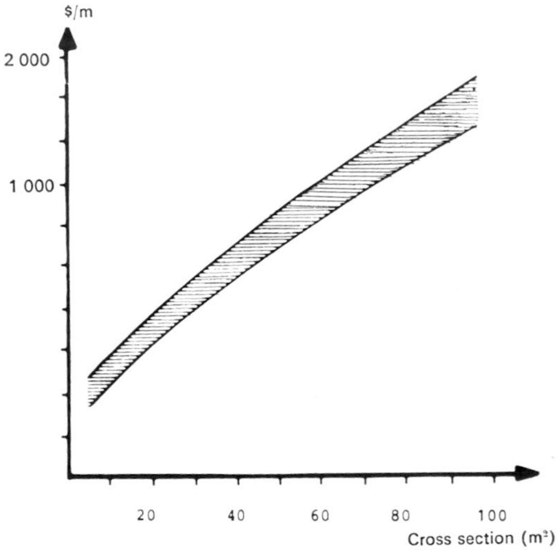

al cost for tunneling

A + B + C	

ect work, administration,
ntrol, etc. +30 %

Total cost	

Cost for blasting per m as a function
of cross section

FIG. 13:16. Model for the calculation of costs in tunneling.

A. Cost for blasting

Object	Volume(m³) Length (m)	Price $	Total
Rock chamber		16 $/m³	
Raise		(2–3) K	
Transport tunnel		1.2 K	
		Total A	

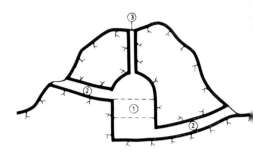

① Rock chamber
② Access and transport tunnels
③ Raise

B. Cost for the treatment of the rock excavated

Object	Volume (m³)	Price $/m³	Total
Truck transport		0.4/km	
Tipp fee		0–2	
Masses sold (income)		0–4	
		Total B	

C. Cost for reinforcement and injection

5–15 % of cost for the
 excavation

Total C

Total cost for the underground chamber

Sum A + B + C

Project work, administration control,
 unforeseen etc. + 25–35 %

Total cost

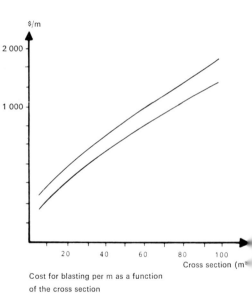

Cost for blasting per m as a function
of the cross section

FIG. 13:17. Model for the calculation of cost for underground chamber.

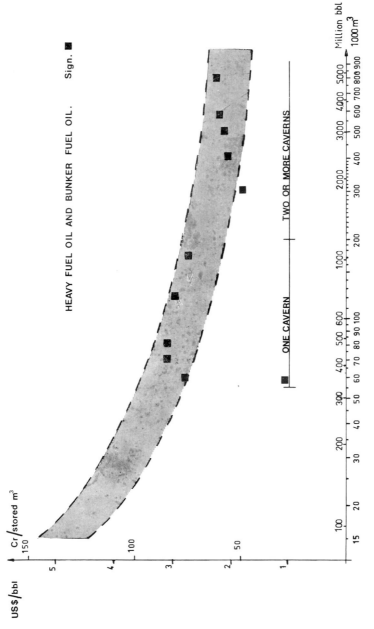

Fig. 13:18. Price level for underground storage in Sweden (Dec. 1975) (Courtesy AB Skånska Cementgjuteriet).

System / Cost Category	Shelter	Production Systems and Industrial Structures	Water Resources	Waste Water Control	Solid Waste	Transportation	Communications	Energy Distribution	Total
Direct Costs	2.8	3.5	1.0	1.0	2.0	2.3	0.5	1.2	14.3
Time	0	4.0	0	0	0	8.0	0	0	12.0
Land	1	0.3	0	0	0	1	0.1	0.1	2.5
Energy	3.8	3.8	0	0	0.1	0	0	0	7.7
Pollution	0	0	0	1.3	0.1	4	0	0	5.4
Safety	3	0	0	0	0	3.1	0	0.1	6.2
Reliability	0	0	0.1	0	0	0.2	0.1	1.0	1.4
Material Resources	3	2.5	0.8	0	0.8	1.0	0.5	0.5	9.1
Total	13.6	14.1	1.9	2.3	3.0	19.6	1.2	2.9	58.6

FIG. 13:19. Annual economic benefits accruing from transferring civil works functions to sub-surface space in billions of dollars.

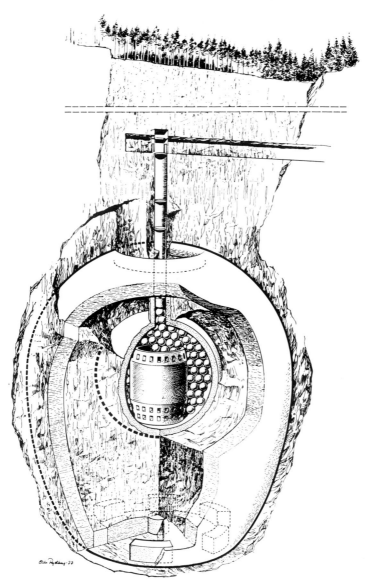

FIG. 13:20. Rock cavern with clay barrier for disposal of nuclear wastes (Courtesy WP System AB).

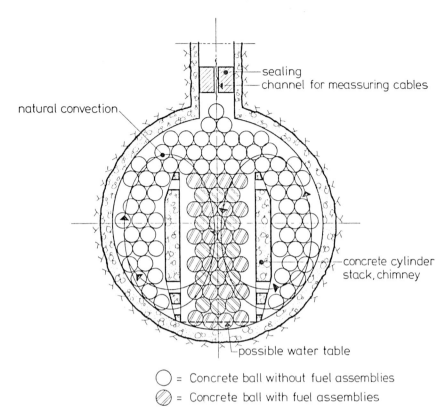

sealing

channel for meassuring cables

natural convection

concrete cylinder
stack, chimney

possible water table

◯ = Concrete ball without fuel assemblies

⬭ = Concrete ball with fuel assemblies

Fig. 13:21. Heat stack is constructed in cavern. In parallel, space outside
stack is filled with empty concrete balls. Inside space (being available from
access shaft also after construction of stack) is filled with loaded balls followed
by empty balls. Ball filling supports cavern walls which may tend to scale
because of thermal compressional stresses. Porosity of ball filling permits free
convection. Cavern can be sealed at any time. It can later be reopened for easy
retrieval of loaded balls. (Courtesy WP System AB).

BIBLIOGRAPHY

ABEL, D., *Channel Underground*. Pall Mall Press, London 1961.

AGNEW, W. G., Blasting Raise Rounds with Millisecond Delays. *Mining Congr. J. 35*, 70–71 April 1949; *35*, 30–32 Oct. 1949.

—— Blasting Cross-cut Rounds with Millisecond Delays and No Cut Holes. *Mining Congr. J. 36*, 85 Oct. 1950.

AHLMAN, H., Blasting with A.N. (Sprängning med AN) Swedish Rock Blasting Committee, p. 167, Stockholm 1960.

ARMSTRONG, E. L., Developing of Tunneling Methods and Controls. *Journal of the Construction Division*, ASCE, Vol. 96, No. C02, Oct., 1970, p. 99.

ATCHISON, T. C., DUVALL, W. I. and OBERT, L., Mobile Laboratory for Recording Blasting and Other Transient Phenomena. *U.S. Bureau of Mines Report of Investigations 5197*, 1956.

ATCHISON, T. C. and TOURNAY, W. E., Comparative Studies of Explosives in Granite. *U.S. Bureau of Mines Report of Investigations 5509*, 1959.

BARKER, J. S., Drilling and Blasting Long Rounds in Tunnels. *Mine and Quarry Eng. 24*, p. 312–321, 350–354, London 1958.

BAUER, A. and COOK, M. A., Observed Detonation Pressures of Blasting Agents. *The Canadian Mining and Metallurgical Bulletin. Jan. 1961.*

BAULE, H., Erschütterungsmessungen beim Schiessen mit Millisekundenzündern und anderen Zündern in Bergwerken und Kalksteinbrüchen. *Nobel Hefte, 21*, p. 132–150, Essen 1955.

BAWA, K. S., Design and Instrumentation of an Underground Station for Washington Metro System. *Proc. in International Symposium on Large Permanent Underground Openings*. Oslo 1969.

Bechtel Corporation for Civil Defense. *Final Study Report for Protective Blast Shelter System Analysis*. April, 1967.

BELIDOR, B. F., *Nouveau cours mathématique à l'usage d'Artillerie et du Genie*. p. 505, Paris 1725.

BENDEL, L., *Ingenieurgeologie II*. Springer-Verlag, Wien 1948.

BERGER, J. and VIARD, J., *Physique des Explosifs Solides*. Dunod, Paris 1962.

BJARNEKULL, T., Bench Blasting on Construction Projects in Sweden. *Manual on Rock Blasting*, 8:51. Atlas Copco AB and Sandvikens Jernverks AB, Stockholm 1957.

BLAIR, B. E., Physical Properties of Mine Rock. Part III. *U.S. Bureau of Mines Report of Investigations 5130*, 1955.

—— Physical Properties of Mine Rock. Part IV. *U.S. Bureau of Mines Report of Investigations 5244*, 1956.

BLAIR, B. E. and DUVALL, W. I., Evaluation of Gauges for Measuring Displacement, Velocity and Acceleration of Seismic Pulses. *U.S. Bureau of Mines Report of Investigations 5130*, 1955.

BLAKE, F. G., Spherical Wave Propagation in Solid Media. *Jour. Acoustical Soc. America, 24*, No. 2, 1952.

BROBERG, K. B., Studies in Scabbing of Solids Under Explosive Attack. *J. Appl. Mech. 22*, p. 317–323, 1955.

BROWN, F. W., Simplified Methods for Computing Performance Parameters of Explosives. See and annual *Symposium on Mining Research, Bull. Univ. Missouri, School of Mines and Metallurgy, 94*, p. 123–136, 1957.

—— Determination of Basic Performance Properties of Blasting Explosives. *Quart. Col. School of Mines, 51* No 3, p. 169–188, 1956.

BRUZEWSKI, R. F., CLARK, G. B., YANCIK, J. J., and KOHLER, K. M., An Investigation of Some Basic Performance Parameters of Ammonia Nitrate Explosives. *Fourth Annual Symposium on Mining Research, Bull. Univ. Missouri School of Mines and Met., 97*, p. 175–203, 1959.

BRÄNNFORS, S., Drilling and Blasting without Removing the Overburden (in Swedish). *FKO-meddelande nr 30*, p. 273–279, Stockholm 1959.

BULLEN, K. E., *An Introduction to the Theory of Seismology.* Cambridge University Press 1947, p. 103.

BULLOCK, R. L., Fundamental Research on Burn-Cut Drift Rounds. *Third Annual Symposium on Mining Research, Bull. Univ. Missouri, School of Mines and Met., 95*, p. 84–121, 1958.

CARMICHAEL, R. L., *The Application of the Livingston Theory of Rock Failure in Blasting to the Design of a New Type of Drift Round.* Doctor of Science thesis, Colorado School of Mines, February 1952.

CLARK, GEORG B., Mathematics of Explosives Calculations. *Fourth Annual Symposium on Mining Research, Bull. Univ. Missouri, School of Mines and Met., 97*, p. 32–80, 1959.

—— *Mining Research.* Vol. I–II, Pergamon Press, New York 1962.

Colorado School of Mines Quarterly *51* No. 3 1956. *Symposium on Rock Mechanics.* Papers and discussion from the First Annual Symposium on Rock Mechanics held at the Colorado School of Mines, April 23–25 1956 at Golden, Colorado.

COOK, M. A., Large Diameter Blasting with High Ammonium Nitrate Non-Nitroglycerine Explosives. *Third Annual Symposium on Mining Research, Bull. Univ. Missouri, School of Mines and Met., 95*, p. 135–149, 1958.

—— *The Science of High Explosives.* ACS Monograph Series, Reinhold Publ. Corp. New York 1958.

—— Water-Compatible Ammonium Nitrate Explosives for Commercial Blasting. *Fourth Annual Symposium on Mining Research, Bull. Univ. Missouri, School of Mines and Met.*, p. 101–123, 1959.

COOLEY, C. M., Properties and Recommended Practices for Use of Ammonium Nitrate in Field-compounded Explosives. Third Annual *Symposium on Mining Research, Bull. Univ. Missouri, School of Mines and Metallurgy. 95*, p. 123–128, 1958.

CRANDELL, F. J., Ground vibration due to blasting and its effect on structures. *J. Boston Soc. Civil Ingrs. 36*, p. 222–245, 1949.

DALEY, R., *The World beneath the City.* J. B. Lippincott, Philadelphia 1961.

DAMBRUN, Etude sur les Effets des Mines militaires. *Memorial de l'Officier Du Genie, 21*, Paris 1873.

DAVIS, V. C., Taconite Fragmentation. *U. S. Bureau of Mines, R.I. 1948*, 1953.

DAVIS, H. and FLOWERS, A. E., Make-it-yourself explosive shatters blasting costs. *Eng. and Min. J. 156*, p. 96–99, 1955.

DEFFET, L., Some Results of the Work of the Belgian Research Center for the Explosives Industry. Paper presented at the 28th International Congress of Industrial Chemistry, held in Madrid, Spain 1955.

DEFFET, L. and DEMARQUE, J., Resultate recents de travaux en carriére. *Explosifs*, 2, p. 49, 1956.

DIXON, J. F. C., Development of Physical Properties and Technique Suitable for Commercial Applications of Slurry Explosives. *Fourth Annual Symposium on Mining Research, Bull. Univ. Missouri, School of Mines and Met.*, 97, p. 124–139, 1959.

VAN DOLAH, R. W., HANNA, N. E., MURPHY, E. J. and DAMON, G. H., Further studies on AN–FO compositions. *Bureau of Mines Pap. No. 1288 Proj. 787*, Dec. 1959.

DREYER, W., Ladegeräte und ihr Einsatz bei der Rationalisierung von kleinen und mittleren Steinbruchbetrieben.

DuPONT, *Blasters' Handbook*, 1966.

DUVALL, W. I. and ATCHISON, T. C., Vibration Associated with a Spherical Cavity in an Elastic Medium. *U.S. Bureau of Mines Report of Investigations 4692*, 1950.

—— Rock Breakage by Explosives. *U.S. Bureau of Mines, R.I. 5336*, 1957.

—— Rock Breakage with Confined Concentrated Charges. *Mining Eng. 11*, 605–611, 1959.

DUVALL, W. I. and FOGELSON, D. E., Review of criteria for estimating damage to residences from Blasting vibrations. *U.S. Bureau of Mines reports of investigations 5968*, 1962.

ERLER, Bemerkungen zum Abschnitt "Bodenerschütterungswirkungen und Luftschallstoss" der Vorschriften des Staates New Jersey über Steinbruchsprengungen. *Nobel Hefte*, 22, p. 96–100, 1956.

EDMOND, T. W., Blasting Ironstone in Quarries. *Mine and Quarry 23*, p. 162–165, 1957.

—— Breaking Rock at a Prophyry Quarry. *Mine and Quarry 23*, 205–258, 1957.

—— A Theoretical Examination of a Number of Factors Directly Connected with Breaking Ground. *Mine and Quarry 23*, 250–256, 1957.

EDWARDS, A. T., Reduction of hydraulic shock-performance of air curtain. *Report No 61–84, Hydro-Electric Power Commission of Ontario, Canada* 1961.

EDWARDS, A. T. and NORTHWOOD, T. D., Experimental studies of the effects of blasting on structures. *The Engineer*, 210, p. 538–546, 1960.

ENHAMRE, E., Effects of Underwater Explosions on Elastic Structures in Water. *Trans. Royal Inst. of Techn.*, 82, Stockholm 1954.

ENOKSSON, B., Investigations on Measurements of Earth Faults (in Swedish). Swedish Rock Blasting Committee, p. 102. Stockholm 1957.

EWING, W. M., JARDETSKY, W. S. and PRESS, FRANK, *Elastic Waves in Layered Media*. McGraw-Hill Publishing Co., 1957.

FAGERBERG, G. and JOHANSSON, C. H., Use of Shaped Charges for Rock Breaking. *Jernkontorets Ann. 133*, p. 199–232, Stockholm 1949.

FARNAM, H. E. JR., Large Scale Use of Ammonium Nitrate Slurries by Iron Ore Company of Canada. Fourth Annual. Symposium on Mining Research,

Bull. *Univ. of Missouri, School of Mines and Metallurgy, 97*, p. 140–146, 1959.

—— Blasting Slurries. *The Canadian Mining and Metallurgical Bulletin for October*, Montreal 1959.

FISH, B. G., Quarry Heading Blasts; Theory and Practice. *Mine and Quarry Eng. 17*, p. 5–10, 53–58, 1951.

FISH, B. G. and HANCOCK, J., Short Delay Blasting. *Mine and Quarry Eng. 15*, p. 339–44, 1949.

FISH, B. G., Seismic vibrations from blasting. *Mine and Quarry Eng. 17*, p. 111, 145, 189, 217, London 1951.

FRAENKEL, K. H., *Bergsprängningsproblem*. Nordisk Rotogravyr, Stockholm 1944.

—— Factors Influencing Blasting Results. *Manual on Rock Blasting*, 6 : 02. Atlas Copco AB and Sandvikens Jernverks AB, Stockholm 1952.

FOGELSON, D. E., DUVAL, W. I. and ATCHISON, T. C., Strain Energy in Explosion-generated Strain Pulses. *U.S. Bureau of Mines Report of Investigations 5514*, 1959.

FOGELSTRÖM, G. and FRAENKEL, K. H., Drilling Patterns. *Manual on Rock Blasting*, 8 : 20. Atlas Copco AB and Sandvikens Jernverks AB, Stockholm 1961.

FORSBERG, B. and GUSTAVSSON, B., *Connection between specific charge and fragmentation* (in Swedish).

FRIEDE, H., Grosssprengungen zum Offnen des Unterwasserkanals eines Wasserkraftwerkes. *Nobel Hefte 28*, p. 79, Essen 1962.

—— Ein neues Verfahren beim Sprengen eines Behelfsdammes in Venezuela. *Nobel Hefte 27*, p. 14, Essen 1961.

GIES, J., *Adventure Underground: The Story of the World's Great Tunnels*. Doubleday & Co., Garden City, N. Y. 1962.

GILBERT, G. H., WIGHTMAN, L. I. and SANDERS, W. R., *The Subways and Tunnels of New York*. John Wiley, New York 1912.

GIRNAU, G., *Unterirdischer Stadtebau, Planungs-, Konstruktions- und Kostenelemente*. Wilhelm Ernst & Sohn, Berlin 1970.

GRANT, B. F., DUVALL, W. I., OBERT, L., ROUGH, R. L. and ATCHISON, T. C., Use of Explosives in Oil and Gas Wells—1949 Test Results. *U.S. Bureau of Mines Report of Investigations 4714*, 1950.

GROSVENOR, N. E., *A* Method for Determining The True Tensile Strength of Rock. Paper presented to AIME Meeting, New York, February 1960.

GUSTAVSSON, R., *Swedish Rock Blasting Technique*. SPI, Göteborg 1973.

GUTENBERG, B., Theorie der Erdbebenwellen. *Handbuch der Geophysik, Bd 4*, Kap. 3, Berlin 1932.

HABBEL, F., Theory of Blasting with Millisecond Igniters. *Geol. en Mijnbouw 16*, p. 118–119, 1954.

HABBEL, H.-G., When and by what Means is Aggregate of Small Size Achieved by the Use of Millisecond Intervals. *Nobel Hefte 22*, p. 40–44, 1956.

HAHN, E., Die Anwendung der seismischen Magnitude und Seismizität. *Freiberger Forschungshefte, C. 25*, Akad. Verlag, Berlin 1956.

HAHN, L. and CHRISTMANN, W., Investigations on the Mechanism of the Action of Millisecond Shooting. *Nobel Hefte 24*, p. 1–35, 1958.

HARTMAN, H. L., *Mine Ventilation and air conditioning.* The Ronald Press Company, New York 1961.

HAWKES, I., A Study of Stress Waves in Rocks and the Blasting Action of an Explosive Charge. *Colliery Engineering*, p. 186, 1959.

HEIDRICH, A., The State of the Large-Borehole Process for Quarry Operations. *Nobel Hefte 21*, p. 49–88, 1955.

HETTWER, A., Betriebskonzentration als Aufgabe bei der Gewinnung in Kalk- und Dolomitsteinbrüchen. *Nobel Hefte, 28*, p. 93, 1962.

HINO, KUMAO, Fragmentation of Rock Through Blasting. *Jour. of Ind. Expl. Soc., Japan, IV*, No. 1, 1956.

—— Theory of Blasting With Concentrated Charge. *Jour. of Ind. Expl. Soc., Japan, 15*, No. 4, 1954.

—— Concentrated Type of No-Cut Round Blasting. *Jour. Ind. Explosives Soc., Japan, 16*, No. 3, p. 38–51, 1955.

—— *Theory and practice of blasting.* Nippon Kayaku Co, Ltd 1959.

HOBERSTORFER, G. and POUSETTE, I., Developments in AN–FO blasting underground at Boliden. *Mine and Quarry Eng.* June 1960.

Holmes & Narver, Inc. *Materials Handling for Tunneling* 1970.

ITO, I., WAKAZONO, Y., FUJINAKA, Y., and TERADA, M., A Study on the Mechanism of the Millisecond Delay Blasting. *Mem. Fac. Eng., Kyoto Univ. 18*, p. 149–161, 1956.

JACOBSEN, R. C., Air-bubble curtain to cushion blasting. *Research News, Hydro-Electric Power Commission of Ontario, Canada,* April–June 1954.

JANELID, I., Bench Blasting in Swedish Limestone Quarries. *Manual on Rock Blasting,* 8:50 Atlas Copco AB and Sandvikens Jernverks AB, Stockholm 1953.

JANELID, I. and OLSSON, G., Janol Method—A New Mining Concept. *Eng. Mining J. 160*, p. 86–89, 1959.

JOHANSSON, C. H., Charging of Drill Holes. *Mine and Quarry Eng. 21*, p. 236–244, 1955; or *Canadian Mining Journal, 76*, p. 55–61, 1955.

—— The Use of the pneumatic cartridge loader for Rock Blasting. *International Symposium on Mining Research, Univ. Missouri*, 1961.

—— The Breaking Mechanism in the Blasting of Rock. IVA *23*, 293, Stockholm 1952.

JOHANSSON, C. H. and LANGEFORS, U., Short Delay Blasting in Sweden. *Mine and Quarry Eng. 17*, London 1951, p. 287.

JOHANSSON, C. H. and LUNDBORG, N., Firing Electrically. *Manual on Rock Blasting 16: 10*, Atlas Copco and Sandvikens Jernverks AB, Stockholm 1958.

JONES, R. J., *Effects of a Reverse Order of Firing Using Millisecond Delay Electric Blasting Caps in a Quarry Operation.* Master's Degree Thesis, Dept. Min. Eng., School of Mines and Metallurgy, Univ. Missouri, 1950.

KALLIN, Å., The Planning of Rock Blasting Operations in Civil Engineering. *Manual on Rock Blasting* 8:01, Atlas Copco and Sandvikens Jernverks AB, Stockholm 1952.

KANATANI, A., *Undersea Tunnels in Japan.* Japanese Railway Engineering, 1960.

KEUCHEL, Das Abtun grosser Serien elektrischer Brückenzünder. *Nobel Hefte, 27*, p. 30–33, 1961.

KIHLSTRÖM, B., Measuring the Insulation in an Underwater Round (in Swedish). *Swedish Rock Blasting Committee, Örebro-Gyttorp*, p. 107, Stockholm 1957.

KNOTT, C. G., Reflexion and refraction of elastic waves with seismological application. *Phil. Mag.*, *48*, p. 64–97, 1899.

KOCHANOWSKY, B. J., *Layout and calculation of coyote tunnel blasts as a contribution to the determination of size of explosive charges in quarrying hard rock.* Doctoral thesis, Univ. Clausthal, Germany 1955.

—— Blasting Research Leads to New Theories and Reductions in Blasting Costs. *Mng. Engr.*, Sept. 1955.

—— Principles of Blasting. Second Annual Symposium on Mining Research. *Bull. Univ. Missouri, School and Metallurgy, 94*, p. 138–48, 1956.

KÖHLER, R., Beurteilung der Erschütterungswirkung von Sprengungen. *Nobel Hefte 21*, p. 127–131, 1955.

KOLSKY, H., *Stress Waves in Solids*. The Clarendon Press, Oxford 1953.

KOLSKY, H. and SHEARMAN, A. C., Investigation of Fractures Produced by Transient Stress Waves. *Res. Suppl. 2*, p. 384–389, 1949.

KUESEL, T. K., BART Subway Construction: Planning and Costs. *Civil Engineering*, March, 1969, p. 60.

LAMPE, H., Chamber Explosions in Hard Rock. *Nobel Hefte 25*, 17–49, 1959.

LANGEFORS, U., Reduction of Ground Vibrations in Blasting. (Minskning av markskakningarna vid sprängning.) *Teknisk Tidskrift 79*, p. 141–147, Stockholm 1949.

—— Short Delay Blasting. *Manual of Rock Blasting 16: 11*, AB Atlas Copco and Sandvikens Jernverk, Stockholm 1952.

—— *Calculation of Charges in Rock Blasting* (in Swedish). NAB, Stockholm, Sweden 1953.

—— The Calculation of Charges for Bench Blasting and Stoping. *Manual on Rock Blasting*, 6:05, AB Atlas Diesel and Sandvikens Jernverks AB, Stockholm 1954.

—— Principles of Tunnel Blasting. *Water Power*, *7*, pp. 9–14, 63–66, London 1955.

—— New Methods for Calculating Explosive Charge. *Highway Research Abstracts, 26*, p. 34–40, Washington, D.C. 1956.

—— New Methods for Driving of Raises. *Mine and Quarry Eng. 23*, London 1957.

—— Smooth Blasting. *Water Power 11*, p. 189, London 1959.

—— Model Scale Tests in Rock Blasting. *Quarterly of the Colorado School of Mines, 54*, 219, Golden, Col., 1959.

—— Underwater Blasting and Blasting through Overburden. *Water Power, 12*, p. 141–146, 189–197, London 1960.

—— Bench Blasting with AN-explosives. *International Symposium on Mining Research, Univ. Missouri*, p. 249–272, 1961.

—— Bench blasting with ammonium nitrate explosives. *Mine and Quarry Eng. 28*, p. 19–28, London 1962.

—— Tunnel- und Streckenvortrieb mit Parallelbohrloch-Einbrüchen. *Nobel Hefte 28*, p. 116, 1962.

LANGEFORS, U., WESTERBERG, H. and KIHLSTRÖM, B., Schadenwirkungen von Bodenerschütterungen durch Sprengungen. *Nobel Hefte 23*, p. 220, 1957.

403

—— Ground vibrations in blasting. *Water Power 10*, p. 335–338, 390 395, 420–424, 1958.

LARBY, W. G., Short Delay Blasting. *Indian Min. J. 2*, p. 9–16, 22, 1954.

LARES, H., Zur Frage der genauen Berechnung bergmännischer Sprengladungen. *Zeitschrift f. d. ges. Schiess- u. Sprengstoffwesen, 28*, p. 105, 146, 181, 206, 250, 1933.

LAUTMANN, K.-F., *The Coromant-Cut*, Sandvikens Jernverks AB, Sandviken, Sweden 1959.

LEET, L. D., Vibrations from blasting. *Explosives Eng. 24*, p. 41–44, 55–59, 85–89, 1946.

LIVINGSTON, C. W., Explosions Research Applied to Mine and Quarry Blasting. *Mining. Engr.*, Jan. (1960).

LJUNGBERG, S., Charging of Drill Holes (in Swedish). *FKO-meddelande nr 30*, p. 87–91, *IVA*, Stockholm 1959.

LOVE, A. E. H., *A Treatise on the Mathematical Theory of Elasticity.* Cambridge Univ. Press, 4th Edit., 1934.

LUSEBRINK, W., The Technique of Blasting and Explosives in Quarries. *Steinbruch und Sandgrube (Susa) 45*, p. 211–213, 1952.

MACELWANE, J. B., *Introduction to Theoretical Seismology, Part I. Geodynamics.* John Wiley & Sons, New York 1936.

MALAN, H., Use of Explosives in Quarries. *Explosifs 11*, p. 29–38, 1958.

MANIFEST, A., Moving Overburden with Explosives. *Mng. Congr. Jour.* April 1960.

MAY, A. N., The Measurement of Rock Pressures Induced by Mineral Extraction. The Canadian Mining and Metallurgical Bulletin, p. 747–753, 1960.

McADAM, R. and WESTWATER, R., *Mining Explosives.* Oliver and Boyd, London, 1958.

McCORMICK, R. F. and HANCOCK, J., Millisecond Delay Blasting in Rippings and Drifts. Meeting of Mining Institute of Scotland at the Royal Institute, Glasgow, September 1953.

MECIR, R. and VALEK, D., Contribution to the problem of optimum delay period of MS delay... *Int. Symp. on Mining Research, Univ. Missouri* p. 273. Editor G. B. Clark, Pergamon Press, New York 1962.

MEISTER, F. J., Die Empfindlichkeit des Menschen gegen Erschütterungen. *Forschung auf dem Gebiet des Ingenieur-Wesens, 6*, p. 116–120, 1935.

MERCER, A., Estimation of the Number of Holes and Charge per Round in Tunnel Drivage. *Operational Research Quart. 8*, p. 22–29, March 1957.

MITTERER, A. B. and SCOTT, S. A., Ammonium Nitrate Blasting in Potash Mining. AIME Meeting, San Francisco, Calif., Febr. 1959.

MORRIS, G., Vibrations due to blasting and their effects on building structures. *The Engineer, 190*, p. 394, 414, 1950.

MURATA, T. and TANAKA, K., Mechanical Theory of Rock Blasting. *Jour. Ind. explosives Soc. Japan, 15*, No. 4, p. 99–101 1954,.

NORDBERG, B., How Producers Use Millisecond Delay Blasting. *Rock Prods. 56*, p. 134–146, 1953.

NOREN, C. M., The Ripple Rock Blast. *Fourth Annual Symposium on Mining Research, Bull. Univ. Missouri, School of Mines and Met., 97*, p. 3–15, 1959.

OBERT, L., WINDES, S. L. and DUVALL, W. I., Standardized Tests for Determining the Physical Properties of Mine Rock. *U.S. Bureau of Mines Report of Investigations 3891*, 1946.

OBERT, L. and DUVALL, W. I., Generation and Propagation of Strain Waves in Rock. *U.S. Bureau of Mines, R.I. 4663*, 1950.

—— A Gauge and Recording Equipment of Measuring Dynamic Strain in Rock. *U.S. Bureau of Mines Report of Investigations 4581*, 1949.

OHNESORGE, A., Fundamentals for Planning and Calculating Rounds for Drifting Tunnels. *Explosivstoffe 5*, p. 46–52, 1957.

OTTO, G., Ausbreitung der Erschütterung um bergmännische Ladungen. *Bergbautechnik, 6*, p. 659–662, 1956.

PAINE, R. S., HOLMES, D. K. and CLARK, H. E., Controlling overbreak by presplitting. Int. Symp. Mining Research, Univ. Missouri, p. 179–210, *1961*. Pergamon Press, 1962.

PAINE, R. S., HOLMES, D. K. and CLARK, G. B., Presplit Blasting at the Niagara Power Project. *Expl. Eng.*, p. 72–92, 1961.

PARKER, A. D., *Planning and Estimating Underground Construction.* 1970.

PARROTT, F. W., Use of Ammonium Nitrate Blasting Agents in Strip Mines Operations. Third Annual *Symposium on Mining Research, Bull. Univ. Missouri, School of Mines and Metallurgy. 95*, p. 129, 1958.

PATTERSSON, E. M., Photography Applied to the Study of Rock Blasting". *J. Phot. Sci. 5*, p. 137–142, 1957.

—— The Photography of Moving Rock During Blasting. *Mine and Quarry Eng. 22*, p. 234–240, 1956.

PEARCE, G. E., Rock Blasting: Some Aspects on the Theory and Practice. *Mine and Quarry Eng. 21*, p. 25–30, 1955.

PEELE, R. and CHURCH, J. A., *Mining Engineers Handbook.* Two Volumes, Third Edition, Second Printing, John Wiley & Sons, New York, 1944.

PEHOSKI, W. L., Application of Tunnel Mining in Hard Rock. Second Annual Symposium on Mining Research. Bull. Univ. *Missouri, School Mines* and Met. *94*, p. 17–24, 1956.

POULET, M., Charging of Boreholes. *Explosifs 11*, p. 22–28, 1958.

REID, T. J., The Efficient Use of High Velocity Explosives. *Mine and Quarry Eng. 26*, p. 298, 1960.

REIHER, H. and MEISTER, F. J., Die Empfindlichkeit des Menschen gegen Erschütterungen. *Forschung auf dem Gebiet des Ingenieurwesens, 2*, p. 381–386, 1931.

RICHARDSON-MAYO. *Practical Tunnel Driving.* McGraw-Hill, New York 1941.

RILEY, G. G. and WESTWATER, R., Blasting with AN-fuel Mixtures. *Mine and Quarry Eng. 25*, p. 211, 1959.

RINEHART, J. S., Scabbing of Metals Under Explosives Attack-Multiple Scabbing. *Jour. Appl. Phys. 23*, p. 1229–1233, 1952.

—— Some Quantitative Data Bearing on the Scabbing of Metals Under Explosives Attack. *Jour. Appl. Phys. 22*, p. 555–560, 1951.

—— Fracturing Under Impulsive Loading. *Third Ann. Symp. Mining Res., Bull. Univ. Missouri, School of Mines and Met., 95*, p. 46–83, 1958.

ROMIG, J. L., Practical Studies of Field Blasting. Second am. *Symposium on Mining Research, Bull. Univ. of Missouri, School of Mines and Metallurgy. 94*, p. 149, 1957.

RUFF, A. W., The Seismic Wave from Plaster and Drill-Hole Explosive Charges. AIME Meeting, San Fransisco, Calif. Febr. 1959.

RUMPF, K., Blasting Methods in Gallery or Tunnel Structures, as well as in Shaft Levels". *Sprengpraxis 5*, p. 103–107, 1956.

RUSSEL, P. L. and AGNEW, W. G., Blasting No-Cut-Hole Raise Rounds Using MS Delays. *U.S. Bureau of Mines, R.I. 4902*, 1953.

RYON, J. L. JR., Underground Use of Ammonium Nitrate for Blasting at the Detroit Mine of the International Salt Company. *Fifth Annual Symposium on Mining Research, Bull. Univ. Missouri, School of Mines and Met., 98*, p. 136–142, 1960.

SANDER, A., Das Problem der seismischen Grensschichtwelle. *Freiberger Forschungshefte, C. 26*, Akad. Verlag, Berlin 1956.

SANDSTROM, G. E., *Tunnels*. Holt Reinhart & Winston, New York 1963.

SCHAFFER, L. E. and Noren, C. H., The Influence of Cartridge Diameter on the Effectiveness of Dynamite. *Bull. Univ. Missouri, School of Mines and Met., 19*, No. 1, 1948.

SHARPE, J. A., The Production of Elastic Waves by Explosion Pressure—I. Theory and Empirical Observations. *Geophysics 7*, p. 144, 1942.

SIEBERG, A., Die Erdbeben. *Handbuch der Geophysik, Bd. 4*, kap. 14, Berlin 1932.

SIMMONS, R. L. BODDORFF, R. D. and LAWRENCE, R. W., Ammonium Nitrate Blasting Agents—Properties and Performance. *Fourth Ann. Symp. Mining Res., Bull. Univ. Missouri, School of Mines and Met., 97*, p. 208–217, 1959.

SLATER, H. and BARNETT, C., *The Channel Tunnel*. Allan Wingate, London 1958.

SPAETH, G. L., Formula for Proper Blasthole Spacing. *Engr. News Rec.* April 1960.

STEIDLE, E., *Some Practical and Theoretical Aspects of the Burn Cut*. Joy Manufacturing Co., Pittsburg, Pa.

STETTBACHER, A., *Spreng- und Schiessstoff*. Racher Verlag, Zürich 1948.

STOBBE, D., Berechnung von Minenladungen für grössere Tiefen. *Zeitschrift f. d. ges. Schiess- u. Sprengstoffwesen, 22*, p. 161, 206, 1927.

STUBBS, F. W., *Handbook of Heavy Construction*, 1959.

STÜSSI, F., Sprengtechnik und Minentheorie. *Techn. Mitt. f. Sappeure*, p. 216–221, 1943, p. 65–75, 1944, p. 150–161, 1945.

—— Theorie der Minen. Techn. Mitt. f. Sappeure p. 209–231, 1942.

SUTHERLAND, H. and BOWMAN, M. L., *Structural Theory*. John Wiley & Sons, Inc. Third Edition, Second Printing, January 1944.

SZECHY, K., *The Art of Tunneling*. Tankonyvkiado, Budapest 1961.

TEICHMANN, A. and HANCOCK, J., Blasting vibrations and the householder. *Quarry Managers' J.* Jan. 1951.

THOENEN, J. R. and WINDES, S. L., Seismic effects of quarry blasting. *U.S. Bureau of Mines, Bull. No. 442*, 1942.

—— House movement caused by ground vibrations. *U.S. Bureau of Mines, Rep. Invest. 3431*, 1938.

THOENEN, J. R., WINDES, S. L. and IRELAND, A. T., House movement induced by mechanical agitation and quarry blasting. *U.S. Bureau of Mines, Rep. Invest. 3542*, 1940.

WEICHELT, F., *Handbuch d. gew. Sprengtechnik*. Carl Marhold Verlagsbuchh. Halle 1949.

WETTERHOLM, A., Explosives for Rock Blasting. *Manual on Rock Blasting,* *16: 01,* Atlas Copco AB and Sandvikens Jernverks AB, Stockholm 1959.

WHITE, H. H.. Measurements of Blasting Efficiency. *Rock Prods. 56,* p. 133, 204 –205, January 1953.

WILD, H. W., The Driving of Tunnels Through Rock in Mines in Northern France, with Particular Regard to Blasting. *Nobel Hefte, 24,* p. 97–124, 1958.

WINDES, S. L., Physical Properties of Mine Rock, Part II. *U.S. Bureau of Mines, Report of Investigations 4727,* 1950.

WRIGHT, F. D., Millisecond-Delay Blasting of Bench Rounds. *Mining Congr. J. 39,* p. 51–53, 108 June 1953.

—— Research on Blasting by the U.S. Bureau of Mines. *S. African Mining Eng. J. 64,* Part I, p. 361–365, 1953.

WUERKER, R. G., The Shear Strength of Rocks. *Mng. Engr.* October 1959.

YANCIK, J. J. JR. and CLARK, G. B., Experimentation with large hole burn cut drift rounds. *Bull. Univ. Missouri, School Mines and Metallurgy, Techn. Ser. No. 83,* 1956.

Appendix 1

NOTES ON THE PROBLEMS

Chapter 2

1. $Q = 0.075$ kg (0.17 lb).
2. $q_e = 0.050$ kg/m³ (0.17 lb/cu yd) according to table 2:2.
   $Q_1 = 0.056$ kg (0.12 lb).
   $c_0 = 0.3$.
3. a) $l_b = 1.4$ kg/m (0.9 lb/ft), $V = 1.7$ m (5.6 ft or 1.87 yd).
   b) $l_b = 1.9$ kg/m (1.25 lb/ft), $V = 2.0$ m (6.6 ft or 2.2 yd).
   c) The depth of a hole is $H = K + 0.3\,V = 3.3\,V$, the rock volume VEK $=$ 3 V^3. For a) the drilled length per m³ (cu.yd) is then $1.1\,V/V^3 = 0.38$ m/m³ ($1.1 \cdot 5.6$ ft/6.5 cu.yd = 0.93 ft/cu.yd) for b) 0.27 m/m³ ($1.1 \cdot 5.5$ ft/10.7 cu.yd $= 0.57$ ft/cu.yd).
4. a) $V = 4.2$ m (14 ft).
   b) $U = 2.9$ m (9.7 ft).
5. a) $l_b = 7.5$ kg/m (5 lb/ft) of dynamite.
   b) As the weight strength is $s = 0.9$ the charge *per m* needed must be $l_b = 1.5/0.9 = 8.4$ kg/m ($5/0.9 = 5.6$ lb/ft).
   c) $d = 110$ mm (4.4 in).
   d) $d = 82$ mm (3.3 in).
6. For $K/V \geqslant 1.8$ or $V \leqslant 3$ m (10 ft) Diagram 2:1 should be used. For greater values for the maximum burden take Diagram 2.9. Even in this case the left hand part of Diagram 1 can be used to get the connection between diameter d, degree of packing P and weight of explosive per m (vertical scale).
7. Take Diagram 2:5. The shovel volume should at least equal the volume L^3 of the boulders. However, it is generally advisable to try to get a fragmentation where the L^3-value is less than $\frac{1}{4}$ of the shovel as this may considerably reduce the repair-cost and -time for the equipment and increase the capacity.
8. Compare table 2:1 For a hole in a row of several holes the conditions are as in bench blasting, $V = 2.25$ m (7.5 ft). For the reliefers that are to be fired one by one the degree of fixation can be greater. Take $c = 0.4 \cdot 1.45 = 0.6$ (instead of 0.4) to compensate for this effect as suggested in table 2:1. The maximum burden $V = 1.85$ m (6.2 ft). (Compensation for deviations in the drilling has to be made according to 3.3.)
9. d depends on the drilling equipment.
   Ex: $d = 32$ mm (1.28 in), $l_b = 1.0$ kg/m (0.7 lb/ft) $V = 1.05$ m (3.5 ft).
   $H = 1.12 \cdot K + 0.3\,V = 1.66$ m (5.5 ft).
10. As the charge should extend up to $h = (H - V)$ in the hole, the charge is $(H - V) \cdot L = 0.61$ kg (1.35 lb). Compare diagram 2:3 ($K' = 1.34$) which gives $Q_b = 0.6$ kg (1.3 lb).

11. The volume excavated per hole is $VE \cdot (1.12 \, K)$. The spacing is $E = 1.25 \, V$, the bench height is increased by 12% to give the volume (due to the slope 2:1) $q = 0.6/(1.05 \cdot 1.3 \cdot 1.34) = 0.33$ kg/m³.
    ($q = 1.35/3.5 \cdot 4.4 \cdot 4.5 = 0.02$ lb/cu ft $= 0.55$ lb/cy.yd).
12. Congratulations!

Chapter 3

1. $V = 3.4$ m (11.3 ft), $E = 1.25 \, V$ (table 3:1).
   $VE = 14.5$ m² (160 sq.ft or 17 sq.yds).
2. a) $VE = 14.5/1.27 = 11.5$ m³.
   b) $VE = 14.5/1.27 \cdot 1.5 = 17$ m².
3. $V = 3.4$, 3.0 and 3.7 m should be reduced 0.50 m each.
   $V_1 = 2.9$, 2.5 and 3.2 m.
   $V_1 E_1 = 10.5$, 7.8 and 12.8 m².
4. In table 3:1 the maximum burden for $d = 50$ mm (2 in) is 2.3 m (7.6 ft) but that value corresponds to a slope 3:1 in relation to the desired bottom. In the problem the degree of fixation should be assumed to be 1.0 if a breakage perpendicular to the hole is desired at the bottom (table 2:2). The $V \cdot E \cdot f$-value should be constant, and thus both V and E should be multiplied by $f = \sqrt{0.9}$.
   One gets $V = 2.2$ m, $E = 2.7$ m, $VE = 6$ m².
   $V_1 = 2.2 - 4.8 \cdot 0.05 = 1.95$ m, $V_1 E_1 = 4.7$ m².
5. a) If the spot where the collaring is to be made deviates 0.5 m *and* if the length of the hole is not increased, 0.5 m of the rock at the bottom may be left as a stump. The following hole must then hava a bottom charge that extends V m (ft) above this stump. For the first hole, however, no extra charge is needed.
   b) In tunnel blasting there is much less restriction on the throw. The column charge can extend to some 0.5 m (2 ft) from the free face of the rock.
6. It can be increased $1/0.8 = 1.25$ (25%) compared with problem 4 and $0.9/0.8 = 1.12$ compared with the figures obtained from table 3:1.
7. For $f = 1.0$ and $d = 50$ mm (2 in) as in problem 4 $V = 2.2$ m. The relievers mentioned are not in a row, but what are called single holes in table 2:2. $f = 1.45$. V is to be reduced to $V = 1.8$ m and the practical burden is $V_1 = 1.5$ m (5 ft).
8. $V_1 = 0.95$ m (3.2 ft), $E_1 = 1.20$ m (4 ft), $Q = 0.41$ kg (0.9 lb) (table 3:5).
9. $V = 0.86$ m (table 3:2), $V_1 = 0.83$ m. $l_b = 2.5$ kg/m
   $H = 1.1 \, (K + 0.3 \, V) = 0.95$ m. Height of the charge $h = H - V = 0.09$ m.
10. Use the column for $d = 50$ mm and increase all geometrical linear quantities 50%. Thus $d = 75$; $K = 0.45$, 0.675, 0.9 m and so on in the left hand column, and the corresponding figures for V are 0.85, 1.06, 1.3 m and so on. Note that the sublevel drilling should be increased in the same proportion and not be less than 0.375 m (1.5 ft).
11. With $q = 0.22$ kg/m³ and $V = 1.5$ m the side of the boulders can be expected to be about $L = 1$ m. The biggest ones must come from the top of the bench that is uncharged.
    With a bigger charge in the column part, $l = 1.3$ kg/m, the specific charge for this part of the bench will be $q = 0.5$ kg/m³ (0.8 lb/ft) and the cor-

responding side length $L = 0.4$ m, for instance a volume of $\frac{1}{16}$ of the previous problem. The conditions at the bottom are unchanged with $q = 0.4$ kg/m³. Even here the big boulders will come from the top of the bench that, however is not unaffected by the heavier column charge.

In fact the influence from the charge below is great enough to loosen the top of the bench and may correspond in the diagram to q being about 0.2 kg/m³. With some cartridges, say 0.5 kg (1 lb), distributed between 1.5 and 0.8 m from the top, the specific charge in that part increases about 0.1 kg/m³. This reduces the mean side length of the boulders by a factor 0.5. The volume is reduced to $0.6^3 = 0.2$ m³, which is a considerable gain. It should be kept in mind that different kinds of rock can give essentially different fragmentation. The diagram 2:5 is given more as a basis for an estimation and for further studies of the connections involved. It has not been investigated thoroughly enough for exact calculations.

Chapter 4

1. a) 1.0 dm³/m, b) 2.0 dm³/m, c) 4.5 dm³/m.
2. a) volume 0.38 dm³/m   weight $0.38 \cdot 1.5 = 0.56$ kg/m
     b)   ,,   0.5   dm³/m   ,,   0.75 kg/m
     c)   ,,   0.66 dm³/m   ,,   1.0   kg/m
     d)   ,,   1.25 dm³/m   ,,   1.9   kg/m
   The figures are a little lower if the diameter include the paper.
3. a) 0.44 kg/m, b) 0.58 kg/m, c) 0.76 kg/m, d) 1.44 kg/m.
4. Fig. 4:22 gives the degree of packing $P \simeq 1.0$ kg/dm³. This value is *defined* as $P = 1.27 \cdot 10^3 \cdot l/d^2$ which means that $l = d \cdot 10^{-3}/1.27 = 1.6/1.27 = 1.26$ kg/m. (According to the definition the P-value is identical with the average density in the hole only if the volume of the hole is exactly the same as that of a cylinder with a diameter d. If the volume per m of the hole is 6 % greater the P-value remains unchanged if l does, but the density is 6 % lower, or $\varrho \simeq 0.94$ kg/dm³.)
5. $P = 1.27 \cdot 10^3 \cdot 2/(40)^2 = 1.58$ kg/dm³.
   The density is 6 % lower, $\varrho = 1.49$ kg/dm³.
6. The breaking capacity is proportional to d^2 and to P.s. Thus the diameter should be large enough for the equality $1.4 \cdot 1.0 \cdot (50)^2 = 0.9 \cdot 0.9 \cdot d^2$, which gives $d = 66$ mm.
7. $1.55 \cdot 0.78(150)^2 = 0.9 \cdot 0.9 \cdot d^2$.
   $d = 183$ mm.

Chapter 5

1. a) 0.071, b) 0.048, c) 0.025.
2. Take for instance drill hole cost 0.4 \$/dm³. (Compare table 5:6.)
   a) 0.33–0.26, b) 0.31–0.24, c) 0.285–0.215.
3. The relative cost of the explosive increases from 0.10 to 0.12.
4. The relative cost of the drilling is 0.153 (at the cost 0.4 \$/dm³).
5. 0.253. There are still many other cheaper alternatives.
   For the highest relative cost of the drilling (1.20 \$/dm³) in table 5:6 the cost will be 0.50.

6. The $P \cdot s$ value here is 1.35 instead of 1.50. The drilling cost has to be 10% higher, or 0.44, 0.22, 0.14 and 0.10 for alternative E in table 5:5. The final cost in tables 5:6 is 0.53, 0.31, 0.23 and 0.19. The alternatives C and D are also changed accordingly. Observe that the discussion on alternatives giving the lowest cost is not influenced in principle by this somewhat lower degree of packing $P = 1.35$ kg/dm³.

7. a) The height of the bottom charge is V and that of the column charge 2.3 V. The concentration of the charges is $l = Pd^2 \cdot 10^{-3}/1.27$, $l_b = 4.7$ kg/m and $l_p = 3.8$ kg/m (table 5:3) and the cost per kg 0.50 and 0.13 respectively table 5:1). The total cost of explosive per hole is: $4.7 \cdot 0.5 V + 3.8 \cdot 0.13 \cdot 2.3 V$
$= 2.35 V + 1.2 V = 3.55 V$.

With Nabit cartridges in the column $l = 1.0 \cdot 50^2 \cdot 10^{-3}/1.27 = 2.0$ kg/m at a cost (table 5:1) of 0.38 \$/kg the total cost of the explosive in a hole is: 2.35 V + 1.75 V = 4.10 V or about 16% more than in alternative E.

In table 5:5 this means that the figure 0.09 is also increased by 0.015 (16%) to 1.05. In table 5:6 the total cost of explosive and drilling together is increased 0.015 units. At a drilling cost of 0.40 \$/dm³ the relative figure will be 0.235. The use of a column charge in cartridges evidently does not change the total cost of the blasting much compared with the difference between the various alternatives A–I.

8. The maximum burden with $Ps = 1.27$ is (table 3:1) $V = 7.0$ m (23 ft). In A we have $Ps = 1.08$ for crystalline AN-oil and $V = 6.5$ m (21 ft). The practical burden allowing for deviations in the position of the drill holes at the bottom of 1.0 m (3.3 ft) is $V_1 = 5.5$ m. Thus $V \simeq K/2$ and we have the same conditions as in table 5:8. Suppose the drilling cost to be 0.28 \$/dm³. The relative cost of drilling and explosive together is 0.22 (the higher figure 0.28 applies with prills).

The mucking cost is 0.11.

If the diameter is reduced to $d = 75$ mm the cost of drilling is supposed to increase in proportion to $\sqrt{d}$ to 0.40 \$/dm³ and the burden is reduced to $V_1 = K/4$. The cost of drilling and explosive (table 5:8) is then 0.27. The cost of the mucking is reduced to 0.055. The summarized cost is 0.325 as compared with the original value 0.33. Thus the two diameters discussed give the same total costs and not lower for the big diameter as often supposed.

Furthermore, with 75 mm (3 in) holes it is possible to reduce the cost of drilling and explosive from 0.27 to 0.24. This makes the whole operation more than 10% cheaper with smaller ground vibrations and other advantages with the 75 mm (3 in) holes.

Chapter 6

1. a) In one series, a blasting machine CI 50 or bigger.

b) Blasting machine CI 600, 4 series with 60 detonators in each, or fewer detonators and more series (at most 10 series).

CI 2400 or CI 6000 could of course also be used with at least 2 series in a parallel connection.

c) CI 2400 or CI 6000 with at most 120 detonators in each series.

The number of delays does not influence the connection of the round.

2. The rock moves 1 cm/ms (0.4 in/s).

   a) With the Swedish detonators the scattering in delay time is $\Delta\tau = \pm 4$ and the time difference between two is less than 8 ms and the distance thus less than 8 cm (3.2 in).

   In the same way about 30 ms and 30 cm (6 in) are obtained with the American detonators, 16 cm (6.4 in) with the British and 3 or 8 cm (1.2 or 3.2 in) with the German ones.

   b) The scattering is $\pm \Delta\tau = 70\text{--}150$ ms and the distance may be 1.40–3.00 m (56–120 in, or 4.7–10 ft).

3. a) The delay between no 5 and 6 is 14–40 ms depending on the make of detonators. 14–40 cm (5.6–16 in).

   b) About 170–600 cm (68–240 in or 5.6–20 ft). Note the increased risk of throw in this case.

4. The resistance of a series is 500 ohm. According to Lundborg (6.1 c) an earth-fault of 1000 ohm = 1 kohm can then be permitted. In this case with a great overcapacity (and no higher voltage) of the blasting machine the 0.8 kohm and 0.5 kohm could also be accepted presupposing that the round can be blasted immediately. If the resistance is declining successively which is rather often the case, and if the round cannot be blasted untill one or some days after checking, an exchange of detonators with the lowest earth resistance is recommended.

5. —

6. —

7. "Burden" $V = 1.2$ m (4 ft), three holes in a "row" with a bottom charge of 0.80 kg are to be used. The middle hole has to be ignited first and is placed one third of the "burden", 0.4 m (1.3 ft) ahead of the other two. With 20 delays available 10 such "rows" and 12 m (40 ft) can be blasted in a round. With a detonating fuse or LEDC (low energy detonating cord) there is no limitation to the length of a round.

8. $V = 0.8$ m (2.7 ft), the middle hole 0.3 m (1 ft) ahead of the others, the same type of connection as above.

9. The VA detonators could be used all the way, the HU-detonators to about 20 m (70 ft) from the power line and the ordinary detonators to 30 m (100 ft) presupposing that the note of table 6:9 is strictly observed (The regulations of local authorities may differ and must in such a case be observed.)

10. About 3000 m (10,000 ft). Ordinary detonators should not be used. The round can safely be connected and blasted with VA as well as HU detonators.

Chapter 7

1. The burden at the bottom should be less than $V = 1.3$ m (4.5 ft) acc. to table 3:1. $V = 1.3 - H \cdot 0.05 = 1.15$ m (3.8 ft) when blasting a row of holes. In stoping hole by hole it must be kept in mind that the angle of breakage for some of the holes may be smaller than expected, leaving a greater burden (up to 20 % more) for adjacent holes. The V_1-value has to be reduced correspondingly.

412

2. $V = 2.3$ m (7.8 ft), $V_1 = 2.3 - 5.4 \cdot 0.1 = 1.75$ m ($\simeq 6$ ft).
   When the burden has to be reduced a further 20 % as mentioned above
   we get $V_1 = 1.4$ m (4.7 ft).
3. a) $V = 0.30$ m (1 ft), b) $V = 0.50$ m (1.7 ft), c) $V = 0.65$ m (2.2 ft).
4. $2\,H \geqslant 0.8\,\sqrt{2^2 + 2^2 H^2 + 0.6^2 \cdot H^3}$
   $6.25\,H^2 \geqslant 4 + 4\,H^2 + 0.36\,H^3$
   $2.25\,H^2 \geqslant 4 + 0.36\,H^3$
   $1.5 \leqslant H \leqslant 6$.
   At depths 1.5–6 m (3–20 ft) the linear deviation is in itself more than 80 %
   of the total value.
5. Use table 7:2. For $d = 30$ mm the concentration of charge can be $1 = 1.0$ kg/m.
   $V = 0.60$ m (2 ft), $V_1 = 0.40$ for 4 holes creating a new square opening of
   about $0.9 \cdot 0.9$ m^2 ($3 \cdot 3$ sq.ft). The 4 next holes can have $V \simeq 0.8$ m (2.7 ft)
   and $V_1 = 0.6$ m (2 ft) which gives an opening of $1.5 \cdot 1.5$ m^2 ($5 \cdot 5$ sq.ft).
6. —
7. $q_0 = 7$ kg/m^3
   $q_1 = 0.6 - 1.0$ kg/m^3
   a) With $q_1 = 1.0$ kg/m^3
        $q = 6/4 + 1.0 = 2.5$ kg/m^3
   b) With $q_1 = 1.0$ kg/m^3
        $q = 6/60 + 0.8 = 0.9$ kg/m^3
8. —
9. The greatest advance with V-cut and a small deviation in the drilling is
   0.6 $B = 3.6$ m (12 ft), ordinary value at this depth 0.5 $B = 3.0$ m (10 ft).
   With a double spiral cut 2.95–6.0 m (10–20 ft) can be obtained when drilling to a depth of $H = 3.1$–6.3 m (10.3–21 ft) depending on the size of the
   empty hole $\varnothing = 2 \cdot 57$–200 mm ($2 \cdot 2.5$–8 in). See following chapter, table
   8:4.
10. a) $N = 32$ holes, b) $N = 56$ holes, c) $N = 83$ holes.
    With $d = 30$ mm the figures are 32, 72 and 125 holes.

Chapter 8

1. The a, b, c and d-values are given in table 8:6 (120, 140, 175, 290 mm)
   $H = 3.3$ m (11 ft).
2. The $\varkappa$-value is reduced from 1.05 to 0.97 when the distances is increased
   20 %. An increase of 10 % can be supposed to correspond to at least $\varkappa = 1.01$.
   (The practical variation seems to be ± 10 %). The same drilling depth as
   in table 8:4 can be adopted.
3. $l_1 = 0.35$ kg/m semi-gelatine explosive ($s = 1.0$).
   $l_1 = 0.39$ kg/m powder explosive ($s = 0.9$) in holes of a diameter $d = 32$ mm.
4. The semi-gelatine explosive in 22 mm cartridges gives a concentration of
   $l = 0.5$ kg/m loosely piled in the hole. This concentration can be used when
   $d = 45$ mm (table 8:1).
   For the powder explosive $l = 0.38$ kg/m in 22 mm (7/8 in) cartridges; just
   the concentration recommended for a 32 mm hole.
5. a) In table 8:3 a four-section cut has an advance of 95 % at a drilled depth
   of $H_{95} = 3.6$ m.

b) At this depth the drilled hole deviation is about $\sigma = 1.96$ cm/m (estimated in table 8:2). This is 2% less (better) than supposed in table 8:3 and the drilling can be 2% deeper $H_{95} = 3.67$ m.

6. $H_{95} = 4.7$ m for $\varnothing = 125$ mm and $\sigma = 2.0$ cm/m (table 8:3).

a) $H_{95} = 4.7 \cdot 83/125 \cdot 2.0/\sigma$. As $\sigma > 2.0$ cm/m, $H_{95} < 3.1$ m. Try some lower H-values. ($\sigma = (4 + 0.64\,H)^{1/2}$) for which $H = H_{95}$.

$H = 2.0$ m	2.5 m	2.6 m
$\sigma = 2.30$ cm/m	2.37 cm/m	2.40 cm/m
$H_{95} = 2.7$ m	2.63 m	2.6 m

$H_{95} = 2.6$ m is the value sought.

b) 2×57 mm hole gives the same advance as one 78 mm hole. $H_{95} = 47 \cdot 78/125$. $H_{95} = 47 \cdot 78/125 \cdot 2.0/\sigma < 2.9$ m as $\sigma > 2.0$ cm/m. $H_{95} = 2.5$ m.

c) $H_{95} < 4.7$ m, $H_{95} \simeq 3.7$ m.

d) $H_{95} < 5.65$ m, $H_{95} = 4.3$ m.

e) $H_{95} < 7.5$ m, $H_{95} = 5.5$ m.

7. a) Drilled depth $H = 4.5$ m. At this depth $\sigma = 2.64$ cm/m and the drilling deviation are the same at the bottom as for a cut with $\sigma = 2.0$ cm/m and drilled 32% deeper to $H = 5.95$ m. This will give a mean advance 86% of the drilled depth, as will (approximately) the cut in question.

b) Drilled depth $H = 3.0$ m, $\sigma = 2.43$ cm/m. The table 8:3 can be remade for $\varnothing = 75$ mm simply by reducing the H-values in proportion 75/125. With this table the same discussion as above gives about 87%.

c) Drilled depth $H = 5.1$ m, $\sigma = 2.7$ cm/m. 83%.

8. —

9. a) The four inner holes should have $l_1 = 0.35$ kg/m the four outer holes $l_2 = 0.75$ kg/m.

b) $l_1 = 0.6$ kg/m, $l_2 = 1.3$ kg/m.

10. a) $A_m = 71/24$, $\varkappa = 0.96$. $H_{95} = 3.3$ m.

b) $\alpha = 0.98$, $H_{95} = 3.65$ m.

11. The values of the first group in table 8:9 are the only ones of interest in this connection. They are.

Site No	1	2	8	9	15	19	20	21–27
$\varkappa$	1.00	1.28	1.07	1.08	1.18	0.96	1.03	0.94–1.23
Ignition	fuse	HS	MS	fuse	MS	HS	fuse	MS

The ignition with fuse seems to have been quite successful and done about as well as the MS-delay detonators. Both ignition systems as well as HS delay could be used.

Chapter 9

In the International System of Units, the so-called MKSA-system (m.kg.s.A), a force giving 1 kg the acceleration of 1 m/s^2 is denoted as N (Newton). The force exerted by gravity on 1 kg is thus $g \cdot N$, or about 10 N.

1. a) The gravity acts on the system with the force $(4.1 + 2.4)gN = 65$ N. This force can give the unsuspended body an acceleration of $6.5/4.2\ g = 1.6\ g$.

When the ground for a short moment moves downward at a greater acceleration the vibrograph will lose contact with the ground.

b) The suspended mass (M_s) and the unsuspended one (M_u) must together
be 8 times as great as the unsuspended. Thus $(M_s + M_u)/M_u = 8$ gives
$M_s = 7\ M_u$.

2. $2A = 7$ mm after amplification $50 \times$. The real amplitude is 0.14 mm (140 μ).
1 sec. in the vibrogram $\simeq 720$ mm. Half a period is about 2 mm. This gives
a frequency $f \simeq 720/4 = 180$ c/s.
Vibration velocity is, according to equ. (9:8), $v = 2\pi \cdot 180 \cdot 0.14 = 159$ mm/s.
Acceleration is, according to equ. (9.7), $a = 4\pi \cdot 180^2 \cdot 0.14 = 180,000$ mm/s² =
180 m/s² $= 180\ N = 18\ g$.
Yes, $v = 180$ mm/s.

3. a) $v = 100$ mm/s.
b) $v = 250$ mm/s (approximately).
In the logarithmic scale 10 mm is a factor 2, one third of it, 3.3 mm, is a

factor $\sqrt{2} = 1.25$.

4. 2.0 kg, 3.7 kg and 5.6 kg (2.2, 8.2 and 12.5 lb).

5. $\gamma = \frac{1}{2} \cdot Q_{red} = \frac{1}{2} \cdot 10 \cdot 3 = 15$ kg (33 lb).

6. $R = 24$ m.

7. The reduction factor is then $\gamma = 1/6$. Three times as many charges can be
ignited in this interval.

8. a) 70 kg (155 lb) in one single hole or in several smaller ones.
b) With $\tau = 25$ ms practically all charges in the two intervals detonate
within half a period (50 ms) of the vibrations in question and they can
consequently all cooperate. Both intervals make together 70 kg (155 lb).
With $\tau = 50$ ms the interference condition (9:19) is fulfilled with $n = 2$ and
$k = 1$ and the optimum reduction effect can be counted on. At least 70 kg
in each interval can be used as this case is more favorable then with one
single interval.

9. Take for instance the big curve for $\tau = 0$ in fig. 9:20 and change the time
scale so that the frequency is $f = 5$ c/s and thus $T = 200$ ms. Show that 8
charges (delays) in this special instance give much lower vibration values
than 4.

10. $Q = 180$ kg (400 lb). $T = 50$ ms. Charges ignited at time delays $\tau \geqslant 3 \cdot 50 = 150$
ms can be regarded as not cooperating.
One possibility is for instance to take for the Atlas detonators the delay
Nos 3, 9, 12 (50, 200, 350 ms) and so on with intervals of 150 ms and use
180 kg in each delay. In this way we can pick out 26 intervals with delays
of 125–150 ms. In all $180 \times 26 = 4700$ kg (10,000 lb) can be detonated in
one round.
Another way is to try to get delays of $\tau = T/2 = 25$ ms for which time the
Swedish VA-detonators have 20 intervals. The ground vibrations will then
get a "negative" cooperation as they successively reduce one another.
The same charge as above can be used in each interval. The total charge is
$180 \times 20 = 3600$ kg (8000 lb).
(No general conclusion regarding the data for the various types of detonators
should be drawn from special examples. The consumer who can choose the
best timing and scattering for his special purpose is best off.)

Chapter 10

1. $V_1 = 1.1$ m (3.7 ft), $E = 0.8$ m (2.7 ft).
2. The burden at the bottom can amount to $V = 1.1 + 7.5 \cdot 0.05 \simeq 1.5$ m if the drill hole deviation is 0.05 m/m.

   The bottom charge given in table 3:1 is $Q_b = 2.0$ kg (4.4 lb) when the concentration at the bottom is $l_b = 1.0$ kg/m (0.7 lb/ft). 30 mm (1.2 in) cartridges can be loosely piled up to a height of 2.0 m (6.7 ft). The rest of the hole can be loaded with: two qurit strings aside one another, *or* a qurit cartridge and a 0.1 kg dynamite cartridge alternating, *or* 0.25 kg dynamite per m of the hole taped on a detonating fuse.

   The ignition should be made with detonating fuse (preferably) or with short delay detonators of one and the same delay. (Even with detonating fuse it may be advisable to use the short delay detonators in every charge if there is a risk for a break in the fuse from the first holes in the round).
3. The practical burden for the round as a whole is $V_1 = 2.0$ m (6.7 ft). Acc. to table 3:1 this requires a column charge of $l = 0.8$ kg/m. The first question is: can this charge be used and the row in question thus have the same burden and spacing as the main part of the round?

   The extended charge corresponds to a concentrated single charge of about the size $2\,lR$ at the distance R. We obtain the following relation between the concentration of charge and the distance $R_{0.5}$ and $R_{1.0}$ for the charge levels $Q/R^{3/2} = 0.5$ and 1.0 respectively.

$l = 0.25$	0.50	0.75	1.00	kg/m
(0.17	(0.38)	(0.5)	(0.6)	lb/ft
$R_{0.5} = 1$	4	9	16	m
$R_{1.0} = 0.25$	1.0	2.25	4	m

   The regions for shattering at the levels greater than 1.0 may get cracks from the blasting. For the contour row this region is only 0.25 m, from the row at a distance of 1.1 m, the region does not extend into the rock to remain after the blasting for $l = 0.50$ kg/m, and $2.25 - 1.1 = 1.15$ m for $l = 0.75$ kg/m.

   As a general conclusion it can be said that a charge concentration of twice that in the smooth blasting can in any case be used. Even 3 times this value could often be accepted. (The ordinary concentration in the holes of the round is about 10 times that in the smooth blasting).
4. a) Smooth blasting: $E_1 = 0.6$ m (2 ft), $V = 0.9$ m (3 ft).

   b) Presplitting: $E_1 = 0.3$ m (1 ft).
5. If the presplitting is done before the round is charged, there is no problem as some of the presplit holes can be loaded with the required bottom charge and be blasted in the main round.
6. $E = 1.6$ m (6 ft) in smooth blasting.

   $E = 0.8$ m (3 ft) in presplitting. In both cases a charge concentration of about $l = 0.9$ kg/m (0.6 lb/ft) of dynamite could be used.
7. Suppose that the burden is $V_1 = 4.0$ m (13 ft) and the maximum value due to the deviations is $V_{max} = 4.7$ m (16 ft). Table 3:1 gives a bottom charge of $Q_b = 60$ kg (130 lb) at a spacing of 1.25 $V_1 = 5.0$ m (17 ft). In order to avoid such heavy charges they could be distributed in 6 (6 or 7) holes with the spacing 0.8 m instead of in 1 hole with the distance 5.0 m to the next one. This gives $Q_b = 9.5$ kg (20 lb) in every hole.

   If this is also too much the burden must be reduced to 3 m or 2 m.

Appendix 2
LIST OF WORDS

Spanish[1]	English	Swedish
apoyo, estribo	abutment	anfang, stöd
aceleración	acceleration	acceleration
acumulador	accumulator	ackumulator
túnel de acceso	access tunnel	tillfartstunnel
exactitud	accuracy	noggrannhet
aumentativo, aditivo	additive	additiv
ajustar	adjust	justera
avance	advance	framdrift
avance por ciclo	advance per round	indrift
convenio, contrato	agreement	avtal
presión de aire	air pressure	lufttryck
tolerancia, permiso	allowance	tolerans
corriente alterna	alternating current (AC)	växelström
nitrato de amonio	ammonium nitrate (AN)	ammoniumnitrat
amplitud	amplitude, A	amplitud, utslag
perno de anclaje	anchor bolt	ankarbult
ángulo	angle	vinkel
ángulo de reposo	angle of repose	rasvinkel
angular	angular	vinkel
antena	antenna (pl. -ae)	antenn
apatita	apatite	apatit
aproximación	approximation	tillnärmelse
arco	arch	valv
cargador de tipo de arco	arch-type loader	kastlastningsmaskin
área, zona, superficie	area	område, yta
colocación de barrenas (perforaciones)	arrangement of holes	hålsättning
fractura de coceo	backbreak	bakbrytning
barra, palanca	bar	stång, spett
capa, estrato	bed	skikt
estratificación	bedding	skiktning
lecho de roca	bedrock	berggrund
banco, escalón	bench	avsats, pall
voladura de banco	bench blasting	pallsprängning
altura de banco	bench height, K	pallhöjd
excavación en bancos	benching	pallbrytning

[1] The Spanish vocabulary has been kindly revised by Mr G. Landberg, Mr P. Warming and by Sr M. Castillo and Sr A. Barocio.

Spanish	English	Swedish
resistencia a la flexión	bending strength	böjhållfasthet
broca, corona	bit	borrskär, borrkrona
volar; voladura	blast	spränga; sprängskott
barreno, agujero	blast hole	spränghål
voladura, pega	blasting	sprängning
cápsula, detonador	blasting cap	stubinsprängkapsel, sprängkapsel
dinamita gelatina } gelatina explosiva }	blasting gelatine	gummidynamit, spräng-gelatin
máquina explosora, explosor	blasting machine	tändapparat
estera para voladuras	blasting mat	sprängmatta
trabajo de voladura	blasting operation	sprängningsarbete
minería por socavación y derrumbe	block caving	blockrasbrytning
bloque	block	block
fractura en bloques, roca suelta	blocky	bomt (berg)
golpe; soplar	blow	slag, blåsa, slå
ventilador	blower	fläkt
perno	bolt	bult
apernado, empernado	bolting	bultning
tiro fallido	bootleg, misfire	dola
fondo, suelo, piso	bottom	botten, sula
banqueo de piso	bottom bench	bottenpall
carga de fondo	bottom charge	bottenladdning
diámetro de fondo	bottom diameter	bottendiameter
galería de avance de piso	bottom heading	sulort
nivel de fondo	bottom level	bottennivå
piedra grande, bloque	boulder	skut
tamaño de piedra grande	boulder size	skutstorlek
anchura, ancho	breadth, width, B	bredd
demoler, quebrantar	break up	slå sönder
fractura, rotura	breakage	brott, lossbrytning
resistencia a la rotura	breaking strength	brotthållfasthet
carga de rotura (fractura)	breaking stress	brottbelastning
roca quebradiza (frágil)	brittle rock	skört berg
roca quebrada, piedra de voladura	broken rock	sprängsten
cubo, cuchara, cucharón	bucket	skopa
cargador de cucharon	bucket loader	lastmaskin
región poblada	built up area	bebyggt område
volumen, tamaño	bulk	volym
potencia de volumen	bulk strength	volymstyrka
peso, espesor	burden, V	försättning
quemar	burn	bränna

Spanish	English	Swedish
cuele (cuña) quemada, "burn cut"	burn cut	brännkil
cable	cable	kabel
barrena batidora de cable	cable churn drill	linstötborr
galería de cables	cable shaft	kabelschakt
jaula	cage	hisskorg
mineral calcáreo	calciferous ore	kalkhaltig malm
calcular	calculate	beräkna
fulminante, detonador	cap	sprängkapsel
capacidad	capacity	kapacitet, kapacitans
capacidad de cuchara (cucharón)	capacity of bucket	skopvolym
costo del capital	capital costs	kapitalkostnad
poner el detonador a la mecha	capping fuse	aptera stubin
cartucho	cartridge	patron
oruga, "caterpillar"	caterpillar	
derrumbe, corrimiento de roca	cave	bergras
material derrumbado	caved material	rasmassor
explotación de mina	caving	rasbrytning, brytning
corrimiento de mina	caving-in of a mine	gruvras
cavidad	cavity	hålrum
central	central	central, centrum
cemento	cement	cement
barreno central	central hole	centrumhål
greda, creta	chalk	krita
barrenación de cámara	chamber blasting, springing	grytsprängning
cargar; carga	charge	ladda; laddning
cargar barrenos	charging	laddning
esquema (plan) de carga	charging scheme	laddningsschema
control	check	kontroll
barato	cheep	billig
barrena de cincel	chisel bit	mejselskär
barrenación batidora	churn drilling	stötborrning
coladero, canalón	chute	bergtapp
circuito	circuit	krets
control de circuito	circuit control	ledningskontroll
probador de circuito	circuit tester	ledningsprovare
cucharón de almeja	clamshell bucket, grab	gripskopa
arcilla	clay	lera
pizarra arcillosa	clay slate	lerskiffer
limpiar (barreno)	clean (hole)	rensa (hål)
voladura limpia	clean blasting	rensprängning
limpieza	cleaning	rensning
tubo soplador	cleaning pipe	blåsrör

Spanish	English	Swedish
grueso	coarse	grovt
coeficiento	coefficient	koefficient
principiar el barrenado; cuello	collar	ansätta; krage
columna	column	pelare, (borr-)pipa
carga distribuida de barreno	column charge	pipladdning
aire comprimido	compressed air	tryckluft
compresión	compression	tryck, kompression
resistencia a la compresión	compression strength	tryckhållfasthet
compresor (de aire)	(air-)compressor	kompressor
concentración de carga	concentration of charge	laddningskoncentration
conclusión	conclusion	slutsats
hormigón, concreto	concrete	betong
revistimiento de hormigón	concrete liming	betonginklädnad
explosor de condensadores	condenser blasting machine	kondensatortändapparat
uniforme, conforme	conformal	likformig
conformidad	conformity	likformighet
conglomerado	conglomerate	omvandlat berg
diagrama de conecciones	connecting pattern	kopplingsschema
cable de conexión	connecting wire	tändkabel
conexión	connection	koppling
conexión en series	connection in series	seriekoppling
constante	constant	konstant
fondo constreñido	constricted bottom	inspänd botten
constricción	constriction	inspänning
construir	construct	konstruera, bygga
consumo de explosivos	consumption of explosives	sprängämnesförbrukning
contenido	content	innehåll, halt
continuo	continuous	kontinuerlig
contrato	contract	avta
trabajo (obra) a destajo	contract work, piece work	ackordsarbete
barreno (agujero) de borde	contour hole	randhål, konturhål
corazón	core	kärna
perforación con corazón	core drilling	kärnborrning
barreno (agujero) de angulo	corner hole	hörnhål
cuele (cuña) coromant	coromant-cut	coromant-kil
corregir	correct	korrigera
costo de voladura	cost of blasting	sprängningskostnad
curso, progreso	course	förlopp
cubrir, sobrecapa	cover	täcka; överliggande berg

Spanish	English	Swedish
cubierta, capa de protección	covering	fördämning
voladura por túneles (coyotera)	coyote blasting	kammarsprängning
grieta	crack	spricka
formación de grietas	crack formation	sprickbildning
grúa	crane	kran
cráter	crater	krater
oruga	crawler	larvfötter
equipo, cuadrilla	crew	arbetslag
galería transversal	crosscut	tvärort
sección transversal	cross-section	area, tvärsnitt
corona, bóveda	crown	hjässa
machacar	crush	krossa
roca quebrada (machacada)	crushed rock	makadam
planta quebradora, planta de machaqueo	crushing plant	krossanläggning
cristalino	crystalline	kristallin
metro cúbico	cu.meter	kubikmeter
corriente	current	ström
pérdida de corriente	current leakage	läckström
cuele, cuña, corte; cortar	cut	kil, skärning; skära
corte y relleno	cut-and-fill	takbrytning med igensättning
barreno del cuele (cuña)	cut hole	kilhål
barreno cortado (robado)	cut-off hole	ryckare
corte, trinchera	cutting	förskärning, skärning
recortes, polvo	cuttings	borr-kax
ciclo, período	cycle	period
cuele cilíndrico	cylinder-cut	cylinder-kil
presa, embalse	dam	damm
compuerta de presa	dam gate	dammlucka
corazón impermeable	dam core	dammkärna
corona de la presa	dam crest	dammkrön
daño	damage	skada
datos (pl.)	data (-s, pl.)	uppgift (-er, pl.)
minería de roca estéril	dead rock mining	gråbergsbrytning
roca decompuesta	decomposed rock	vittrat berg
perforación profunda	deep drilling	djupborrning
defecto	defect	fel
deflexión	deflection	avvikning
grado de compactación	degree of packing, P	packningsgrad
voladura de retardo	delay blasting	intervallsprängning
detonador de retardo	delay detonator	intervallsprängkapsel
elemento retardador	delay element	fördröjningssats
ignición retardada	delay ignition	intervalltändning

Spanish	English	Swedish
densidad de carga	density of charge	laddningstäthet
profundidad	depth	djup, mäktighet
profundidad de barreno	depth of hole, *H*	håldjup
proyecto, plan, diseño	design	konstruktion, ritning
desintegración	desintegration	sönderbrytning
broca (corona) desmontable	detachable bit	löst skär
determinar	determine	bestämma
detonar	detonate	detonera
mecha detonante	detonating cord	detonerande stubin
detonador	detonator	sprängkapsel
desviación	deviation	avvikning, avvikelse
diabasa	diabase	diabas
diametro de barreno	diameter of drill hole	borrhålsdiameter
diametro en el fondo	diameter at the bottom	bottendiameter
barrenado de diamante	diamond drill	diamantborr
excavadora, pala	digger	grävmaskin
lampara indicadora	dim glowing lamp	glimlampa
dimensión	dimension	dimension
corriente continua	direct current (D.C.)	likström
dirección de barrenos	direction of holes	hålriktning
dirección de estratos	direction of streta	slagriktning
desventaja	disadventage	nackdel
volcamiento, descarga	discharging	tippning
dislocación	dislocation	dislokation
desarreglo, desplazamiento, amplitud	displacement	förskjutning, förkastning, amplitud
distancia, espaciamiento	distance, spacing	avstånd
tubo múltiple	distribution manifold	tubgrenar
distribución de carga	distribution of charge	laddningsfördelning
zanja	ditch	dike, rörgrav
túnel de desvío	diversion tunnel	omloppstunnel
dolomía, dolomita	dolomite	dolomit
corte de doble espiral	double-spiral cut	dubbelspiral-kil
abajo	down	ned
barrenación con máquina dentro del barreno	down-the-hole rock drill	sänkborrmaskin
tubo de aspiración	draft tube	sugrör
cubo de arrastre, trailla	drag scraper bucket	släpskopa
desaguar, drenar	drain	dränera
agujero de desagüe	drainage hole	dräneringshål
túnel de desagüe	drainage tunnel	dränagetunnel
dibujar proyectar	draw	rita, konstruera
desventaja	drawback	nackdel
plano, dibujo	drawing	ritning
galería	drift	ort
minero	drifter	ortdrivare

Spanish	English	Swedish
excavation de galerías	drifting	ortdrivning
perforar; perforadora	drill	borr; borra
corona cortante, broca	drill bit	borrkrona, borrskär
polvo de barrenado	drill dust	borrdamm, borrkax
barreno	drill hole	borrhål
plataforma de barrenación	drill jumbo	borrställning
escalera de perforadora (guía)	drill ladder	borrstege
barras de barrenación	drill steel	borrstål
metro de barrenado	drilled meter	borrmeter
barrendador, perforista	driller	borrare
perforación, barrenado	drilling	borrning
metraje de barrenado	drilling meterage (footage)	borrmeter
perforadora	drilling machine	borrmaskin
diagrama (plano) de barrenación	drilling pattern	borrplan
plataforma de perforación	drilling platform	borrställning
velocidad de barrenado	drilling rate	borrhastighet, borrsjunkning
pedestal de barrenación	drilling stand	borrstativ
polvo de barrenado	drillings	borrkax
motorista	driver	förare
martinete de caida	drop hammer	fallhammare
prueba de impacto	drop weight test	fallhammarprov
pila seca	dry cellar battery	torrbatteri
descargar; escombrera	dump	tippa; tipplats
dumptorista	dumptor operator	dumptor förare
polvo	dust	damm
región poblada	dwelled area	bebyggt område
dinamita	dynamite	dynamit
relleno de tierra, terraplén	earth filling	jordfyllnad
efecto	effect	effekt
eficiencia	efficiency	verkningsgrad
elasticidad	elasticity	elasticitet
límite de elasticidad	elastic limit	elasticitetsgräns
circuito eléctrico	electric circuit	strömkrets
detonador eléctrico	electric detonator	elsprängkapsel
ignición eléctrica	electric ignition (firing)	elektrisk tändning
red de eléctricidad	electric network	elnät
electricidad	electricity	elektricitet, elkraft
ascensor, elevador	elevator	hiss
factor de energía	energy factor	arbetsfaktor
maquinista	engine operator	maskinist

Spanish	English	Swedish
jefe de trabajo	engineer in charge	driftschef
agrandar, ampliar	enlarge	upprymma
montar, levantar, eregir	erect	resa (upp)
montaje	erection	montage
erróneo, defectuoso	erroneous	felaktig
error, falta, defecto	error	fel
planta, establecimiento	establishment	anläggning
excavar	excavate	schakta
excavadora, pala	excavator	grävmaskin
pala excavadora	excavator shovel	grävskopa
sobrecarga	excess charge	överladdning
gases de escape	exhaust gases	avgaser
expansión	expansion	utvidgning
perno de expansión	expansion-shell bolt	expanderbult
caro	expensive	dyr
explotar, estallar, volar	explode	explodera
agujero de exploración	exploratory drill hole	undersökningshål
gases de explosión	explosion gases	spränggaser
explosivo	explosive	sprängämne
cartucho explosivo	explosive cartridge	sprängpatron
carga explosiva	explosive charge	sprängladdning
desencapado, expuesta	exposed	frilagd, blottad
unir, extender	extend (connect)	förlänga, skarva
extensión	extension	utvidgning
frente de avance (galería)	face of heading	ortgavel
derrumbe de roca	fall of rock	bergras
soterramiento de mina	falling-in of a mine	gruvras
cuele de abanico	fan cut	solfjäderkil
limite de fatiga	fatigue limit	utmattningsgräns
error; falla	fault	fel; förkastning
perforación errónea	faulty drilling	felborrning
roca fallada	faulty rock	slagrikt berg
zona de falla	fault zone	sprickzon
alimentación	feeding	matning
rellenado	filled	igensatt
relleno, terraplén (en una mina)	filling	fyllning
fino	fine	fint
finos, tierra	fines	mull
rgnición, disparo	firing	tändning
orden de ignición (disparo)	firing sequence	tändningsföljd
fisura	fissure	spricka
roca grietosa	fissured rock	sprickigt berg
constricción, fijación	fixation	inspänning
agujero de lavado	flashing hole	spolhål
capacidad de transmisión	flash-over capacity	överslagsförmåga

Spanish	English	Swedish
flexible	flexible	böjlig
piso, suelo	floor	sula
placa de soporte	foot-plate	sula
capataz, sobrestante	foreman	förman
formación	formation	bildning
formula	formula	formel
corte de cuatro secciones	four-section cut	fyrsektionskil
fractura, rotura	fracture	brott
fragmentación	fragmentation	sönderslagning, styckefall
peso (V) sin constricción	free burden, V	fri försättning
frecuencia	frequency, f	frekvens
fricción	friction	friktion
combustible, fuel	fuel	bränsle
petrol	fuel oil	brännolja
avance a sección completa	full-face driving	gavelsprängning
mecha fusible	fuse	stubin
cabeza de fusible	fuse head	tändpärla
galería	gallery	galleri, tunnel
arenisca	ganister	sandsten
empaquetadura	gasket	packning
lumbrera de compuertas	gate shaft	luckschakt
generador	generator	generator
atascar	get stuck	fastborrning
gneis	gneiss	gnejs
cucharón de almeja	grab	gripskopa
nivel de suelo (piso), fondo	grade, grade level	bottennivå, pallbotten
granito	granite	granit
grafito	graphite	grafit
piedra machacada, grava	gravel	grus
afilar	grind	slipa
vibración de suelo	ground vibration	markskakning
nivel freático	ground-water level	grundvattennivå
lechadear, inyectar; lechada de cemento	grout	injektera, injektering
perno cementado	grouted bolt	ingjuten bult
barreno de guía	guide hole	ledhål
guías	guides	styrskenor
hormigón a soplete	gunite	sprutbetong(-era)
gurita	gurit (explosive)	gurit
martillo	hammer	hammare
carga a mano, rezaga a mano	hand mucking	handlastning
manija	handle	handtag
ignición involuntaria	haphazard ignition	oavsiktlig tändning
metal duro, widia	hard metal	hårdmetall
broca con inserto, widia	hard metal bit (tip)	hårdmetallskär
transportar, acarrear	haul	transportera

Spanish	English	Swedish
galería para transporte	haulage drift	transportort
galería de avance, frente	heading	galleri
calor de explosión	heat of explosion	explosionsvärme
voladura grande	heavy round	storsalva
altura de carga	height of charge, h	laddningshöjd
explosivo de alta potencia	heigh explosive	brisant sprängämne
alta presion	high pressure	högtryck
levantar; malacate	hoist	hissa; hiss
profundidad del barreno	hole depth	håldjup
distancia de barrenos, espaciamiento	hole distance, spacing, E	hålavstånd
inclinación de barreno	hole inclination	hållutning
tolva	hopper	tappficka
horizontal	horizontal	horisontell
corte y relleno horizontal	horizontal cut-and-filling	horisontal takbrytning med igensättning
barreno horizontal	horizontal hole	liggare
manguera	hose	slang
casa	house	hus
cimiento, fundación de la casa	house foundation	husgrund
iniciar, encender	ignite	tända
ignición	ignition	tändning
cable detonador	ignition cable	tändkabel
plano (diagrama) de ignición	ignition pattern	tändplan
inclinar	incline	luta
inclinación	inclination	lutning
inclinación de barreno	inclination (slope) of hole	hållutning
inclusión	inclusion	inneslutning
indicación de fractura	indication of fracture	brottanvisning
irregularidad	irregularity	ojämnhet
iniciar	initiate	initiera
iniciación	initiation	initiering
control, inspección	inspection	kontroll
cuele (cuña, corte) instantanea	instantaneous cut	momentkil
detonador instantáneo	instantaneous detonator	momentsprängkapsel
ignición instantánea	instantaneous firing	momenttändning
aislamiento	insulation	isolering
toma, entrada	intake	intag
barrena con broca integral	integral drill steel	borrstål med fast skär
esfuerzo interno	internal stress	inre spänning
intervalo (de tiempo)	interval (-time)	intervalltid
hierro	iron	järn
empujador de pierna, pierna neumática	jack leg, air leg	knämatare

426

Spanish	English	Swedish
perforadora con pierna neumática	jackhammer with air-leg	knämatad borrmaskin
perforación térmica	jet piercing	brännborrning
caolin	kaoline	kaolin
clase de roca	kind of rock	bergart
interruptor de cuchillas	knife switch	knivströmbrytare
laboratorio	laboratory	laboratorium
obrero	labourer	arbetare
perforación de escalera guía	ladder drilling	stegmatning
retraso; retardo	lag	eftersläpning; fördröjning
cuele (cuña) de barreno grande	large hole cut	storhålskil
voladura de barrenos grandes	large hole blasting	storhålssprängning
voladura de grandes bancos	large scale benching	storpallssprängning
ley	law	lag
capa	layer	skiva
indicación de barrenos	laying out	utsättning
plan, estudio	layout	plan, projektering
prueba al bloque de plomo	leadblock test	blyblockprov
longitud, largo	length	längd
cable de conexión	leg wire	kopplingstråd
leptita	leptite	leptit
nivel, elevación	level	nivå
nivelar (allanar) por voladura	levelling	plansprängning
ligero	light	lätt
rayo	lightning	blixt
cal	lime	kalk
caliza	limestone	kalksten
limite	limit	gräns, begränsning
carga limite	limite charge	gränsladdning
revestimiento	lining	infordring, inklädnad
cargar, descombrar; carga	load	ladda, lasta; belastning;
cargar sin retaque (atacar)	load without packing	laddning
carga con tacos de madera intermedios	load with intermediate wood sticks	stapla löst pinnladdning
cargadora, pala rezagadora	loader, mucking machine	lastmaskin
cargador neumático para explosivos	loader (pneumatic)	laddapparat (pneumatisk)
descombrado	loading	upplastning
capacidad de carga	loading capacity	laddnings-, lastningskapacitet
relevo (turno) de carga	loading shift	utlastarskift

427

Spanish	English	Swedish
largo	long	lång
barrenación profunda	long-hole drilling	långhålsborrning
roca suelta (floja)	loose rock	bomt berg, löst berg
aflojar, desprender; mullir	loosening	lossbrytning; uppluckra
pérdida	loss	förlust
voladura de banco bajo	low bench blasting	plansprängning
madera	lumber	virke
sala de máquinas, central	machine hall	maskinhall
descombrado (rezagado) a maquina	machine mucking	maskinlastning
roca magmática	magma rock	magmberg
magnesita	magnesite	magnesit
magnetita	magnetite	magnetit, svartmalm
nivel principal	main level	huvudnivå
mantenimiento	maintenance	underhåll
mármol	marble	marmor
margen	margin	marginal
masa de roca	mass of rock	bergmassa
medida	means	medel
medida; medir	measure	mäta; mått, åtgärd
medida	measurement	mätning
excavadora	mechanical shovel	grävmaskin, grävskopa
mecánica	mechanics	mekanik
método de voladura	method of blasting	sprängmetod
milisegundo	millisecond, ms	millisekund, ms
mina	mine	gruva
voladura de mina	mine adit, mine blasting	minsprängning
roca volada (extraida)	mined rock	utbruten bergmassa
minero	miner	gruvarbetare
minería, explotación de minas	mining	bergshantering, gruv-brytning
tiro (voladura) fallado (-a)	misfire (missed round)	bomskott, dola (bom-salva)
barreno quedado (fallado)	missed hole	dola
mezclar	mix	blanda
mezcladora	mixer	blandare
ensayo de modelo (a escala)	model (scale) test	modellförsök
morena	moraine	morän
sierra	mountain chain	berkgskedja
deslizamiento de la mon-tana	mountain slide	bergras
movimiento	movement	rörelse
descombrado, desescom-bro	mucking	utlastning
fango, lodo	mud	slam
banqueo múltiple	multibench blasting	flerpallsprängning
voladura de filas múltiples	multiple-row blasting	flerradssprängning

428

Spanish	English	Swedish
red, malla	net	nät
voltaje de linea (v. de la red)	net voltage	nätspänning
nitroglicerina	nitroglycerin(e), NGL	nitroglycerin
nomograma	nomogram	nomogram
normal	normal	normal, vanlig
aceite, petróleo	oil	olja
minería a cielo abierto	open cast	dagbrott
banqueo a cielo abierto	open cut bench	dagpall
minería a cielo abierto	open-cut mining	pallbrytning i dagbrott
cantera a cielo abierto	open pit	dagbrott
gastos de explotación	operating costs (pl.)	driftskostnad
maquinista, motorista	operator	förare
normal,corriente,commún	ordinary	vanlig, normal
mineral	ore	malm
area de mineral	ore area	malmarea
cuerpo mineral	ore body	malmkropp
roca metalífera	ore-carrying rock	fyndigt berg
yacimiento de mineral	ore deposit	malmfyndighet
exploración de minas, cateo de minas	ore-prospecting	malmletning
filón, vena	ore vein	malmåder
afloramiento	outcrop	berg i dagen
salida	outlet	utlopp, avlopp
croquís	outline	skiss
sobreexcavación	overbreak	överberg
sobrecapa, encape, sobrecarga	overburden	täckande lager
barrenación sin desencapar	overburden drilling (O.D.)	borrning utan avtäckning
ensanche de techo	overhand stope	takstross
roca pendiente	over hang	överhäng
retacar	pack	packa
perforadora de percusión	percussion rock drill	hammarborrmaskin
perímetro, contorno, perfil	perimeter	kontur
período	period	period
explosivos permisibles	permitted explosive	säkerhetssprängämne
perpendicular	perpendicular	vinkelrät, lodrät
obra a destajo	piece work	ackord
pilar, columna	pillar	pelare
galería de avance	pilot heading	riktort
tubo, caño	pipe	rör
pozo	pit	schakt, grop
disposición de barrenos	placing of holes	hålplacering
proyectar	plan	planera
plano	plane	yta, plan

Spanish	English	Swedish
plano de debilitamiento	plane of weakness	svaghetsplan
planta	plant	anläggning
voladura de pegadura	plaster-shooting	skutsprängning
pegadura	plaster shot	bulldosa
explosivo plástico	plastic explosive	plastiskt sprängämne
cuele (corte) arado, cuele de cuña	plough cut	plogkil
cargador neumático	pneumatic loader	pneumatisk laddapparat
empujador neumático	pneumatic pusher	knämätare
punto de ataque	point of attack	angreppspunkt
barra, poste	pole	stång
petardo	pop shot	bulldosa
fuerza, potencia	power	kraft
sala de máquinas	power house	maskinhall
series de potencia	power series	potensserie
excavadora, pala	power shovel	grävmaskin
planta de energía, central	power plant	kraftstation
precisión, exactitud	precision	noggrannhet
ignición involuntaria	premature ignition	oavsiktlig tändning
voladura de corte previo, precorte	presplit blasting	förspräckning
cortar previamente	presplitting	förspräckning
presión	pressure	tryck
lumbrera de presión	pressure shaft	trycktub
roca primaria	primary rock, bed rock	urberg
procedimiento, proceso	procedure, process	process
propagación	propagation	utbredning
proporcional	proportional	proportionell
protocolo	protocol, report	protokoll
pozo de bomba, cárcamo	pump, pit	pumpgrop
empujador	pusher	matare
cuele (cuña) piramidal	pyramid cut	pyramidkil
cantidad de carga	quantity of explosive	laddningsmängd
cantera	quarry	stenbrott
cuarzo	quartz	kvarts
cuarcita	quartzite	kvartsit
radiofrecuencia	radio frequency	radiofrekvens
radio	radius	radie
riel, carril	rail	räls
contracielo	raise	stigort
rampa	ramp	ramp
región, extensión	range	område
relación	ratio	förhållande
escariador, "reamer"	reamer (hole)	upprymmare
reducir	reduce	minska
reducción	reduction	reduktion, minskning
factor de reducción	reduction factor	reduktionsfaktor

Spanish	English	Swedish
area, región	region	område
reforzar	reinforce	förstärka
hormigón armado	reinforced concrete	armerad betong
refuerzos	reinforcement	förstärkning
relación	relation(-ship)	förhållande
remover, quitar	remove	avlägsna
reparación	repairs (pl.)	reparation
investigación, búsqueda	research	forskning
resistencia	resistance	motstånd
cable de retorno	return wire	återledare
laboreo rítmico	rhythmic driving	rytmisk drift
camino	road	väg
roca	rock	berg
minero	rock blaster	bergsprängare
voladura de roca	rock blasting	bergsprängning
anclaje de rocas	rock bolting	bergbultning
estallido de roca	rock burst	bergskott
cámara en roca	rock cavity	bergrum
constante de la roca	rock constant, c	bergkonstant
barrena, perforadora de roca	rock drill	bergborr
perforación de roca	rock drilling	bergborrning
fragmentación de roca	rock fall	styckefall
suelo rocoso	rock ground	berggrund
presión de roca	rock pressure	bergtryck
estrato de roca, capa	rock stratum	bergskikt
lanzamiento de la roca	rock throw	utkastning
barra, varilla	rod	stång, käpp
perno de techo	roof bolt	bergbult
labor de anchurón y pilar	room and pillar mining	rum- och pelarbrytning
rosa de grietas	rose of cracks	sprickros
perforadora rotatoria	rotary drill	rotationsborrmaskin
voladura	round	salva
fila (hilera) de barrenos	row of holes	hålrad
ruptura, fractura	rupture	brott
límite de ruptura	rupture limit	brottgräns
seguridad	safety	säkerhet
mecha (fusible) de seguridad	safety fuse	krutstubin
arena	sand	sand
piedra arenisca	sandstone	sandsten
andamiaje	scaffolding	ställning
sanear, quitar partes sueltas a la roca	scale	skrota
ensayo a escala	scaled-down test	modellförsök
sanéo de roca, amacizando	scaling	skrotning, bergrensning

431

Spanish	English	Swedish
lanzamiento, dispersión	scattering	kastning, spridning
esquema	scheme	schema
esquisto, pizarra	schist	skiffer
rasguño	scratch	rits
disparo	shot	sprängskott
cuele "costura", cuele de sierra	seam cut	sömkil
voladura secundaria	secondary blasting	skutsprängning
sección de túnel, area	section, area of a tunnel	(tunnel)-area
barrena de extension	sectional drill steel	skarvstål
sedimento	sediment	sediment
roca sedimentaria	sedimentary rock	sedimentärt berg
explosivo semiplastico	seim-plastic explosive	halvplastiskt sprängämne
sensibilidad	sensitivity	känslighet
serie	series	serie
juego	set	uppsättning
pozo	shaft	schakt
excavación de pozo hacia abajo	shaft sinking	schaktsänkning
pizarra arcillosa	shale	lerskiffer
cortante	shearing	skjuvning
cuña (corte) en V	shearing V-cut	saxad plogkil
sobrestante de turno	shift forman	skiftförman
ripia	shingle	singel
corto circuito	short circuit	överledning
barreno	shothole	borrhål
falta, escasez	shortage	brist
voladura de retardo corto	short delay blasting	kortintervallsprängning
detonador (estopino) de tiempo MS	short delay detonator	kortintervallsprängkapsel
pala mecánica	shovel excavator	grävmaskin
cargador de cucharón	shovel loader	grävmaskin
minería de grada	shrinkage stoping	magasinbrytning
encofrado	shuttering	form, formsätta
banco lateral	side bench	sidostross
ensanche lateral	side stoping	väggstrossning
pozo hacia abajo	sink shaft	sänke
forma sinusoidal	sinus shape (form)	sinusform
tamaño, medida	size	storlek
casco de protección	skull-guard	skyddshjälm
voladura de losa	slab blasting	skivsprängning
pizarra	slate	skiffer
altura de tajado	slice height	skivhöjd
derrumbe, deslizamiento	slide	ras
regla de cálculo	slide rule	räknesticka
barreno inclinado	sloping hole	lutande hål
escrepa de mina	slusher	skrapspel

432

Spanish	English	Swedish
voladura tersa, recorte	smooth blasting, smooth-ing	slätsprängning
explosivo sólido	solid explosive	fast sprängämne
roca sólida	solid rock	fast berg
solución	solution	lösning
roca sana	sound rock	bra berg
fuente de corriente	source of current	strömkälla
espaciamento, distancia (de barrenos)	spacing (of hole), E	(hål-)avstånd
carga específica	specific charge, q	specifik laddning
peso específico	specific weight	specifik vikt
túnel vertedor (alivia-dero)	spillway tunnel	spillvatten tunnel
espiral	spiral	spiral
carga alternada con tacos de madera	splint charging	pinnladdning
fragmentos, astillas	splinters	splitter
dispersión	spread	spridning
ampliación del fondo de barreno	springing	grytsprängning
estable	stable	stabil
pedestal	stand	stativ
norma, normal	standard	standard, normal
escarpado, pino	steep	brant
atacadura	stemming	förladdning
depósito; acopiar	stock pile; pile	tipp; lagra
ensanchar, destroza	stope	strossa; stross
ensanche	stoping	strossning
relleno	stowing	igensättning
(ex-)tensión	strain	töjning
estratificación	stratification	skiktning
estratificado	stratified	lagrad
estrato, capa	stratum (pl. strata)	lager, skikt
fuerza (por peso)	strength (of explosive), s	(vikt-) styrka
peso, tensión	stress	tryck, spänning
desencapado; descapar	stripping	(jord-) avtäckning
construcción, estrutura	structure	konstruktion
cliente, "tronco"	stump	gadd
banqueo de sub-nível	sub-level benching	skivpallbrytning
excavación (socavación) de sub-nível	sub-level caving	skivrasbrytning
minería vertical desde sub-nível	sub-level stoping	skivpallbrytning
asiento	subsidence, settlement	sättning
succión, aspiración	suction	sugning
tubo de aspiración	suchon pipe	sugrör
pozo	sink shaft	sänkschakt

Spanish	English	Swedish
reforzamiento, adema de protección	supporting	förstärkning
entibación, soporte de roca	supporting of rock	bergförstärkning
superficie, área	surface, face	yta
minería de cielo abierto	surface mining	dagbrytning
camara de equilibrio, pozo de oscilación	surge chamber	svallbassäng
hinchamiento	swelling	svällning
detonación simpática	symphatetic detonation, flash-over	överlag
sistema de grietas (fracturas)	system of cracks	spricksystem
tabla, cuadro	table	tabell
cuele (cuña) de Täby	Täby-cut	Täby-kil
túnel de descarga (desagüe)	tailrace tunnel	avloppstunne
atacar	tamp	ladda, packa
atacador	tamping pole (rod, stick)	laddkäpp
tangencial	tangential	tangentiell
colado	tapping	tappning
técnica	technique	teknik
temperatura	temperature	temperatur
resistencia a la tensión	tensile strength	draghållfasthet
(ex-)tensión, dilatación	tension	töjning
prueba, ensayo	test	prov
probador	tester	(lednings-)provare
perforación térmica	thermal piercing	termisk borrning
espesor, grosor	thickness	mäktighet, tjocklek
cuña (corte) de tres secciones	three-section cut	tresektionskil
lanzamiento	throw, throwing	kastning
entibar	timber	timra, stötta
intervalo de tiempo	time interval	tidsintervall
descargar, volcar; escombrera	tip	tippa; tipp
cubierta	tire	däck
barreno inferior, zapatera	"toe hole"	liggare
tolerancia	tolerance	tolerans
avance por galería superior	top heading	takort
rebaje de techo	top slicing	skivbrytning
grúa de torre	tower crane	kabelkran
vía, carrilera	track	spår
túnel de transporte	transport tunnel	transporttunnel
vibración, agitación	trembling	skakning
zanja	trench	dike

434

Spanish	English	Swedish
prueba, ensayo	trial	försök
voladura de prueba	trial blasting	provsprängning
cuele (cuña) triangular	triangular cut	triangelkil
perfilar, peine, recorte	trim	profilera, trimma
peinador, barrenos de re-corte	trimmer	konturhål
camión	truck	lastbil
tubo, tubería	tube	rör
carburo de tungsteno, wi-dia, metal duro	tungsten carbide	hårdmetall
barrena de metal duro	tungsten carbide drill	hårdmetallborr
túnel	tunnel	tunnel
fondo (piso) de túnel, so-lera	tunnel bottom	tunnelbotten
excavación de túnel	tunnel driving	tunneldrivning
frente (del túnel)	tunnelface,tunnelheading	tunnelfront
	tunnelling	tunneldrivning
techo de túnel, bovéda	tunnel roof	tunneltak
turbina	turbin	turbin
turno, jornada, relevo	turn	skift
libre	unconfined	liggande
roca suelta (floja), roca no consolidada	unconsolidated rock	löst berg
desencapado, descubierto	uncovering	avtäckning
socavación de guía in-ferior	undercut caving	blockrasbrytning
banqueo subterráneo	underground benching	pallbrytning under jord
voladura subterránea	underground blasting	underjordssprängning
cámara subterránea, ca-verna	underground chamber	bergrum
minería subterránea	underground mining	underjordsbrytning
laboreo (trabajo) sub-terráneo	underground work	underjordsarbete
banco de fondo	underhand bench (slope)	bottenpall
desigualdad, rugosidad	unevenness	ojämnhet
unidad	unit	enhet, aggregat
no cargado, sin cargar	unloaded	oladdad
hacia arriba	up grade	uppför
válvula, llave, grifo	valve	ventil, kran
cámara de válvulas	valve, chamber	ventilkammare
variable	variable	variabel
variación	variation	variation
cuele (cuña) en V	V-cut	plogkil
vena	vein	stråk
velocidad	velocity	hastighet
pozo (lumbrera) de ven-tilación	ventilation shaft	luftschakt

Spanish	English	Swedish
vertical	vertical	vertical
vibración	vibration	skakning
velocidad de vibración	vibration velocity	svängningshastighet
vibrograma	vibrogram	vibrogram
vibrógrafo	vibrograph	vibrograf
voltaje	voltage	el. spänning
volumen	volume	volym
perforadora de carretilla "waggon drill"	waggon drill	vagnborrmaskin
pared	wall	vägg
desperdicio, escombro	waste	avfall
roca estéril, ganga	waste rock	gråberg
suministro (aprovisionamiento) de agua	water supply	vattentillförsel
aguja de agua de limpieza	water tube	spolrör
tubo de agua	water pipe	vattenledning
longitud de onda	wave length	våglängd
eflorescerse, desgastarse a la intemperie	wither	vittra
cuele en cuña	wedge cut	kil (med snedstuckna hål)
peso	weight	vikt, belastning
peso de carga	weight of charge, Q	laddningsvikt
potencia por peso	weight strength, s	viktstyrka
soldar	weld	svetsa
soldadora	welding machine	svetsaggregat
rueda	wheel	hjul
ancho	width	bredd
anchura (longitud) de banqueo	width of bench, B	pallbredd, pallängd
espersor del filón, anchura de veta	width of ore body	malmbredd
cable, alambre	wire	tråd, ledning (el.)
diagrama de circuito	wiring layout	kopplingsschema
penetrar, calar, romper	work through	bryta igenom
obrero, operario, trabajador	worker	arbetare
explotación, laboreo	working	brytning
plan de trabajo	working scheme	arbetsplan
obrero	workman	arbetare
taller	workshop	verkstad
envoltura	wrapper	omslagspapper
límite de elasticidad	yield point	sträckgräns
zona	zone	stråk

Appendix 3

TABLE OF UNITS

The International System of Units (MKSA-system) is extremely convenient in technical application, and has been recommended by the International Union of Pure and Applied Physics and the General Conference on Weights and Measures (1960).[1]

This system is based on four basic units for length, mass. time and electric current intensity:

meter	m
kilogram	kg
second	S
ampere	A

which units are used throughout this book together, with their decimal fractions: mm (10^{-3} m), cm (10^{-2} m), dm (10^{-1} m), μ (10^{-6} m), g (10^{-3} kg) and ms (10^{-3} s). The MKSA or m-kg-s-A system is enlarged with three other basic units for temperature, luminous intensity and amount of substance (in chemical physics). They are not used here but are given for the sake of completeness

degree Kelvin	°K
candela	cd
mole	mol

The Kelvin temperature scale has its zero point at -273.15°C, its temperature interval, named degree, equals that of the centigrade scale (°C) being used here.

In the text the English units are also given in () after the metric units so that the reader familiar with this system will have no difficulties in the reading.

---

[1] Report published with financial support of UNESCO.

	English	Metric
length	1 inch (in) = 25.3 mm	1 m = 39.4 in
	1 inch = 2.53 cm	1 m = 3.28 ft
		1 m = 1.09 yd
	1 foot (ft) = 304.8 mm	1 cm = 0.39 in
	1 foot = 30.5 cm	
	1 foot = 0.305 m	1 mm = 39 mils
	1 yard (yd) = 0.91 m	1 μ = 39.10^{-3} mils
	1 mil (10^{-3} in) = 0.025 mm	
surface	1 sq in = 6.45 cm²	1 m² = 1550 sq in
	1 sq ft = 0.093 m²	1 cm² = 0.15 sq in
	1 sq ft = 930 cm²	
	1 sq yd = 0.836 m²	
volume	1 cu yd = 0.765 m³	1 m³ = 1.31 cu yd
	1 cu ft = 0.028 m³	1 cm³ = 0.06 cu in
	1 cu in = 16.39 cm³	
time	1 sec (s)	1 second (s)
	1 millisec (ms, MS) = 0.001 s	1 ms = 0.001 s
	1 microsec (μs) = 10^{-6} s	1 μs = 10^{-6} s
frequency	cycles per sec (c/s)	cycles (periods) per sec. (c/s, p/s)
velocity	1 ft/s = 0.31 m/s	1 m/s = 3.28 ft/s
	1 in/s = 2.5 cm/s	1 cm/s = 0.39 in/s
	1 mile/s = 1.61 km/s	1 km/s = 0.62 mile/s
mass	1 lb = 0.45 kg	1 kg = 2.20 lb
density	1 lb/cu ft = 16.0 kg/m³	1 kg/m³ = 0.063 lb/cu ft
	= 0.016 g/cm³	1 kg/dm³ = 62.5 lb/cu ft
	1 lb/cu in = 27.68 g/cm³	1 g/cm³ = 62.5 lb/cu ft
spec. charge	1 lb/cu yd = 0.59 kg/m³	1 kg/m³ = 1.69 lb/cu yd
konc. of charge	1 lb/ft = 1.48 kg/m	1 kg/m = 0.67 lb/ft
acceleration force	1 ft/s² = 0.31 m/s²	1 m/s² = 3.28 ft/s²
		1 kp = g N
		1 N = 1 kgm/s²
pressure	1 p. s. i = 0.07 Atm	1 Atm = 760 mm Hg at 0°C
		1 at = 1 kp/cm²1 $\cong$ Atm
		1 at = 14.2 p. s. i (lb/sq in)
temperature	°F = 32 + 9/5°C	°C = 5/9 (°F − 32)
energy		1 Nm = 0.238 cal
		1 Nm = 1 joule
effect	1 horse power (HP)	1 watt (W) = 1 Nm/s
	= 746 watt	
	1 HP = 0.746 kW	1 kW = 1.34 HP
el. current	1 ampere (A)	1 ampere (A)
el. resistance	1 ohm (Ω)	1 ohm (Ω)
el. tension	1 volt (V)	1 volt (V)
capacity	1 farad (F)	1 farad (F)